D062177

GEOLOGY OF THE OLDUVAI GORGE

GEOLOGY OF THE OLDUVAI GORGE

A STUDY OF SEDIMENTATION IN A SEMIARID BASIN

BY

RICHARD L. HAY

UNIVERSITY OF CALIFORNIA PRESS

BERKELEY LOS ANGELES LONDON

University of California Press
Berkeley and Los Angeles, California

University of California Press, Ltd.
London, England

Copyright © 1976, by
The Regents of the University of California
ISBN 0-520-02963-1
Library of Congress Catalog Card Number: 74-29804
Printed in the United States of America

Designed by Michael Bass

CONTENTS

v

FIGURES

TABLES

PLATES

(following page 174)

FOREWORD

BY

MARY D. LEAKEY, F.S.A., F.B.A.

Over the past twelve years Richard Hay has studied the stratigraphy of the Olduvai basin, and I have been fortunate to have worked there myself while the story of the geology was being unravelled; particularly memorable were those occasions when some new piece of evidence provided the answer to a long-standing problem.

This monograph gives us the data, meticulously collected and critically evaluated, on which Richard Hay has based his interpretation of the record, and it is unique as a detailed study of Pleistocene paleogeography. The picture now seems so complete and is presented so clearly and precisely that one gets the impression that pieces of the stratigraphic column fitted neatly and effortlessly into the right places and at the right times, but this was seldom the case and the occasions when it happened still stand out as red-letter days. More often, results were obtained by unremitting hard work and patient measuring of detailed sections up and down the length of the gorge. Sometimes the sections could be matched and the deposits correlated, but on other occasions there appeared to be insurmountable difficulties, particularly in the complex history of Bed II. In many instances correlations could only be achieved by mineralogic analysis in the laboratory after the field seasons were over.

Before this stratigraphic record was established, the positions of many sites where hominid, faunal and cultural remains had been found were uncertain or even unknown, but it is now possible to fit nearly all the important sites into the sequence.

The original terminology for the Olduvai deposits, set up by Reck in 1913, has been retained in the present work except for the higher part of the sequence where some alterations have become necessary and new terms have had to be added to distinguish subdivisions not previously recognised as significant.

The deposits in the Olduvai basin reach a maximum depth of 100 metres, of which Bed I (the lowest unit) has the greatest thickness. It spans the period from 2.1 to 1.7 m.y. ago. The division between Bed I and Bed II is easy to identify only in areas where the marker bed at the top of Bed I occurs, and it is evident that the conditions of deposition in lower Bed II were closely similar to those of Bed I. During the whole of this period there was a permanent, alkaline lake with a known diameter of 25 km. Camp sites of the early hominids are predominantly along the southeastern shoreline where freshwater streams from the volcanic highlands drained into the lake.

It is estimated that Bed II was deposited from 1.7 to 1.15 m.y. ago. The onset of faulting at about 1.6 m.y. substantially changed the topography of the Olduvai basin. It reduced the size of the existing lake and gave rise to an increase of open grassland, which encouraged the plains animals to occupy the area. The realisation that major earth movements began at this time and were responsible for significant changes in the paleogeography was one of the most dramatic discoveries in the geological study.

In the past, Beds III and IV have been considered as two discrete units. It has now been shown that they can only be subdivided in the eastern part of the gorge where the red Bed III forms such a prominent feature. To the northwest, where the red deposits are lacking, these beds are referred to as III-IV undivided. The paleogeography of this period has been interpreted as open grassland with meandering rivers and streams draining into a main channel in which faunal and cultural remains are more common than elsewhere.

The latest deposits to be laid down before the cutting of the gorge, (formerly included in Bed IV and termed IVB) are now known as the Masek Beds. They mostly consist of aeolian material and are disconformable to the underlying beds. They are succeeded by the Ndutu Beds which were laid down after the gorge had been partially cut.

Unfortunately, there is only scanty evidence for events in the Olduvai basin at this time. Richard Hay has subdivided the Ndutu Beds into lower and upper units, the lower consisting of residual patches of conglomerates and sandstones which still adhere to the sides of the gorge in certain areas, while the upper unit consists mostly of aeolian tuff. It partially filled the gorge but was later largely scoured out by renewal of drainage following the subsidence of a fault at the mouth of the gorge.

Richard Hay has estimated ages of 0.6 to 0.4 m.y. ago for the Masek Beds, 0.4 m.y. to 60,000 years for the lower Ndutu unit and 60,000 to 32,000 years for the upper unit. These last deposits contain artifacts of Middle Stone age facies and their date agrees closely with other Middle Stone Age occurrences in Africa.

Apart from a recent eruption of black ash from Oldoinyo Lengai dated at about 1,400 yr. B.P. the last deposit of significance is the Naisiusiu Beds which were formerly included in Bed V. These were laid down between 22,000 and 15,000 B.P. They contain a microlithic industry which is almost certainly contemporary with the skeleton of *Homo sapiens* found by Reck in 1913. For a time, this was considered to be of Middle Pleistocene age but has now been dated by radiocarbon to about 17,000 yr. B.P.

Two of the most important results to emerge from this detailed stratigraphic study have been to show the contemporaneity of certain hominid taxa and of two cultural traditions, and to provide an explanation for the change in fauna between lower and middle Bed II.

We now know that *Australopithecus boisei* continued through from Bed I times to the top of Bed II. He was contemporary with *Homo habilis* in Bed I and in lower Bed II. *Homo habilis* is not found higher than the lower part of middle Bed II, but in upper Bed II *Australopithecus boisei* was co-existent for a while with *Homo erectus.*

A single stone industry occurs in Bed I, that is the Oldowan. It is characterised by a variety of chopper and other "heavy duty" tools as well as small flake tools for lighter work. This industry then continues in a more evolved form with an expanded tool kit through Beds II and III and even into upper Bed IV. I have termed it the Developed Oldowan when it contains a more complex tool kit than in Bed I.

The Acheulean or handaxe culture is contemporary with the Developed Oldowan from middle Bed II upwards. Whether or not the roots of the Acheulean are to be found at Olduvai is uncertain, although I am now inclined to believe that some of the assemblages of tools from middle Bed II which were previously classed as Developed Oldowan may be earliest Acheulean. These industries have a low handaxe ratio but those that do occur appear to have been made with Acheulean expertise, unlike those in the later stages of the Developed Oldowan.

Even an archaeologist such as myself can appreciate the firm basic data and sound reasoning on which Richard Hay has based his conclusions. The text is clearly written and a multitude of excellent diagrams illustrate the changes through time in the Olduvai basin. The book will long stand as a model for future workers in this field.

The following lines from Emerson truly reflect Richard Hay's contribution: "What is originality? It is being one's self and reporting accurately what we see."

ACKNOWLEDGMENTS

I am particularly indebted to M. D. Leakey and the late L. S. B. Leakey for the use of camp facilities and assistance in many phases of my work from 1962 to the present. Until his death in 1972, L. S. B. Leakey unfailingly supported and encouraged my work despite occasional disagreements over geologic interpretations. Mary Leakey has not only supported and encouraged my work but has been deeply interested and involved in all aspects of field work and has been a continual and rigorous testing ground for new ideas. She aided in the naming of stratigraphic units, and she participated in various of the geologic investigations, including in particular the Laetolil Beds, the geologic setting of archeologic sites, the sources of raw materials for artifacts, and the dunes and trailing sand-ridges on the Serengeti Plain. Many times she collected and sent samples to fill gaps in my field collection. The African staff of the Olduvai camp deserves my thanks for many and varied services. George Karonga and Lucas Kioko were invaluable as field assistants, and F. K. Lili is largely responsible for measuring rainfall, evaporation, and movement of an active barchan dune. The cooperation of the United Republic of Tanzania is gratefully acknowledged, and I particularly wish to thank A. A. Mturi, Conservator of Antiquities.

The National Geographic Society and the National Science Foundation (grants G-22094 and P2A-0375) generously provided funds for both field and laboratory work and preparation of this volume. Research professorships in the Miller Institute for Basic Research (Berkeley) provided the time for extensive laboratory studies during 1962/63 and 1969/70, and this work was essential in establishing the relationship between mineral assemblages and the chemical nature of depositional environments at Olduvai. The University of California (Berkeley) provided funds for miscellaneous expenses, and the Associates of Tropical Biogeography at Berkeley defrayed the cost of transportation for one trip to Olduvai Gorge. Finally, a grant from the L. S. B. Leakey Foundation covered the cost of drafting, typing, and other expenses incurred in preparing the manuscript.

Many scientists have helped in this work, either directly or indirectly. I am grateful to B. H. Baker and J. Walsh, both formerly of the Kenya Mines and Geological Department, for the loan of equipment, assistance in shipping specimens, and for profitable discussions of Rift-Valley geology. Age determinations on the Olduvai Beds are the work of many hands: of J. L. Bada, R. Berger, A. Brock, A. Cox, G. H. Curtis, J. F. Evernden, C. S. Grommé, R. Protsch, and D. L. Thurber. Mineralogic conclusions depended to a considerable extent on X-ray diffractometer analyses of A. Iijima, R. N. Jack, and R. J. Moiola. Professor A. Pabst generously contributed his mineralogic expertise, and he translated one of Hans Reck's early papers for me. For chemical analyses I depended on R. N. Jack, H. Onuki, J. C. Stormer, and W. A. Wright. Fission-track studies were attempted by D. MacDougall, and although these were unsuccessful for dating, they provided important information regarding the low-temperature annealing of calcite. I am indebted to J. R. O'Neil for isotopic measurements, and to R. A. Sheppard for permission to give an unpublished analysis of chabazite from Olduvai Gorge. Stephen G. Custer and John Wakely are responsible for grain-size and heavy-mineral data on the Holocene sediments. L. H. Robbins served as my field assistant in 1964. I have profited from discussions with others too numerous to mention, although special thanks are due to G. L. Isaac, C. M. Gilbert, Kay Behrensmeyer, L. B. Leopold, Alan Gentry, and J. J.

Jaeger. The manuscript was reviewed by G. L. Isaac, C. M. Gilbert, C. V. Haynes, and Kay Behrensmeyer.

R. I. M. Campbell is responsible for most of the photographic work. He took many high-quality photographs of the sides of the gorge, which were useful both in working out the stratigraphy in the field and in illustrating it for this volume (Plates 1-6). He also took a series of low-altitude aerial photographs that were used to make a small-scale aerial mosaic of a part of the gorge, and this mosaic proved essential to a detailed paleogeographic reconstruction of Bed II (for example, Fig. 27). For photomicrographs (Plates 9, 10), I am indebted to J. Hampel. Permission to publish aerial photographs of the plain to the north of the gorge (Plates 11, 12) was granted by the Commissioner for Surveys and Mapping of the United Republic of Tanzania. Plate 8 was reprinted from Plate 7 in *Contributions to Mineralogy and Petrology* 17, pp. 255-274, by permission of the publisher.

Stanley J. Chebul, L. Luedke, and R. Laniz were responsible for preparing the thin sections used in microscopic study. Leonard J. Vigus made the equipment used in sampling sediments of modern saline lakes, which were essential to interpreting the Pleistocene lake deposits. Nearly all of the illustrations were drafted by Judith Ogden, and the manuscript was typed, in its various stages, by Debbie Aoki and Fanchon Lewis. The manuscript was edited by my wife Lynn.

INTRODUCTION

Although known principally for its content of artifacts and hominid remains, Olduvai Gorge for more than a decade has been a focus of varied scientific work centering on geology. Contributions have been made in paleontology, geochronology, geomagnetism, pedology, Rift-Valley tectonism, geochemistry, and mineralogy. These studies provide a wealth of detail concerning the nature and development of the Olduvai basin and its fauna, including man, over the past two million years.

I have worked on the stratigraphy of Olduvai Gorge for nine periods between 1962 and 1974. The first and most essential aim has been to work out the geometrical framework, that is, to subdivide the vertical succession and find marker beds for correlating. This was relatively easy for Bed I, which contains several widespread, mineralogically distinctive tuffs. The stratigraphic framework has been more difficult to establish in the overlying deposits, both because widespread marker beds are relatively rare and because the major subdivisions commonly change in thickness and lithology over short distances. A disproportionately large amount of field work has been devoted to Bed II, which is the most laterally variable of the Olduvai Beds.

Dating of the deposits has been a second goal of the stratigraphic work, and this has relied heavily on the application of geophysical methods by specialists. The initial dating, by the potassium-argon (K-Ar) method, was by my colleagues G. H. Curtis and J. F. Everden. They collected and dated their samples before the stratigraphy had been worked out in any detail, and much of my earlier stratigraphic work was directed to finding out the stratigraphic position of dated samples. Geomagnetic polarity, fission-track dating, radiocarbon dating, and amino-acid racemization have been used more recently to date the Olduvai Beds.

The third and major goal has been to interpret the various stratigraphic units in terms of the environment(s) which prevailed during deposition. The Olduvai Beds afford extraordinary possibilities for environmental interpretation, as they contain a rich and varied fauna, and their deposits represent a wide variety of geographic environments. The deposits are generally well exposed, and sediments deposited in different environments (= lithofacies) can, in most cases, be correlated in time, thus permitting a detailed environmental reconstruction for the Olduvai basin during most of the Pleistocene. Limited exposure outside of the gorge is the principal obstacle to a fully satisfactory paleogeographic reconstruction.

The initial laboratory work showed that the Olduvai Beds contain a wealth of mineralogic and textural features diagnostic of relatively specific chemical environments and climatic regimes. Samples from all environmental subdivisions have now been studied in the laboratory, and the number of man-hours spent in this work almost certainly exceeds the time spent in field work.

Eolian sediments occur widely in the Olduvai sequence and are emphasized in environmental interpretations. Most common are tuffs redeposited by wind and altered to an unusual assemblage of minerals at the land surface before burial. These deposits of eolian tuff appear thus far to be unique to the Olduvai region, and modern eolian sedimentation of volcanic ash in the Olduvai region is discussed in some detail to help explain the origin of the older deposits.

These geologic results have been applied to the archeology, both to date the archeologic sites and to determine the paleogeography and environments of hominid activities. In addition, the raw materials of artifacts were studied and their sources identified for the purpose of showing the degree to which different rock types were selected by hominids for particular tool types and the distance to which artifacts were transported from their sources.

In conclusion, the present volume can be viewed as a case history in the stratigraphic development of a small sedimentary basin in a semiarid climate. Several sections are included to make the main points understandable to the reader with a minimal geologic background. These sections are a glossary of geologic terms, a discussion of the principles and methods of environmental interpretation, and a nontechnical summary of the stratigraphy.

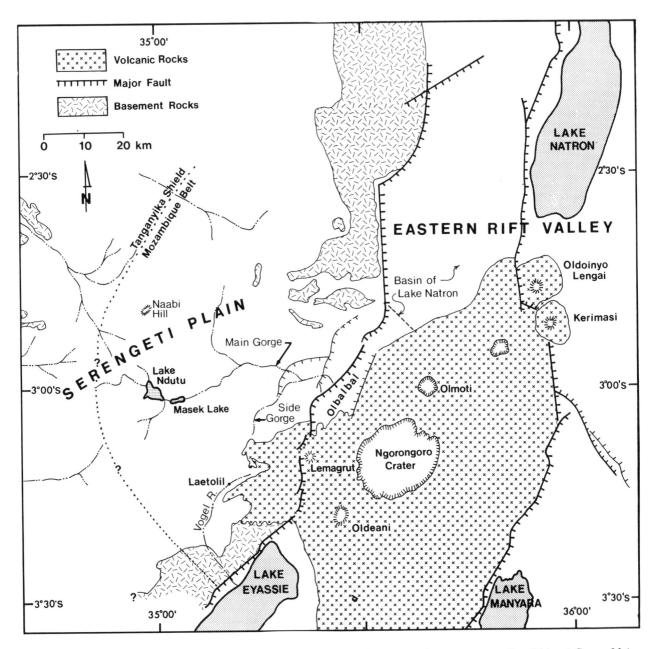

Figure 1. Regional map showing major geologic and topographic features in the area surrounding Olduvai Gorge. Major faults are shown by heavy hachured lines and lesser faults by thinner hachured lines. Data are taken from quarter-degree geologic maps (Pickering, 1958, 1960b, 1964, 1965; Guest et al., 1961) and from the regional synthesis of Baker et al. (1972).

1

THE SETTING OF THE GORGE AND HISTORY OF GEOLOGICAL INVESTIGATIONS

Olduvai Gorge is a valley at the western margin of the Eastern Rift Valley in northern Tanzania (Fig. 1). The valley is cut in the Serengeti Plain, which extends about 110 km northeast toward Lake Victoria, 260 km distant. The plain has an elevation of 1,360 to 1,520 m in the vicinity of the gorge, and it slopes gently eastward and is offset in a series of steps. Several hills and inselbergs of metamorphic rock break the even surface of the plain in the vicinity of the gorge, and extensive highlands of metamorphic rock bound the plain about 20 km to the north. The main branch of the gorge, known as the Main Gorge, has its origin in lakes Masek and Ndutu, and it flows 46 km eastward to empty into Olbalbal, a depression within a fault graben. The divide between drainage flowing toward Lake Victoria and toward Olbalbal lies 10 to 15 km west of Lake Ndutu. The graben with Olbalbal is bordered on the west and south by several large volcanoes, collectively termed the volcanic highlands.

The Main Gorge is broad, shallow, and underlain by a basement complex of metamorphic rocks in the western 20 km of its course. It deepens abruptly in a series of rapids and falls known as Granite Falls. These falls represent the western margin of the Olduvai basin, which is filled with Pleistocene sedimentary deposits termed the Olduvai Beds. In the eastern 26 km of its course, the gorge is steep-sided, 46 to 90 m deep, and 0.5 to 1.5 km wide. It has an average gradient of 7.8 m/km (1:128), which is about the same as the average gradient of the plain (1:137). About 9 km upstream from its mouth, the Main Gorge is joined by a smaller, southern branch, named the Side Gorge, which originates on the western slopes of Lemagrut volcano (Figs. 2, 3).

The Olduvai basin appears to have originated in growth of the volcanic highlands on an eastward-sloping surface of metamorphic basement rocks. Lavas and ignimbrites both underlie and interfinger with the basin sediments. Eruptive rocks beneath the Olduvai Beds presumably thin from east to west, and a single welded tuff, the Naabi Ignimbrite, separates the Olduvai Beds from metamorphic basement rocks near the western margin of the basin.

The basin has been extensively modified by faulting, and Olbalbal, its present drainage sump, lies considerably east of the lowest part of the basin during early and middle Pleistocene time. The steplike topography of the Serengeti Plain is a reflection of fault offsets, with downward displacements on the eastern side of most faults (Fig. 4). Measured displacements range from a few centimeters to about 40 m, but the inferred offset at the western margin of Olbalbal is at least 100 m and may be considerably more. Faulting began early in the deposition of the Olduvai Beds and has continued into relatively recent times. Individual fault blocks were deformed to varying extents as reflected both in obvious folds and in variation in thickness of stratigraphic units within a given fault block.

Geological investigation of the gorge was begun with the arrival, in 1913, of a German expedition led by Professor Hans Reck. The expedition was sent to collect fossils, following up on discovery of the gorge and remains of extinct mammals by Professor Kattwinkel in 1911. Reck's work was concentrated in the lower 9 km of the gorge, where fossils were collected and the geology studied over a period of three months. World War I ended German exploration of the gorge, and geologic results of this first expedition were published by Reck (1914a, 1914b, 1933). In his first paper (1914a), Reck subdivided the sedimentary sequence overlying basaltic lavas at Olduvai into five major units, termed *Schichtkomplexes* and numbered from 1 to 5. The term *Schichtkomplex* was later replaced by the English term *Bed*, but the numbering system was continued without modification until recently. As we will see, the further work at Olduvai shows the weakness of a numerical system for designating stratigraphic units. Reck himself raises the problem, by implication, in referring to a series of beds high in the section which he was unable to assign either to Schichtkomplex 4 or 5 (1914a). Four of the major faults were noted, and these were numbered from 1 to 4.

He believed that most of the sequence was deposited in a fresh-water lake, and he correctly inferred an eolian origin for the uppermost unit, Schichtkomplex 5, which is said to consist of dusty, loesslike clay and eolian dust tuff (äolischen Staubtuffe).

Fossils collected on the expedition were the subject of numerous papers, by Dietrich and others, over a period of many years. These faunal reports are referenced by L. S. B. Leakey (1951, 1965). Particular interest was aroused by the nearly complete human skeleton, which Reck believed to come from Schichtkomplex 2 (= Bed II).

Reck was a member of L. S. B. Leakey's first archeological expedition to Olduvai, in 1931, which remained for about two months. As in the 1913 expedition, work was largely confined to the lower 9-10 km of the gorge. Reck's observations are presented in L. S. B. Leakey's first volume on Olduvai Gorge (1951). In this report, the term Bed replaces Schichtkomplex, but the stratigraphic subdivisions remain unchanged, and a fifth fault was added to the numbered sequence. The stratigraphy is interpreted in terms of pluvial and interpluvial stages, with Beds II and IV representing expansion of the lake for a lengthy period of time. Bed III was interpreted as the deposit of an interpluvial period.

In August 1932, Leakey revisited the gorge with E. J. Wayland, director of the Geological Survey of Uganda. In a short letter to Nature, Wayland (1932) favored assignment of the human skeleton to the base of Bed V, a view later shared by most participants in the controversy (L. S. B. Leakey et al., 1933). While at Olduvai, Wayland made other observations of a geologic nature, which are given in a mimeographed but unpublished report entitled "The Age of the Oldoway Human Skeleton" and dated September 2, 1932. I received a copy of this report from Mr. Wayland in 1964. It in-

cludes a description of stream gravels deposited during erosion of the gorge and eolian tuffs of Bed V deposited after the gorge was eroded. The gravels are termed Bed Va, and the eolian tuffs are termed Bed Vb. This subdivision was followed by L. S. B. Leakey, who adopted the term Bed Va for the gravels but preferred to use Bed V for the tuffs. Wayland notes that the gorge exhibits two cycles of erosion, an earlier, pre-Bed V valley with comparatively gentle slopes, and a later valley with steep to precipitous slopes. Although this observation is not strictly correct, it shows some awareness of the complex history of erosion in the gorge. His environmental interpretation of the Olduvai deposits is remarkably accurate in view of his short visit and is quoted in full:

> The expedition is in the habit of speaking of the "Oldoway Lake" as though the flat-lying deposits were laid down beneath a permanent sheet of water. I am not prepared to dispute that such a lake probably existed, but I contend that the Oldoway beds themselves were not deposited upon its floor, for they indicate rather the gradual filling in – to a considerable extent by windborne volcanic materials – of a swampy depression, which may have been, and probably was, connected with a more permanent water area. It must be admitted that the remains of fish occur, but they are not common (as they are, for example, in the bone-beds at Kaiso, Uganda, where they are the dominant fossils), and though the relics of semi-aquatic creatures (e.g., Deinotherium and Crocodiles) are to be found, the fauna is essentially terrestrial, as a perusal of the lists on Fig. 3 will show. Nor were these remains carried down by rivers to be buried in deltaic deposits for these, so far as I could discover, are generally wanting. The one-time living creatures died, it would seem, pretty much where their remains are to be found this day. Some, no doubt, were killed by carnivores, and in any case it is probable that the corpses were generally dismembered after death by scavengers. Completely articulated skeletons would therefore be of a rarest occurrence; and none, I understand, has been found – except, of course that of the Oldoway man [i.e., the human skeleton found by Reck].

In 1934, L. S. B. Leakey was accompanied by P. G. H. Boswell, who had been the first to clearly demonstrate that the Olduvai human skeleton postdated Bed IV (Boswell, 1932). At intervals from 1935 onward, L. S. B. and M. D. Leakey studied the archeology and paleontology of the gorge. L.S.B. Leakey's geological interpretations are summarized in his volumes on Olduvai Gorge (1951, 1965), in which he views the Olduvai stratigraphy in terms of pluvial and interpluvial periods. Papers by Cooke (1957) and Flint (1959) cast serious doubt on his pluvial-interpluvial interpretation of the Olduvai sequence.

The region including Olduvai Gorge was mapped geologically by R. Pickering in 1956 and 1957. The results were published in a geologic map (Pickering, 1958) and a short report commenting on the stratigraphy and including a longitudinal section of the gorge (Pickering, 1960a). He infers that the Olduvai sequence was laid down in a lake about 32 km from east to west and 11 to 16 km from north to south, and he uses Olbalbal as a modern example of the type of lake basin in which Beds II and IV were deposited.

A new era in geological investigation of the gorge began with K-Ar dating of Bed I by G. H. Curtis and J. F. Evernden at 1.75 m.y.a. (million years ago). Evernden first sampled the tuffs of Bed I in 1958, and the results were published in 1961 (L. S. B. Leakey et al., 1961). The result was highly controversial at the time (see Koenigswald et al., 1961; Straus and Hunt, 1962), and both additional dating and more detailed geological studies were undertaken. In 1961, R. Pickering spent some months at Olduvai, and he made a geologic map of the entire gorge and a detailed map of the FLK area. He also measured a number of stratigraphic sections and wrote a report on the stratigraphy. These have not been published, but were provided to me in manuscript form by the Tanganyika Geological Survey. In this he identifies for the first time the ignimbrite which underlies Bed I in the western part of the gorge. He was unable to recognize Reck's stratigraphic units in the western part of the gorge, and consequently he replaced Reck's subdivision into Beds I through V with a system emphasizing lithology and lateral facies changes.

M. R. Kleindienst, an archeologist, excavated site JK 2 in 1961/62, and in a series of trenches she showed that deposits of Bed III interfinger with deposits of a type previously considered diagnostic of Bed IV (Kleindienst, 1964; in press). This lateral change is one of the many which are vital to an understanding of the stratigraphy and paleogeography.

In 1962, I spent nearly three months at the gorge, principally working on the stratigraphy of the Bed I tuffs (Hay, 1963a). Reck's subdivisions were modified only to the extent of including the basaltic lavas within Bed I. An important development was the recognition of

extensive deposits of eolian tuffs within Beds II and IV. The unique mineralogy of the tuffs clearly pointed to a saline, alkaline, soil environment in a dry climate much like that of the present (Hay, 1963b). Paleomagnetic measurements by C. S. Grommé demonstrated normal polarity for the lavas of Bed I (Grommé and Hay, 1963, 1967), which was a significant discovery at this early stage in development of the magnetic polarity time scale. My 1964 season, devoted principally to Beds II, III and IV, showed that faulting began in the Olduvai basin during the deposition of Bed II and continued through deposition of Bed IV (Hay, 1965, 1967a).

Laboratory study of samples collected in 1962 and 1964 showed that the Olduvai Beds were divisible into a series of mineralogic zones or facies, which could be correlated with various depositional and postdepositional environments (Hay, 1966). The mineralogic evidence left no doubt that the lake of Beds I and II was saline, alkaline, and essentially similar to lakes in the same region today. Work in 1967 and 1968, of limited extent, was concerned primarily with lake-margin deposits of Bed II and the origin of chert in these sediments (Hay, 1968). Discordant radiocarbon dates for Bed V (L. S. B. Leakey, 1969) led to detailed work on the upper part of the sequence in 1969, and Bed V was found to include two quite different units, an older termed the Ndutu Beds, and a younger termed the Naisiusiu Beds (Hay, 1971). With the elimination of Bed V from stratigraphic usage, the term Masek Beds was applied to the widespread unit, chiefly of eolian tuffs, which previously had been included within Bed IV and named Bed IVb (Hay, 1971). The Masek Beds are probably the deposits which Reck was unable to assign either to Schichtkomplex 4 or 5 (that is, Bed IV or V). I worked at Olduvai Gorge again in 1970, 1972, 1973, and 1974, and the principal new geologic result of this work was the detailed stratigraphy of Bed II. Collaboration with M. D. Leakey resulted in locating sources for raw materials used to make artifacts, and in working out the origin of the belt of sand dunes on the plain to the north of the gorge. The magnetic stratigraphy of the gorge now has been established, at least roughly, on the basis of laboratory work by C. S. Grommé, A. Brock, and A. Cox (see Grommé and Hay,

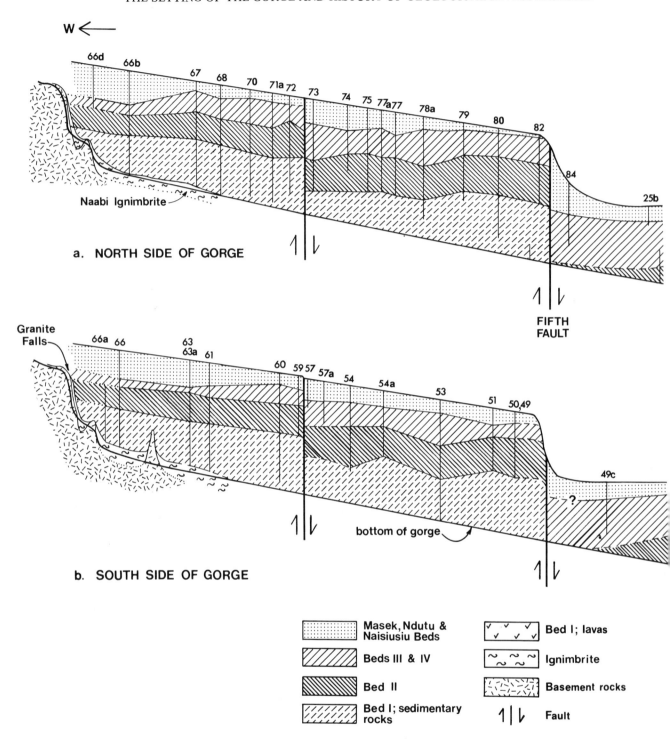

a. NORTH SIDE OF GORGE

b. SOUTH SIDE OF GORGE

Masek, Ndutu & Naisiusiu Beds

Beds III & IV

Bed II

Bed I; sedimentary rocks

Bed I; lavas

Ignimbrite

Basement rocks

Fault

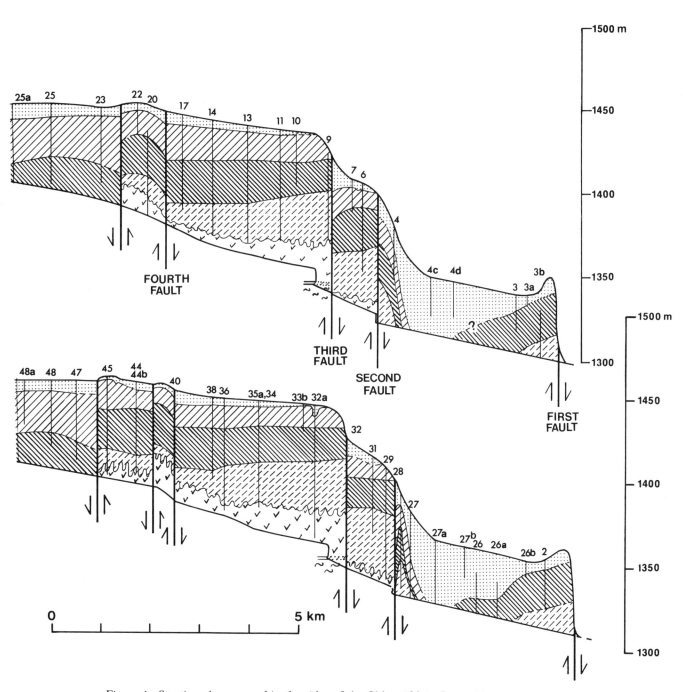

Figure 4. Stratigraphy exposed in the sides of the Olduvai Main Gorge. Numbered vertical lines represent measured sections. Vertical scales represent elevation.

1971; Brock et al., 1972). F. H. Brown contributed materially both to the physical stratigraphy of Bed IV and to the magnetic stratigraphy of Bed II through the Masek Beds during a visit in 1972. Oxygen-isotope analyses were made in 1971/72 by J. R. O'Neil on chert nodules from Olduvai and other East African lake deposits (O'Neil and Hay, 1973). This work provided a new line of evidence concerning the nature of the saline lake of Beds I and II. Another new approach, only recently utilized, is the extent of racemization of amino acids in fossil bones. Racemization is a function of both age and temperature, and this technique has been used to date the upper unit of the Ndutu Beds (Bada and Protsch, 1973) and to determine the late Pleistocene temperature (Schroeder and Bada, 1973).

Between 1914 and the present, many paleontologists have studied the Olduvai fauna, and some of the more recent studies, by specialists, are of immense importance in environmental interpretations. A few selected examples are the work of J. J. Jaeger (in press) on the rodents of Bed I, of Anthea and Alan Gentry on the bovids of Beds I through IV (in press), Greenwood and Todd on the fishes of Beds I and II (1970), and P. Brodkorb on the avifauna of Beds I and II.

STRATIGRAPHIC NOMENCLATURE

The term Olduvai Beds, originally applied to Beds I through V, now comprises the sequence of Bed I through the Naisiusiu Beds. In modern stratigraphic usage, the various beds would be considered formations, or mappable rock units, each containing many deposition units, or strata. Reck's original subdivision into Beds I through IV is continued here, with some modification (Fig. 5). Bed I is redefined to include all rocks between the Naabi Ignimbrite and the base of Bed II (Fig. 4). By this definition Bed I includes the lava flows in the bottom of the gorge and is the thickest of the Olduvai Beds. Bed II is still used now as Reck defined it, although the Lemuta Member is established within it. Beds III and IV are continued in their original sense in the eastern part of the gorge where they are mappable units; farther west they cannot be separated and are termed Beds III-IV (und., or undivided). Thus Beds III and IV can be viewed as members of a single formation, Beds III-IV

(und.). The principal marker tuffs in Beds I-IV are designated Tuff, followed by the Bed in which they occur and by a letter indicating their position relative to other named tuffs (Hay, 1971). Tuff IA, for example, is the lowermost marker tuff of Bed I. Other marker beds are named descriptively (for example, bird-print tuff or lower augitic sandstone of Bed II). The Masek Beds are a widespread unit, characterized by eolian tuffs, which disconformably overlies Beds III and IV. It may have been referred to in Reck's first Olduvai paper (1914a), but it was not mentioned in his final report (1951). Beds V and Va are obsolete and are replaced by two units, the older of which is the Ndutu Beds and the younger of which is the Naisiusiu Beds. The Ndutu Beds comprise eolian tuffs and detrital sediments deposited in the gorge and over the adjacent plain during a lengthy period of intermittent faulting and erosion. The Ndutu Beds are subdivided into a lower and an upper unit. The Naisiusiu Beds are chiefly eolian tuffs deposited in the gorge and over the adjacent plain after the gorge had been eroded to nearly its present depth.

The youngest deposits are ash and alluvium of Holocene age. Unconsolidated alluvium as much as 4 m thick is exposed in the bottom of the gorge, and a much greater thickness presumably underlies the locus of modern sedimentation in the northeastern part of Olbalbal. A single ash deposit, termed the Namorod Ash, is found throughout the Olduvai region. It was earlier named Bed VI by L. S. B. Leakey (1965).

Each major vertical subdivision of the Olduvai Beds varies laterally in lithology, reflecting either different depositional environments or sediment sources, or both. These lateral variations, or lithofacies, are for the most part areally restricted and can be delineated in the stratigraphic succession. The various lithofacies are named for the environments in which they were deposited, as for example the lake deposits of Bed I, but it should be emphasized that such names designate only the principal depositional environment(s) involved, for most of the facies contain sediments deposited in two or more environments. Contemporaneous facies deposited in the same type of environment but differing in source area and mineral composition are named separately, as for example the eastern and western fluvial facies of Beds III-IV.

Reck, 1951; L.S.B. Leakey, 1951, 1965	Hay, 1967a		Hay, 1971		Hay, this volume	
Formation	Formation	Member	Formation	Member	Formation	Member
Bed VI					Namorod Ash	
Bed V	Bed V		Naisiusiu Beds		Naisiusiu Beds	
Bed Va	Bed Va		Ndutu Beds		Ndutu Beds	Upper unit Lower unit
	Bed IV	Eolian Tuff Mbr.	Bed IVb	Norkilili Mbr.	Masek Beds	Norkilili Mbr. Lower unit
Bed IV	Bed IV		Bed IVa / Beds III-IVa (und.)		Bed IV / Beds III-IV (und.)	
Bed III	Bed III		Bed III / Beds III-IVa (und.)		Bed III / Beds III-IV (und.)	
Bed II	Bed II	Eolian Tuff Mbr.	Bed II	Eolian Tuff Mbr.	Bed II	Lemuta Mbr.
Bed I	Bed I	Upper Mbr.		Upper Mbr.	Bed I	
lavas	Bed I	Basalt Mbr.		Basalt Mbr.	Bed I	lavas
				Lower Mbr.		

Figure 5. Stratigraphic subdivision and nomenclature of the Olduvai Beds.

The system for designating geographic localities in the Olduvai Gorge area deserves some comment. L. S. B. and M. D. Leakey generally designated sites with significant faunal remains or artifacts with the initials of the person making the first discovery, followed by *K* (for karonga, the Swahili term for gully) or *C*, for cliff (Table 1). Thus MNK refers to Mary Nicol Karonga. A few other localities were named for some feature, as for example Handaxe *C*, or Elephant *K*. These sites are located by M. D. Leakey (1971*a*). Localities of measured stratigraphic sections or particular geologic features were numbered (Hay, 1971), and additional localities in the vicinity of numbered sections have been denoted by the letters *a, b, c,* and so forth.

Table 1. *Archeologic and Faunal Sites with Corresponding Geologic Localities in Olduvai Gorge*[1]

Site	Geologic Localities
BK	94
"Capsian site"	Near 5
Castle Rock	44
CK	6, 6*a*, 7, 8
CMK	96
Croc. K	48
DC	93
DK	13
DK-EE	12*a*
EF-HR	12
Elephant K	32
FC	89
FK	25

Site	Geologic Localities
FLK	45
FLK-Maiko Gully	45c
FLK-N	45a
FLK-NN	45b
FLK-S	46a, 46b, 46c
GTC	95
HEB	Near 44
HK	24
Hoopoe Gully	74
HWK	44
HWK-E	43, 43a
HWK-EE	42, 42a, 42b
ISK	35a
JK	14
Kar K	77
Kestrel K	84
Kit K	Near 3
KK	41
LK	10
LLK	46
Long K	38, 38a, 38b, 39
MCK	40, 40a, 40b
MK	11
MLK-East	52
MLK-West	53
MNK	88
MRC	95a
NGC	97
PDK	35
PEK	88a, 88b
PLK	23
RHC	80
Rhino K	25a
RK	9
SC	90a
SHK-East	90
SHK-West	91, 92
THC	32a
TK	15-19
TK, Fish Gully	18
VEK	85-87
VFK (Vth Fault K)	83
WK	36

NOTE: Sites are described further by M. D. Leakey (1971a).

Thus localities 45a, 45b, and 45c are in the vicinity of locality 45. Most of the numbered geologic localities are shown in Figures 2 and 3, and a larger scale map (Fig. 6) covers the area surrounding the junction of the Main and Side gorges, which contains numerous archaeologic sites and geologic localities.

CLIMATE AND VEGETATION

The Serengeti Plain is covered by grass with *Commiphora* scrub and scattered *Acacia* (thorn trees). This vegetation is insufficient to prevent considerable movement of sediment by the wind during the dry season. The bottom of the gorge contains denser vegetation, which includes much wild sisal or *Sansevieria*. The name Olduvai means "the place of the sisal" in Masai (*Ol* = the place of; *duvai* = sisal).

The climate is hot and dry. Annual rainfall ranges from 27.6 to 105.2 cm per year, averaging 56.6 cm over the eight years that records have been kept systematically, principally by F. K. Lili. Nearly all of the rain falls in the months of December through May (Table 2). Water flows in the gorge only intermittently during the rainy season, and in some years no water flows at all. Evaporation records, by F. K. Lili, although incomplete, strongly suggest a daily average of slightly more than 1 cm (Table 3), vastly in excess of precipitation during all months in most years.

The temperature ranges between limits of about 12°C and 33°C and varies rather little on a seasonal basis (Table 4). The mean annual air temperature is approximately 22.8°C. This figure is an average of maximum and minimum

Table 2. *Monthly Rainfall at Olduvai Gorge, 1966-1973   (in cm)*

	Jan.	Feb.	Mar.	Apr.	May	June	July	Aug.	Sept.	Oct.	Nov.	Dec.	Total
1966	24.0	31.8	12.7	22.8	0	0	0	0	6.3	0	0	7.6	105.2
1967	6.3	5.1	29.2	7.6	11.4	0	0	0	4.4	1.3	12.7	11.4	89.4
1968	33.0	17.7	28.0	7.6	1.9	0	0	0	6.3	0	0	3.8	98.3
1969	6.4	9.9	7.2	0	6.3	0	0	0	4.5	0	2.5	5.5	42.3
1970	6.1	3.1	5.1	8.1	2.1	0.3	0	0	0	0	1.5	1.3	27.6
1971	5.0	2.0	2.5	3.6	1.7	0	2.1	4.1	0	0	1.2	6.0	28.2
1972	6.4	7.6	2.5	0.7	4.1	2.0	0	0	0	1.2	2.2	7.6	34.3
1973	9.3	3.7	0.6	3.0	2.0	0	0	0	2.1	2.2	1.4	3.4	27.7
Average	12.1	10.1	11.0	6.7	3.7	0.3	0.3	0.5	3.0	0.6	2.7	5.8	56.6

temperatures recorded in the camp between the Main and Side gorges in 1972 and 1973. Approximately the same average is given by averaging the maximum and minimum readings recorded by F. K. Lili at the camp on the north

Table 3. *Evaporation at Olduvai Gorge, 1971-1972*

1971	January	12.4 mm (aver./day)
	October	15.5
	November	14.1
	December	9.4
1972	January	12.0
	February	6.6
	March	10.8
	April	11.3
	May	12.0
	June	8.6
	July	10.4
	Average	11.2

side of the Main Gorge. Temperature was also recorded at hourly intervals during several days in June, July, December, and January to show the diurnal fluctuation. These measurements show that the daily average based on hourly readings approximates very closely the average of maximum and minimum readings for the same day. The ground temperature was obtained with a maximum-minimum thermometer placed at the base of a 2-meter-deep auger hole, which was plugged above the level of the thermometer. The temperature was constant at 78°F (25.6°C) for the period of January 9-11, 1974. It is to be expected that the ground temperature will exceed the air temperature because the ground absorbs heat from the sun, reaching tempera-

Table 4. *Mean Monthly Temperature for Olduvai Gorge*

	Month	No. of days recorded	Mean minimum	Mean maximum
1973	January	31	16.8°C	29.3°C
	February	28	17.2	30.3
	March	11	16.2	31.4
	April	0	n.d.	n.d.
	May	17	16.2	28.5
1972	June	10	15.8	27.8
	July	31	15.6	28.4
	August	31	15.9	29.8
	September	30	15.9	30.7
	October	0	n.d.	n.d.
	November	0	n.d.	n.d.
	December	31	17.1	29.9
Mean			16.3	29.3

tures far in excess of the air above. The mean annual ground temperature is important for use in age determination by racemization of amino acids.

Wind blows most of the time, principally from the east. Wind velocity was recorded twice daily (8 *a.m.* and 6 *p.m.*) during five months of 1973 at the airstrip near the Olduvai Camp, principally by M. D. Leakey (Table 5). The average wind velocity for the recorded period is about 12.4 kph.

Table 5. *Average Wind Velocity at Olduvai Gorge, August-December 1973*

	a.m.	p.m.
August	13.4 kph	14.4 kph
September	12.1	10.2
October	12.8	12.9
November	12.8	12.0
December	11.7	11.2
Average	12.6 kph	12.1 kph

METAMORPHIC BASEMENT ROCKS

Metamorphic rocks in the vicinity of Olduvai Gorge are chiefly granite gneisses and quartzites, which vary considerably in lithology and mineral composition. Foliated biotite granite gneiss is exposed in the western part of the gorge, and sodic granite gneiss of distinctive appearance and composition forms the inselbergs of Kelogi (Table 6). The Kelogi gneiss is relatively coarse-grained and equigranular, and it shows at most a weak foliation. It is characterized mineralogically by 10 to 20 percent of aegirine and a sodic amphibole that is strongly pleochroic from pale yellow to deep purple. Quartzite forms most of the other basement hills and inselbergs in the vicinity of the gorge. The quartzite is extremely coarse-grained and commonly micaceous in the northern and eastern parts of the Olduvai region, including Naibor Soit and the hills and highlands to the north of the gorge. Individual crystals of quartz are generally 1 to 2 cm in diameter in these rocks. Most of the quartzite exposed to the south of the gorge is medium-grained and exhibits primary sandstone textures.

Metamorphic rocks of the Olduvai region belong to the Mozambique belt, which contains a great thickness of sediments that was highly

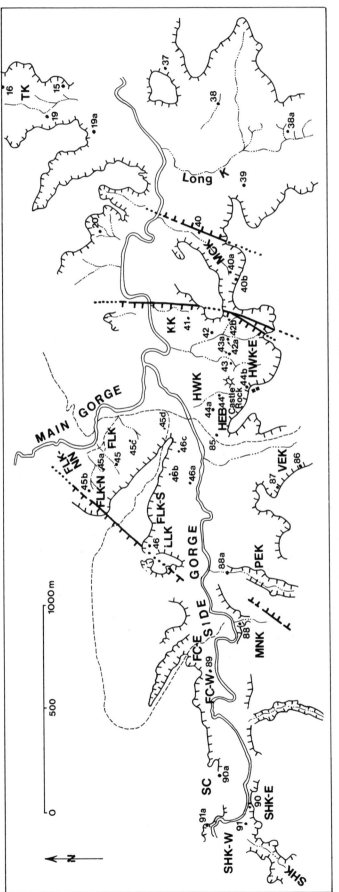

Figure 6. Map showing localities near the junction of the Main and Side gorges.

Table 6. *Composition of Metamorphic Rocks in the Vicinity of Olduvai Gorge*

	Geographic feature and location	Lithologic description	Mineral composition
1.	Granite Falls; in western part of Main Gorge	Granite gneiss; pink and gray, foliated, fine- to medium-grained; has pegmatitic dikes and zones	Quartz, microcline, plagioclase (An_{20}), biotite, muscovite
2.	Naibor Soit; inselberg north of Main Gorge	Quartzite, white or colorless, rarely pale brown or green; generally very coarse-grained; micaceous layers are foliated and lineated	Quartz, muscovite
3.	Hill 4 km northwest of Naibor Soit	Very coarse quartzite similar to Naibor Soit	n.d.
4.	Olongoidjo; E-W ridge 6 km long that lies 12 km north of Olduvai Gorge	Dominantly very coarse quartzite similar to Naibor Soit	n.d.
5.	Kelogi; inselbergs near west edge of the Side Gorge	Granite gneiss, dominantly pale yellow, rarely reddish-brown; medium- to coarse-grained; massive	Quartz, orthoclase, albite, aegirine, alkali amphibole
6.	Kearoni; inselberg east of Side Gorge	Same as Kelogi	Same as Kelogi
7.	Naisiusiu; hill south of Main Gorge near locality 61	Granite gneiss, fine- to coarse-grained; also quartzite, white and medium-grained	Quartz, microcline, plagioclase, aegirine, green hornblende, sphene, biotite, quartz, muscovite, garnet, kyanite, staurolite
8.	Low broad hill 6 km north of Naisiusiu	Poorly exposed medium-grained quartzite; dominantly white?	n.d.
9.	Inselberg 5 km west of Kelogi	Chiefly quartzite, medium-grained, and white, yellow, and dark reddish-brown	Quartz, muscovite, kyanite, biotite
10.	Inselberg 5 km west-southwest of Kearoni	Chiefly quartzite, medium-grained, and white, yellow, pink, dark reddish-brown; some has relict sandstone texture; little amphibolite present	Quartz, altered feldspar, muscovite, rare biotite

deformed and metamorphosed near the beginning of the Paleozoic era, roughly 475 to 650 m.y.a. (Cahen and Snelling, 1966). Metamorphic rocks of the Mozambique belt are bordered to the west by the Tanganyika shield, which has been a stable area during the past 2,500 million years. Unmetamorphosed granites of the shield are unconformably overlain by relatively unmetamorphosed quartzites, sandstones, and shales of the late Precambrian Bukoban System only about 25 km west of Lake Ndutu, which is at the headwaters of Olduvai Gorge (Pickering, 1960b).

VOLCANOES AND VOLCANIC ROCKS

Pleistocene volcanoes and volcanic rocks form mountainous highlands to the south and east of the gorge (Fig. 2). These supplied both detrital and pyroclastic (aerially discharged) sediment to the Olduvai basin. Directly bordering the basin are the volcanoes Olmoti, Ngorongoro, Sadiman, and Lemagrut. Volcanic rocks of Ngorongoro underlie Bed I through much of the Olduvai basin. Olmoti was active during deposition of Bed I and supplied most of the pyroclastic rocks in Bed I. Sadiman is a subsidiary volcano related to Lemagrut, a large volcano to the south of the gorge. The Laetolil Beds are fossiliferous tuffs of Sadiman or Lemagrut that extend into the south edge of the Olduvai basin. Kerimasi and Oldoinyo Lengai are relatively young volcanoes that lie 55 to 65 km east to northeast of the gorge, and they supplied large volumes of ash to the Olduvai region in the latter part of Pleistocene time. One small, isolated volcanic center, named Engelosin, lies within the basin 7 km north of the Main Gorge. These volcanoes and their deposits are briefly described in the following paragraphs.

Engelosin

Engelosin is a steep-sided volcanic neck about 500 m in diameter and 145 m high that is extensively mantled by talus breccia cemented by calcrete. Its age is unknown, but the relatively high degree of erosion suggests that it may well be Pliocene or older. The lava is of dark greenish-gray, fine-grained, slightly porphyritic nepheline phonolite. It has a distinctive mineralogy and texture (Table 7).

Table 7. *Composition of Volcanic Rocks in the Olduvai Region*

	Volcano and locality	Lithology	Mineralogic data[a]
1.	Engelosin	Flow-banded nepheline phonolite lava; slightly porphyritic, dark greenish-gray.	Very few phenocrysts of anorthoclase and sodic augite; common nepheline microphenocrysts 0.1-0.2 mm long; groundmass is nepheline, alkali feldspar, aegirine, analcime, and a rod-shaped rare-earth silicate(?) with 45° extinction and high birefringence. Sphene is rare. Color index is about 20.
2.	Ngorongoro; Lerai road (southern edge of caldera wall)	340 m of interbedded lavas and pyroclastic rocks. Very porphyritic trachyandesite lavas predominate in the lower part and slightly porphyritic trachyandesites in the upper part of the section. Few flows of olivine basalt in upper part.	Phenocrysts of plagioclase, augite, and less commonly olivine. Groundmass feldspar, augite, olivine, hornblende, biotite.
3.	Ngorongoro; Windy Gap road (NW edge of caldera wall)	Trachytic welded tuffs, coarsely devitrified and rich in crystals, and light olive gray to dark greenish gray, interbedded with gray trachyte lava.	Phenocrysts of anorthoclase and less commonly augite; groundmass quartz, anorthoclase, hornblende, and sodic augite.
4.	Ngorongoro; western slopes	At least 25 m of thin flows of porphyritic olivine basalt overlies pale red trachytic welded tuff at least 20 m thick.	Olivine basalt has 5-15% olivine phenocrysts; color index of basalt is approx. 60-70. Welded tuff is 25-40% crystals, mostly of anorthoclase. Quartz forms rare phenocrysts and is common in groundmass.
5.	Ngorongoro; fault scarp north of Lemagrut	Same sequence as (4) overlying lava and greenish-gray trachytic welded tuff.	Greenish-gray trachytic welded tuff has 5-15% phenocrysts of anorthoclase and quartz.
6.	Stream channels at foot of NW slopes of Ngorongoro	Clasts of olivine-rich basalt; pale red welded tuff; purplish-gray and greenish-gray trachyte lava and welded tuff; gray trachyandesite, fine- to coarse-grained.	Same as (4). Same as (4). Same as (3). Same as (2) but commonly coarser-grained.
7.	Naabi Ignimbrite (western part of Main Gorge); erupted from Ngorongoro	Dominantly welded tuff; generally devitrified.	5% phenocrysts of quartz, anorthoclase ($n_z = 1.531$; 5.02% K), augite. Small amount of blue-green hornblende of basement origin. Glass has $n = 1.525$-1.528.
8.	Stream channels at foot of NW slopes of Olmoti	Clasts dominantly medium- to coarse-grained, slightly porphyritic trachyandesite, soda trachyandesite and trachyte lava. Also porphyritic olivine basalt and pale brown coarse trachyte tuff.	One clast of soda trachyandesite has poikilitic hornblende and augite; one of trachyte has nepheline and aegirine. Olivine basalt has large phenocrysts of olivine and augite and a color index of about 75. Groundmass is plagioclase, augite, and olivine with minor biotite. Glass of trachyte tuff is altered to zeolites (principally chabazite).
9.	Lemagrut; stream channels on north side	Clasts are dominantly gray trachyandesite and soda trachyandesite, fine- to coarse-grained, and non-	Trachyandesites and soda trachyandesites have phenocrysts of plagioclase, hornblende, augite, and olivine; groundmass contains in addition anorthoclase and biotite. Olivine basalt has

Table 7. *Composition of Volcanic Rocks in the Olduvai Region* (Continued)

Volcano and locality	Lithology	Mineralogic data[a]
	porphyritic to very porphyritic; common is olivine basalt and rare is trachyte.	phenocrysts of plagioclase, olivine, and augite; same minerals plus anorthoclase are in groundmass. Trachyte has groundmass olivine, augite, hornblende, and biotite.
10. Lemagrut; channel of Side Gorge	Clasts are same as above, with addition of melilitite tuff and greenish-gray porphyritic nephelinite lava.	Nephelinite has phenocrysts of nepheline, pyroxene, and sphene. Melilitite tuff has melilite, nepheline, pyroxene, perovskite, biotite, melanite[a] and sphene.
11. Sadiman; stream channels on north side	Clasts are chiefly porphyritic nephelinite and nepheline phonolite, dark greenish-gray; few are ijolite.	Phenocrysts of nepheline or anorthoclase and pyroxene; microphenocrysts of sphene are common. Groundmass nepheline, anorthoclase, aegirine-augite, and rare perovskite and melanite. Color index is generally 25-50.
12. Laetolil Beds: Laetolil and Silal Artum	Dominantly nepheline melilitite eolian tuffs, light olive gray, massive and cemented by zeolites. Clasts of lava and ijolite.	Principal minerals of tuffs are melilite, nepheline, augite; less common are melanite, perovskite, and biotite; rare are sphene and olivine. Most lava fragments are nephelinite which may have accessory perovskite, melanite, biotite, melilite, and sphene.
13. Kerimasi; exposures in Swallow Crater	Tuff, lapilli tuff, tuff-breccia; dominantly lithic; upper unit also contains limestone clasts.	Dominant minerals are nepheline, melilite, augite (commonly sodic); accessory minerals are perovskite, melanite (abundant in upper unit), ilmenite; relatively rare sphene, olivine, biotite, and apatite.
14. Oldoinyo Lengai; Holocene tuffs on north side	Reworked tuff and lapilli, dominantly lithic.	Dominant minerals are augite (commonly sodic), nepheline, melilite; common accessory minerals are ilmenite, melanite, and perovskite; relatively rare are sphene, biotite, sodalite, olivine.
15. Oldoinyo Lengai; older, yellow tuffs near northern margin of Lake Natron	Tuff.	Dominant minerals are augite (commonly sodic) and nepheline; common accessory minerals are ilmenite, sphene; rare are perovskite and melanite.

[a]*Melanite* is used here to include the related mineral *schorlomite,* which differs from melanite in a higher content of titanium.

Ngorongoro and the Naabi Ignimbrite

Ngorongoro is a broad, deeply eroded volcano noted for the large size of its summit caldera, which is 18 to 22 km in diameter (see Pickering, 1965; Guest et al., 1961). The caldera rim ranges in elevation from approximately 2,100 to 2,400 m, and the floor has an elevation of about 1,700 m. The summit of the volcano may well have had an elevation of 4,500 to 5,500 m prior to the collapse which formed the caldera. The volcano consists largely of lavas, but ignimbrites predominate among the later eruptive rocks. Several relatively small cones on the floor of the caldera represent the latest volcanic activity of Ngorongoro (Pickering, 1965). One of these is Engitati Hill, a relatively flat-topped vitric tuff cone of basaltic composition about 1,100 m in diameter and 60 m high. It is of particular interest because its relatively low, flat top and vitroclastic composition suggest that it was formed by subaqueous eruptions when a lake filled the caldera floor. The caldera floor has no stream outlet, and a saline, alkaline lake occupies the lowest area. Eroded bluffs of horizontal lacustrine sediments as much as 2.5 m thick border the lake on the south and east. These are principally ostracodal clayey limestone or marl, which contains small amounts of biogenic opal and unaltered vitric shards. Rootmarkings of marsh-

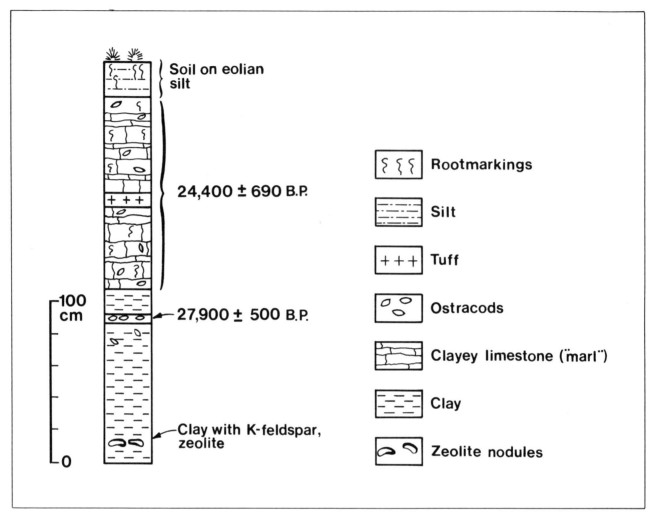

Figure 7. Stratigraphic section and C-14 dates of upper Pleistocene deposits exposed along the eastern margin of the crater lake of Ngorongoro. Exact horizon of sample dated at 24,400 ± 690 yr. B.P. is unknown.

land vegetation are abundant, and there are scattered vertebrate remains and small artifacts. A tuff with unaltered nepheline and glass is interbedded near the middle of these calcareous sediments. These sediments are upper Pleistocene, and ostracodal samples have yielded C-14 dates of 24,400 ± 690 and 27,900 ± 500 yr. B.P. (years before the present) (Appendix A). On the east side of the lake these ostracodal sediments grade down into unfossiliferous claystones with disseminated authigenic zeolites and K-feldspar (Fig. 7). Unusual nodules of zeolite (phillipsite) are found locally in a horizon 50 to 60 cm below the lowermost ostracodal samples. In places the nodule horizon contains broken nodules, fragments of zeolitic tuff, and scattered

artifacts, indicating surface exposure prior to burial by the ostracodal sediments. On the basis of the pattern of alteration and authigenic minerals, the lower part of this stratigraphic section, up to and including the nodule horizon, was deposited in highly saline water, and the overlying ostracodal sediments were deposited in water of lower pH and salinity. Lacustrine tuffs older than these deposits and correlated with the Ndutu Beds crop out at several places on the caldera floor.

Basaltic and trachytic lavas and trachytic welded tuff are the principal rock types exposed on the northwest slopes of Ngorongoro, which supplied sediment to the Olduvai basin. The most widespread of these eruptive rocks is a pale red, crystal-rich quartz trachyte welded tuff (Table 7). It is exposed in valleys on the northwest slope below locality 151 (see Fig. 2) and crops out in fault scarps to the northwest (locs. 152, 153), including the western margin of Olbalbal (loc. 157). It appears to interfinger southward with lavas of Lemagrut. Beneath this welded tuff in a fault scarp at locality 153 are a lava flow and greenish-gray welded tuff. Thin flows of olivine-rich basalt totaling at least 25 m overlie the pale red welded tuff on the northwest slope of Ngorongoro (for example, loc. 150), and similar lava overlies the pale red welded tuff on the fault scarp north of Lemagrut (loc. 157). Stream valleys on the northwest side of Ngorongoro contain trachyte and trachyandesite in addition to olivine basalt and welded tuff.

Potassium-argon dating seems to indicate that the volcano grew to its present size over a period of about 450,000 years, from about 2.45 to 2.0 m.y.a. Rocks exposed in the caldera wall present a difficult problem in dating, for the eighteen dates range from 2.15 to 16.3 m.y.a. (Grommé et al., 1970; Curtis and Hay, 1972). Twelve of the eighteen dates fall between 2.1 and 2.8 m.y.a., and they are considered to be more reasonable than the older dates. Seven of these twelve dates were obtained from plagioclase crystals of a tuff near the base of the section, and the other five are from a fine-grained trachyte lava stratigraphically about 170 m above the tuff (see Fig. 11, Chap. 4). The tuff dates range from 2.23 to 2.81, averaging 2.46 m.y.a.; the dates on the flow range from 2.15 to 2.61, averaging 2.41 m.y.a. The mean of the twelve dates is 2.44 m.y.a. The cause of the variability in these dates is not clear. If the older dates reflect contamination by radiogenic argon, then the younger dates should be the more accurate. If the variability results from heterogeneity in the sample splits used for K and Ar^{40}, the mean should be the more accurate.

The older dates, ranging from 2.9 to 16.3 m.y.a., are all from the plagioclase phenocrysts of thin, porphyritic flows that contain a few inclusions of gneiss. Phenocrysts from a single flow have yielded dates of 2.95, 5.31, and 9.44 m.y.a. (Curtis and Hay, 1972). These variable but old dates most likely reflect variable degassing of phenocrysts that acquired argon in the magma chamber — possibly from the melting of gneiss blocks.

Two additional dates, of 2.0 ± 1 and 2.1 ± 1 m.y.a., were obtained from anorthoclase of the uppermost ignimbrites of Ngorongoro. The date of 2.0 m.y.a. is based on a sample from just beneath Bed I near the Third Fault (Curtis and Hay, 1972); the other, also by G. H. Curtis, is from the pale red, crystal-rich welded tuff at locality 155, near the western margin of Olbalbal (Brock et al., 1972). On the basis of K-Ar dates, a hiatus seems likely between the older eruptive rocks of Ngorongoro, which are chiefly lavas, and the younger ones, which are chiefly ignimbrites.

The magnetic stratigraphy of Ngorongoro is compatible with the K-Ar dating of about 2.45 to 2.0 m.y.a. Rocks with normal polarity are succeeded by rocks with reversed polarity in the caldera wall, and the transition is near the base of the section exposed along the Lerai Road (Grommé et al., 1970). This transition and the K-Ar mean of 2.44 m.y.a. are in reasonably close agreement with the Gauss-Matuyama polarity change at about 2.41 m.y.a. (McDougall and Aziz-ur-Rahman, 1972). Magnetic measurements on rocks of the northwestern slopes of the volcano (loc. 150) appear to bracket a short normal event younger than the reversed welded tuffs. Here a 25-meter-thick sequence of thin basalt flows overlies a reversed ignimbrite, and all of the flows are normal except for the topmost flow, which is reversed. In view of the dates of 2.0 and 2.1 m.y.a. on ignimbrites of Ngorongoro, it is tempting to correlate this normal event with a Reunion event at 2.02 ± .02 m.y.a. (McDougall and Watkins, 1973). It may, however, be correlative with some other short normal event in the lower Matuyama epoch, such as that in the Pribiloff Islands dated at 1.95 to 1.96 m.y.a. (Dalrymple, 1972).

The summit of Ngorongoro probably foundered shortly after discharge of the latest tuffs and lavas (~2.1 to 2.0 m.y.a.), before the present drainage system was established. As evidence, the valleys on Ngorongoro are related to the present shape of the volcano, and none is truncated by the caldera walls. The summit of Ngorongoro appears to have foundered rather quietly, unlike most other volcanoes with large calderas, in which collapse is accompanied by voluminous explosive eruptions. The evidence for this is in the lower part of Bed I to the west of the Fifth Fault, which consists chiefly of Ngorongoro tephra. The only tuff recording a voluminous, unusually explosive eruption is in the lower part of the sequence of Ngorongoro tuffs and presumably antedates the caldera.

Ngorongoro was faulted both early and late in its history. The older faults, which trend northeast-southwest, are found along the northwest side of the caldera rim and nearby on the outer slopes of the volcano. Malanja is a depression on the outer slopes which resulted from this faulting. The older faulting occurred shortly after the volcano had grown to its full size, as based on the subdued, deeply eroded nature of the fault scarp and the fact that stream valleys on the slopes to the northwest of Malanja appear to have developed after faulting. The younger faults, which are dominantly north-south, are found lower on the slopes of the volcano, and at least the lowest of these are related to subsidence of the Olbalbal fault graben. Scarps are little eroded, and the drainage network is disrupted along these younger faults. An example of disrupted drainage is on the uplifted, westward-dipping fault block of Ngorongoro rocks to the north of Lemagrut (Fig. 2, loc. 152). The fault block has deeply incised, presently dry valleys trending northwest that were originally part of the main Ngorongoro network but are presently 50 to 100 m above their former upstream extensions on the main slope of the volcano. Valleys on the main slope now drain into a valley flowing northward along the line of the fault. The uplifted valleys on the fault block north of Lemagrut are important to the present story in showing the drainage pattern which existed for all but the latest of Pleistocene time on this western part of Ngorongoro.

The Naabi Ignimbrite is a quartz trachyte ignimbrite that lies between the basement rocks and Bed I in the western 3 km of the gorge. The name Naabi is taken from Naabi Hill, which lies northwest of Olduvai Gorge at the entrance to the Serengeti National Park. The Ignimbrite is thickest in depressions and thins or pinches out over hills of basement rock, and it laps up over the rise of basement rock that forms the western margin of the Bed I basin (Fig. 4). It must extend even farther to the west, for clasts of the Naabi Ignimbrite are in the riverbed to the west of Granite Falls and the Bed I basin. The maximum exposed thickness is 16 m (loc. 66b), and in a few places the top is eroded beneath Bed I (for example, loc. 66b). In most exposures it is greenish-gray, moderately welded and devitrified; locally the base and top are gray and vitric (n of glass = 1.525-1.528). Minerals of basement origin (principally blue-green hornblende) are in all the samples examined microscopically. Presumably these basement materials were picked up by the deposit in flowing over the land surface.

Ngorongoro is very likely the source of the Naabi Ignimbrite. Quartz-bearing welded tuffs are common among the later eruptives of Ngorongoro, and the Naabi Ignimbrite closely resembles greenish-gray welded tuff exposed in a fault scarp north of Lemagrut (loc. 153). A potassium-argon date of 5.0 m.y.a. was obtained from feldspar of the Naabi Ignimbrite (see Curtis and Hay, 1972). The date seems to argue against Ngorongoro as a source, but blocks of gneiss and crystals of basement origin in the ignimbrite raise doubts about any K-Ar date from this deposit. A. Brock measured reversed polarity on an oriented sample, which fits with the reversed polarity of the uppermost welded tuffs of Ngorongoro.

If, as inferred here, the Naabi Ignimbrite is from Ngorongoro, then it must have flowed uphill at least to some extent in order to lap up over and spread beyond the western margin of the basin later occupied by Bed I. Uphill flow of a large ignimbrite is not unlikely, considering the probable elevation of the crater (~4,500 to 5,500 m) and the relatively small amount of uphill flow required (~100 to 200 m?).

Olmoti and Elanairobi

Olmoti is a large moderately dissected volcano on the northeast flank of Ngorongoro (see Pickering, 1965). It has a summit caldera 6 km across and a maximum elevation of 3,100 m. Stream channels on the northwest slopes contain chiefly lava but also include zeolitic tuff and lapilli tuff (see Table 7). Clasts of basalt are distinguished by large phenocrysts of olivine and augite; at least some of the trachyte is highly sodic and contains nepheline and aegirine. Outcrops of zeolitic lapilli tuff and tuff breccia of trachyte composition are visible in a few places along the western foot of Olmoti. Olmoti erupted trachyandesite lavas and most or all of the coarse pyroclastic deposits in Bed I and the lower part of Bed II. Potassium-argon dates and paleomagnetic measurements from Beds I and II suggest a span on the order of 1.85 to 1.65 m.y.a. for the later of its lavas and for the terminal, explosive eruptive phase of Olmoti. Tuff IB (1.79 m.y.a.) was discharged near the beginning of the explosive phase.

Elanairobi is a large, dissected volcano to the northeast of Olmoti. It has a caldera (Embagai) of about the same size as that of Olmoti, and it appears to be of the same general age as Olmoti and Ngorongoro. Little is known about its composition, although nephelinite tuff and lava are exposed in roadcuts to the east of the caldera.

Lemagrut, Sadiman, and the Laetolil Beds*

Lemagrut is a deeply eroded volcano about 1,500 m high, and it has a maximum elevation of 3,131 m. Its lavas vary widely in composition, although trachyandesite detritus is the most abundant type in stream valleys draining the north half of the volcano. Ignimbrite detritus is apparently absent. The suite of lava types from Lemagrut is readily distinguishable from that of Ngorongoro, both on the relative proportions of different types and on mineralogic differences between the same types (Table 7). Lavas of Lemagrut overlie the pale red, crystal-rich ignimbrite of Ngorongoro in the fault scarp on the north side of Lemagrut. Nephelinite and melilitite pyroclastic rocks are interbedded with lavas on the west and southwest sides of Lemagrut, and clasts eroded from these deposits are in the Side Gorge. Tephra deposits on the southwest side of Lemagrut form at least a major part of the Laetolil Beds, which are discussed in the following paragraphs. Lavas above the Laetolil Beds include trachyandesite and vogesite, and Kent (1941) reports phonolitic agglomerate and dark green olivine nephelinite.

Sadiman is a small, deeply eroded volcano that rises about 450 m above the western flank of Lemagrut. It is formed of tuffs, agglomerates, and subsidiary lava flows of nephelinite and phonolite composition (Pickering, 1964). A stream channel draining the north side of Sadiman contains pebbles, cobbles, and boulders of distinctive dark green and greenish-gray porphyritic nephelinite and nepheline phonolite. Most of these lavas are moderately to highly porphyritic, and they contain large phenocrysts of

either nepheline or anorthoclase, or both. Rounded clasts of Sadiman lava are in the lowest exposed part of Bed I in the eastern part of the Main Gorge, indicating that Sadiman was already present when Bed I was deposited. A K-Ar date of 3.7 m.y.a. was obtained from nepheline crystals of one sample of lava (Curtis and Hay, 1972, K-Ar no. 2238), but this date is not considered acceptable, as nepheline is unreliable for dating, and Ngorongoro appears to be of about the same age as Sadiman.

The Laetolil Beds are fossiliferous pyroclastic deposits, described by Kent (1941), that are interbedded with lavas of Lemagrut on the southwest and west slopes of the volcano. Farther southwest, they overlie basement rocks. Kent gives a thickness of nearly 15 m for exposures at Laetolil, in the valley of the Engarusi River (Fig. 2), which have yielded the bulk of the fossil fauna. I studied and sampled these tuffs and an overlying flow of vogesite, which is at least 8 m thick and has reversed polarity. The tuffs are light olive gray, massive to crudely bedded and contain nodular ledges and anastomosing twiglike structures cemented by calcite. The tuffs consist largely of wind-worked melilitite rock fragments and associated mineral grains of medium-sand size, most of which are coated with clay. Nepheline, melilite, and glass are altered to form a cement of zeolites, as in eolian tuffs of Olduvai Gorge. The accessory minerals form a distinctive suite (Table 7). The coarser clasts include ijolite and nephelinite. Similar but relatively unfossiliferous tuffs as much as 20 m thick crop out at Laitole Spring, known locally as Esére, which lies 9 km south of Laetolil. In 1974 these tuffs were traced 6 km southward to the southern limit of exposure. Southward from Laitole Spring, they are underlain by metamorphic basement rocks.

An estimated 25 to 30 m of reworked tuffs assigned here to the Laetolil Beds are exposed in Silal Artum, a valley on the west side of Lemagrut, near the head of the Side Gorge (Fig. 2). Here they are interbedded near the top of a sequence of lava flows from Lemagrut. The lower one-third of the section consists of massive, medium-grained eolian tuffs lithologically and mineralogically similar to those at Laetolil. The upper part of the section comprises fine- to medium-grained tuffs that include both wind-worked and water-worked deposits. Nephelinite,

*See Note Added in Proof, p. 187.

olivine basalt, and ijolite blocks are present in a few layers. These tuffs are somewhat richer in nepheline and poorer in melilite than the lower one-third of the section. These upper tuffs are extensively altered and cemented by zeolites, with or without calcite. A few bone fragments were noted in the lower part of the section. The tuffs can be traced almost continuously in aerial photographs from Silal Artum to Laetolil.

A 12-meter thickness of Laetolil Beds is exposed near Kelogi (locs. 100, 101a), at the southern margin of the Olduvai basin. These are massive, medium- to coarse-grained zeolitic eolian tuffs of nepheline melilitite composition that are mineralogically similar to the Laetolil Beds in their type area. Angular clasts of ijolite and melilitite lava are as much as 25 cm long. Where Bed I is absent (locs. 101, 101a), the Laetolil Beds underlie the lower part of Bed II. They appear to dip beneath the southernmost exposures of Bed I a short distance to the north. These exposures represent only the upper part of Bed I, but the Laetolil Beds are almost certainly older than all of Bed I. This inference is based on the lack of nephelinite ejecta of Laetolil type in even the lowest part of Bed I.

The Laetolil Beds are composed of tephra erupted either from Lemagrut or Sadiman. As evidence, the large size of many ejecta points to a nearby vent. The Laetolil Beds probably accumulated over a relatively short span of time in view of their highly distinctive and unusual composition and stratigraphic position between lavas of Lemagrut. The vogesite flow which overlies the Laetolil Beds has yielded a K-Ar date of 2.39 ± 0.09 m.y.a. (G. H. Curtis, 1975, personal communication). Very likely the Laetolil Beds are not much older than this lava flow.

A few land snails from the Laetolil Beds (loc. 101) have been identified by B. Verdcourt as follows (1969, personal communication to M. D. Leakey):

OLD/69 Kelogi (3) *Achatina* (*Lissachatina*) very probably a form of *A. fulica* Bowdich. A large range of specimens would be needed to pronounce with certainty.

OLD/69 Kelogi (1) & Kelogi (4) *Burtoa nilotica giraudi* (Bgt.) I have been unable to distinguish from recent shells.

OLD/69 Kelogi (2) *Subulona* sp. unidentifiable without really perfect material.

The Kelogi snails indicate a type of vegetation a good deal drier than the driest evergreen forest — it could in fact be woodland, savanna, or grassland with scattered thickets. The *Burtoa* is common today in the following places — Tabora, Dodoma, Kiboriana Mt., Shinyanga, to mention a few.

Kerimasi and Oldoinyo Lengai

Kerimasi is a slightly eroded volcano about 1,200 m high with a summit elevation of 2,300 m, lying 55 km east of Olduvai Gorge. It was erupted close to the base of a major Rift-Valley scarp, and it buries the escarpment. According to Dawson (1964), the volcano consists of nephelinite pyroclastic rocks overlain by limestone considered to be recrystallized carbonatite ash. I examined these deposits where exposed in Swallow Crater, or Loolmurwak, a maar-type crater 1,000 m in diameter that lies northeast of Kerimasi (Guest et al., 1961; Dawson and Powell, 1969). Deposits of Kerimasi are exposed in the lower 60-90 m of the crater walls, and they are succeeded by 8 to 30 m of tephra erupted from Swallow Crater. The uppermost deposits are eolian tuffs of Oldoinyo Lengai, as much as 5 m thick. The lower three-quarters of the Kerimasi deposits are gray tuffs and tuff-breccias of nepheline melilitite composition. In mineral composition these deposits compare closely with the lower unit of the Masek Beds in Olduvai Gorge, with which they are probably correlative. The uppermost Kerimasi deposits are 15 m of a curious soft, loesslike deposit with abundant blocks of foliated limestone (carbonatite ejecta?) in which calcite replaces a preexisting, bladed carbonate(?) mineral. Nephelinite tuff-breccia is interbedded in the loesslike deposit. Melanite is extremely abundant in the tuff-breccia, suggesting correlation with the Norkilili Member of the Masek Beds. Tephra erupted from Swallow Crater is principally nepheline melilitite tuff and lapilli tuff with abundant coarse crystals of biotite and augite. The overlying eolian tuffs from Oldoinyo Lengai are yellowish-brown massive deposits of nephelinite composition that are correlated with the older, yellow tephra deposits of Oldoinyo Lengai and with the Naisiusiu Beds of Olduvai Gorge.

The Swallow Crater tephra deposits are about 370,000 years old, according to Macintyre et al. (1974). This figure is an average of six K-Ar

18

dates, ranging from 140,000 to 570,000 yr. B.P., that were obtained from lava and tephra deposits of Swallow Crater, and of tephra from nearby, genetically related Kisetey Hill. Kerimasi has not yet been dated radiometrically, but it is bracketed between 0.4 and 1.1 m.y.a. on the basis of its relation to episodes of faulting.

Oldoinyo Lengai is a little-eroded, presently active volcano about 2,100 m high with a summit elevation of 2,887 m that borders Kerimasi on the north. It consists of nephelinite and phonolite pyroclastic rocks interbedded with lavas (Dawson, 1962). Oldoinyo Lengai has erupted sodium-carbonate ash several times since 1917, and sodium carbonate lavas were discharged in 1960. The bulk of the cone consists of yellow pyroclastic deposits of nephelinite and phonolite composition unconformably overlain by dark gray nephelinite tuffs representing the latest eruptive phase. The older, yellow deposits form low hills on the plain north of Oldoinyo Lengai and islands in the southern part of the lake. The younger deposits are represented by an alluvial fan of reworked tuffs on the north side of the island, and by the Namorod ash of the Olduvai region.

Oldoinyo Lengai postdates Swallow Crater, and the older, yellow tephra deposits of Oldoinyo Lengai are mineralogically similar to the Naisiusiu Beds and upper unit of the Ndutu Beds of Olduvai Gorge ($\sim$15,000 to 60,000 yr. B.P.). The Namorod ash was erupted about 1,200 to 1,300 years ago, and calcretes associated with the reworked tuffs to the north of the volcano have given C-14 dates of 1,300 and 2,050 yr. B.P. (Appendix A).

2

PRINCIPLES OF
ENVIRONMENTAL INTERPRETATION

The approach to environmental analysis followed here is based on subdivision of each mappable unit (for example, Bed I) into its lithofacies, which are lithologically different but contemporaneous rock assemblages deposited in a closely related series of environments. Each lithofacies is then analyzed for evidence of (1) paleogeography and physical sedimentation processes, (2) chemical environment of sedimentation and postdepositional processes (for example, weathering), and (3) the fauna and flora. An attempt is also made to estimate the time span represented by major stratal units and paleosols. Data from each facies are then integrated into an environmental synthesis and geologic history for the mappable rock unit over the entire basin.

EVIDENCE FOR PALEOGEOGRAPHY AND SEDIMENTATION PROCESSES

Source areas for clastic sediment vary in their mineral composition (see Tables 6, 7), and thus detrital sediments can generally be assigned to specific sources. Similarly, eruptive sources can be distinguished by the composition of tephra. Sand-size or coarser particles are most diagnostic, but the phyllosilicate, or "clay minerals," of argillaceous rocks may also reflect source areas. Montmorillonite is a characteristic weathering product of mafic volcanic rocks in semiarid climates, and much of that in the Olduvai deposits probably originated in weathering of volcanic rocks, judging from the fact that montmorillonite is commonly associated with coarser detritus of volcanic origin. Much of the illite of the Olduvai deposits was derived from the micaceous basement rocks, as illite is formed readily by weathering of mica, and it is almost invariably the only clay mineral associated with coarser metamorphic detritus in the Olduvai deposits. Clay minerals can, however, be formed

or modified after deposition, hence they cannot be interpreted solely in terms of source rock. Mixed-layer illite-montmorillonite may, for example, be formed by alteration of either illite or montmorillonite, and illite has formed from montmorillonite in some of the Olduvai deposits (Hay, 1970a).

Particle size reflects the strength of waves and currents and the distance from the source area. Clays, for example, are the characteristic detrital sediment of a quiet pond or lake. Within fluviatile and ashfall deposits, particle size normally decreases in a down-current direction. This fact suggests a western sediment source for basement detritus in Beds III and IV, which coarsens from east to west. By contrast, the stream-laid trachytic deposits of Bed I coarsen from west to east, indicating an easterly source.

Sorting of particle size reflects the nature of the transporting medium and the activity of waves and currents. Sediment can be sorted rapidly in aqueous currents, hence the sorting of a sediment commonly reflects the last current action to move and deposit the sediment. Sand deposited on beaches or in shallow water is characteristically better sorted than that deposited in streams, a principle exhibited particularly well in Bed II. Deposits of mass movement, either mudflows or ash flows, are characteristically unsorted and may contain coarse blocks in a fine-grained matrix of clay or volcanic ash.

Pedogenic processes (for example, roots and burrowing organisms) mix materials on the land surface and hence decrease the degree of sorting and stratification. Both ash layers and fluviatile sediments may be "homogenized" by the activity of roots and burrowing organisms. Beds I and II contain massive, poorly sorted tuff beds that were extensively modified by pedogenic processes.

Primary structures are among the most diagnostic features for determining the processes and

20

environment of sedimentation (see Middleton, 1965). Those observed in the Olduvai deposits include cross-bedding, ripple marks, rootmarkings, mud cracks and channeling (cut-and-fill). Channel alignments proved to be highly useful in establishing the stream pattern. A channel orientation alone offers two possible flow directions, as for example a north-south channel, which could represent either a north-flowing or a south-flowing stream. If, however, the source for the detritus in the channel is known, then the flow direction is fixed. Variability of channel alignments may reflect the degree of meandering in a stream.

Deposits of meandering streams can be distinguished from those of braided ones by a combination of textural features and sedimentary structures (see Visher, 1965; Selley, 1970). This is a matter of special importance in interpreting Beds III and IV. The lateral migration in a meandering stream produces a characteristic sequence of grain size and sedimentary structures. The coarsest detritus is at the base, overlying an eroded surface, and the grain sizes decrease upward. Similarly, the scale, or set height, of cross-bedding decreases upward, and at the top micro-cross-laminated and planar-bedded fine sandstones may grade into silts and sands of the floodplain. The thickness of this sequence, if complete, is a measure of the stream depth. Braided river systems consist of an interlaced network of low-sinuosity channels, the deposits of which are typically composed of sand and gravel to the exclusion of silt and clay. It should be noted that the distinction between braided and meandering streams can lose significance from a sedimentologic standpoint in semiarid climates, where a single river may be braided in some places and meandering in others (see Leopold and Wolman, 1957). Beds III-IV appear to contain deposits of rivers of this type.

Eolian sediments of three types can be recognized in the Olduvai Beds: dune deposits, extensive beds of eolian tuff, and localized concentrations of claystone pellets. The large-scale, steeply inclined cross-bedding of dune deposits is exhibited locally in eolian tuffs of the Masek and Ndutu Beds. Laterally extensive beds of massive eolian tuff are the dominant type of eolian deposit in the Olduvai region, and they form a substantial amount of the Laetolil Beds, Bed II, and the Masek, Ndutu, and Naisiusiu Beds. The massive eolian tuffs are deposits of wind-transported tephra, mostly of medium sand size, that accumulated on the land surface, generally as a thin, blanketlike deposit. Claystone pellets of sand size are in many beds, and sand-size mineral grains and rock fragments are commonly coated to a thickness of 10 to 70 μm with claystone or fine-grained tuff that gives the grains a well-rounded pellet shape. The term *pelletoid* is applied here to deposits with a substantial proportion of coated, pellet-shaped clasts. A modern unconsolidated equivalent of the eolian tuffs is found in the surface layer of sediment on the plain to the north of the gorge.

Claystone-pellet aggregates can be massive, thinly bedded, or cross-bedded. The thickest beds are massive and as much as 1 m thick. Claystone-pellet aggregates are often difficult to distinguish from claystones without careful examination, and their common occurrence was not recognized until 1972, ten years after beginning field work at Olduvai. The claystone pellets are generally structureless clasts, rounded to subangular, and 30 to 400 μm in diameter (Plate 10). Detrital sand is commonly associated, and all mixtures of the two can be found. Claystone-pellet aggregates are in Beds I through IV but are most abundant in Bed II. The pellets are almost certainly formed by wind erosion of mudflats, either of a floodplain or at the margin of a lake. Ripples and small dunes of claystone pellets are known from tidal mudflats and intermittently dry lake basins of semiarid regions, where efflorescent salts are important in breaking up of the surface layers of clay into particles small enough to be transported by wind (Price, 1963; Bowler, 1973). At Olduvai, many if not most of the pellet concentrations have fluvial or lacustrine features, indicating that pellets were commonly blown from mudflats into streams or lakes.

EVIDENCE FOR CHEMICAL ENVIRONMENT

Mineralogy can supply much information about the chemical environment of sediments and indirectly about the paleogeography and climate. Saline, alkaline lakes rich in dissolved sodium carbonate-bicarbonate such as lakes Natron and Magadi yield a diagnostic suite of minerals and some very distinctive rock types. Gaylussite

$(Na_2Ca(CO_3)_2.5H_2O)$ and trona $(NaHCO_3.Na_2CO_3)$ are soluble minerals precipitated in sodium-carbonate lakes, and casts and molds of these minerals are in lake deposits of Beds I and II. Sand-size euhedral crystals of calcite are widespread in lake deposits of Beds I and II (Plate 9; Hay, 1973). Although the origin of the calcite is not fully understood, similar crystals of modern age have been found only in saline alkaline lakes, for example, the crater lake of Ngorongoro and of Lake Ndutu to the west of Olduvai Gorge. Beds of fine-grained dolomite also point to saline, alkaline lake water (Friedman and Sanders, 1967). Chert nodules of Beds I and II have unusual shapes and textures (Plate 8) that are diagnostic of an inorganic chemical origin from a sodium-silicate precursor mineral in a saline, alkaline lake (Hay, 1968; Eugster, 1967, 1969). Zeolites, potassium feldspar, searlesite, and fluorite are additional minerals at Olduvai that are characteristically associated with saline, alkaline water (Hay, 1966; Sheppard and Gude, 1968, 1970). All of this mineralogic evidence leaves little room to doubt that the lakes of Beds I and II were chemically similar to saline lakes in the same region today. High salinity of the Pleistocene lakes points to a closed basin in an arid or semiarid climate, as at the present time in the same region.

Stable-isotope ratios (O^{18}/O^{16} and C^{13}/C^{12}) in authigenic lacustrine minerals afford an independent means of estimating the degree of salinity. This salinity determination is based on the principle that the heavier isotopes are concentrated selectively (for example, O^{18} over O^{16}) in the evaporation required to form a brine from dilute meteoric water (rainfall, stream inflow, and so forth). Isotope values of chert nodules from Beds I and II indicate high salinities (O'Neil and Hay, 1973), as do values of euhedral calcite crystals in the same deposits.

Unaltered diatoms, phytoliths, and trachytic glass in lake deposits afford a means of locating areas of relatively fresh water. This is because the opaline silica of diatoms and phytoliths dissolves slowly at a pH of 7 to 9, whereas at a pH of 9.5 or more it dissolves rapidly. The alteration of trachytic glass to zeolites is likewise sensitive to pH, with the rate of reaction increasing rapidly above a pH of 9.

Mineralogic features at Olduvai indicate that many alluvial sediments were exposed to a climate that was at least seasonally dry and in which sodium carbonate was concentrated at the surface by evapotranspiration. Etched quartz grains in the weathered alluvial deposits are evidence for highly alkaline fluids, as quartz is relatively soluble only at a pH above 9.5. Moreover, clay minerals, particularly montmorillonite, have reacted chemically in or near the soil profile to form illite and zeolites, which is the reverse of the usual weathering sequence, and points to a saline, alkaline environment. The zeolites comprise analcime, chabazite, phillipsite, and natrolite, all of which have a high content of alkali ions, especially sodium. Chemically analyzed chabazite from claystones of Bed II is highly unusual in its high content of sodium and low content of silica (see Table 26, Chap. 8). A bright reddish-brown color is produced in most of these deposits by dehydration of brown hydrated ferric oxide to anhydrous ferric oxide (hematite). The end product of this alteration is a reddish-brown, hard, bricklike rock with 20 to 40 percent zeolites (Hay, 1970a). An informative modern example of similar alteration is provided by the deltaic clays of the Peninj River, on the west side of Lake Natron, Tanzania (Hay, 1966, pp. 36-38). The lake is strongly saline and alkaline, and it fluctuates 1 to 2 m in level annually. Modern clays of the delta are reddish-brown, contain zeolites, and are indurated only in the marginal zone flooded at times of high level and dried at low levels. Zeolites are also found in the purely lacustrine clays, which are unconsolidated and gray or brown. Thus the reddish-brown color and induration appear to require either a low or a fluctuating water table in addition to alkaline solutions. Soft brown claystones have been transformed to hard, reddish-brown zeolitic rocks at the surface of the Loboi Plain to the north of Lake Hannington, Kenya (Hay, 1970a). Sodium carbonate concentrated at the surface by evapotranspiration appears to be responsible for the zeolitic alteration, rather than flooding by saline, alkaline lake water. It is not presently clear whether the alteration is late Pleistocene or Holocene. Claystones in both these examples were altered and reddened in a geologically short period of time, unlike the redbeds formed in the Sonoran desert of Baja California (Walker, 1967).

Mineral reactions in land-laid tuffs likewise

document a saline, alkaline soil environment. The more reactive components (glass, nepheline, and so forth) are altered to produce alkali-rich zeolites, calcite, dolomite, among others, which cement the deposit (for example, Table 25, Chap. 7). These altered tuffs are atypical of weathered deposits in their high content of alkali ions (see Table 26, Chap. 8). The rare mineral dawsonite ($NaAl(OH)_2CO_3$) is locally common, and when formed at surface temperatures, dawsonite requires a highly alkaline solution of sodium carbonate (for example, Chesworth, 1971). The clay mineral formed in this alteration is generally a type of mica (illite?) rather than the montmorillonite typically formed in less alkaline environments. Mafic glass in the Masek Beds altered to a mica-type mineral extraordinarily rich in iron (Table 26, no. 8, Chap. 8).

It should be emphasized that authigenic minerals and silicate reactions can be used as evidence for the surface environment only if they can be shown to be penecontemporaneous with deposition. Zeolites, for example, can form long after deposition and provide no information about the environment in which the sediments were deposited. A penecontemporaneous origin can be demonstrated in various ways, but one of the commonest is to find a clast of altered rock (for example, zeolitic tuff) in a deposit (conglomerate) only slightly younger than the altered deposit which supplied the clast. If the authigenic minerals in the clast are abraded at the margin of the clast, then the clast must have been altered before it was eroded and deposited in the conglomerate. Most of the zeolites in Olduvai Gorge were formed penecontemporaneous with deposition, but some others crystallized long after burial. More information about the age and environments of zeolitic alteration at Olduvai Gorge is given elsewhere (Hay, 1966, 1970a).

Caliche soils, or pedogenic concentrations of calcium carbonate ($CaCO_3$), occur throughout the Olduvai sequence and provide another line of evidence for a semiarid climate at Olduvai through the Pleistocene. The caliche layers are of various types, but the most common and distinctive type from the Masek Beds upward is dense, laminated calcrete typically 1 to 3 cm thick. These calcretes are associated with eolian tuffs, and the calcium carbonate of calcretes originated from wind-transported particles of older calcite, from natro-carbonate tephra (see Dawson, 1964), and by chemical reaction of mafic tephra at the land surface (Hay, 1970b). Calcium carbonate was carried downward in solution and precipitated at a horizon of lower permeability — most commonly paleosols and layers of cemented tuff.

FAUNAL, FLORAL, AND ARCHEOLOGIC EVIDENCE

Fossilized remains and organic structures show not only the nature of the biota, but they are among the most sensitive indicators of climate and paleogeography. Faunal remains are both abundant and varied, and most of the bones, even of small fish and rodents, are preserved beautifully. The fauna as a whole can be subdivided into two major units at the disconformity over the Lemuta Member of Bed II. Below this level is a dominantly swamp-dwelling assemblage, whereas above it the proportion of plains animals abruptly increases. The various forms vary greatly in their environmental significance. Large proboscideans, for example, are generally less sensitive environmental indicators than are small rodents. Terrestrial gastropods are among the most sensitive of the faunal elements to climate, with urocyclid slugs, pointing to perennially damp conditions, known from Bed I, and *Limicolaria,* pointing to arid or semiarid conditions, known from the Masek, Ndutu, and Naisiusiu Beds. It should be emphasized that the faunal variations are of limited value in distinguishing purely climatic fluctuations from paleogeographic changes caused by faulting.

Amino acids in fossilized bones have been used to determine the late Pleistocene temperature history. This determination is based on the fact that only the L-amino acids are commonly found in living organisms, whereas the amino acids in fossil shells, bones, and so forth, are partly racemized, which means that some of the L-amino acids have been converted to the D-isomers. The amount of racemization is a function of temperature and age, and if the age of the fossil is known by an independent method (for example, C-14), the mean temperature of the bone can be calculated (Bada and Protsch, 1973). From bones of different ages, changes in temperature can be determined. Where the tem-

23

perature history is known, the racemization can be used as a method of dating (see Schroeder and Bada, 1973).

Fossilized leaves are rare, and little is known of the pollen at Olduvai. Rootmarkings are, however, both widespread and abundant and serve to distinguish savannah grassland from swamp and shore vegetation. Rootmarkings diagnostic of savannah grassland vegetation are narrow, finely textured, and branching; swamp and shore vegetation (for example, *Typha* and some types of grasses and reeds) is indicated by subhorizontal rhizomes and by coarse, generally unbranching vertical channels and casts. Diatoms and opaline encysting cases of chrysophyte algae are in some of the lake-margin deposits (see Hay, 1973). Siliceous phytoliths, or particles of "plant opal," although common in many beds, have not been studied by specialists.

Hominid activities also have a bearing on the geologic interpretation. Undisturbed occupation sites, for example, indicate emergent land surfaces. Hominid transport must be considered in accounting for the location of a single or very small number of clasts which do not fit the inferred paleoglography. As an example, a very few unmodified chert nodules are in the Sandy Conglomerate overlying the disconformity in Bed II at HWK-E. The nodules are of the type formed in saline, alkaline lake water, but all other evidence at HWK points to relatively fresh lake water. Chert tools are abundant in the same bed, and it seems more likely that the unmodified nodules were carried here by hominids rather than formed in this area.

ESTIMATING ELAPSED TIME

Any geophysical method for age determination is potentially useful in determining the interval of time and hence the average sedimentation rate between two horizons. Potassium-argon dates combined with magnetic stratigraphy were used to estimate the time represented by the deposits of Bed I above the lavas ($\sim$75,000 years). A few time-dependent features can be used for time estimates in the absence of geophysical measurements. Silicate reactions are generally rather slow at the land surface, and the extent of zeolitic alteration in eolian tuffs or of clay formed in paleosols can provide a basis for estimating at least the minimum amount of time represented by the observed change. The extent of change is, of course, dependent on the chemical environment as well as time. From study of the Namorod Ash and the Naisiusiu Beds, supplemented by other data (Hay, 1966), it appears that about 10,000 years are required to convert unconsolidated tephra to yellowish-brown, hard zeolite-cemented tuff in the present alkaline soil environment of the Serengeti Plain.

The degree to which mineral grains (for example, quartz or nepheline) are altered can be used as a measure of time. This method is generally satisfactory only for comparing the ages of sediments deposited in the same environment and exposed to the same postdepositional conditions. If one of two such deposits is dated by an independent method (for example, C-14), the age of the other can be estimated from the alteration. As an example, the upper unit of the Ndutu Beds was dated by comparing the alteration of its nepheline crystals with those of the Naisiusiu Beds (M. D. Leakey et al., 1972). Stratal thickness can be used to some extent in estimating time, but its use at Olduvai is complicated by the fact that faulting from Bed II onward has resulted in different sedimentation rates for different areas. Thus valid time estimates require stratal measurements over a large area.

3

STRATIGRAPHIC SUMMARY OF THE OLDUVAI BEDS

The sequence of deposits exposed in Olduvai Gorge, termed the Olduvai Beds, is as much as 100 m thick. This sequence was deposited in a shallow basin about 25 km in diameter which is bordered and underlain on the southeast by late Cenozoic volcanic rocks and on the north and west by metamorphic basement rocks. The Olduvai Beds are subdivided into seven mappable units, or formations, that are listed from oldest to youngest as follows: Bed I, Bed II, Bed III, Bed IV, the Masek Beds, the Ndutu Beds, and the Naisiusiu Beds. These deposits constitute a relatively complete stratigraphic record of the period from about 2.1 m.y.a. to 15,000 yr. B.P. Sediments of Holocene age ($\leqslant$10,000 yr. B.P.) are also present in and near the gorge.

Bed I is the thickest of the major subdivisions. It is as much as 60 m thick in the eastern part of the Main Gorge, where it consists of lavas overlain by sedimentary deposits. The lavas pinch out in a westward direction, and Bed I has a maximum exposed thickness of 43 m in the western part of the basin. Several widespread tuffs serve as marker beds for correlating within Bed I. The oldest deposits of Bed I are found only in the western part of the basin, and are on the order to 2.0 to 2.1 million years old; the top of Bed I has an age of about 1.70 million years. All of the known archeologic sites and nearly all of the faunal remains are in deposits ranging from about 1.85 to 1.70 m.y.a. The basin was slightly warped, and its eastern margin may have faulted during the deposition of Bed I.

Bed I is subdivided into five lithofacies, which comprise the lava flows and deposits of a lake, lake-margin terrain, an alluvial fan, and an alluvial plain (see Fig. 14, Chap. 4). The lake was broad, shallow, and fluctuated considerably in level and extent (see Fig. 18, Chap. 4). Lake-margin sediments were deposited on low-lying terrain intermittently flooded by the lake. Lava flows and alluvial-fan deposits are found along the eastern margin of the basin, and alluvial-plain deposits are exposed only in the western half of the basin, where they underlie the lake- and lake-margin deposits. The lake deposits consist chiefly of claystone and contain a wide variety of authigenic minerals which show that the lake water was highly saline and alkaline. Lake-margin deposits are chiefly tuff and claystone around the southeast margin of the lake and sandstone and claystone around the western margin. Water flooding the southeastern lake-margin terrain was relatively fresh, and deposits of this area contain the vast majority of faunal remains and archeologic sites of Bed I (see Fig. 17, Chap. 4; Table 32, Chap. 12). The lake-margin deposits interfinger eastward with alluvial-fan deposits that consist largely of stream-worked tuffs and related pyroclastic materials originating in explosive eruptions of Olmoti. The lava flows form a thick sequence which underlies the lake-margin and alluvial-fan deposits. The lower of the lava flows are from Olmoti, and the upper flows are from one or more vents to the south of the gorge. The alluvial-plain deposits are principally tuff and claystone. The tuffs originated principally in eruptions of Ngorongoro, and much of the ash was redeposited by wind.

Bed II is 20 to 30 m thick and is found over a slightly larger area than Bed I. It conformably overlies Bed I, and where the marker tuff at the top of Bed I is missing, the contact between the two can be difficult to locate. A regional disconformity subdivides Bed II into two major units differing in lithology and paleogeographic setting. Deposits below the disconformity have many features in common with Bed I, whereas the overlying deposits are in many respects unique within the Olduvai sequence. Many individual beds, particularly tuffs, can be used for correlating, and the Lemuta Member is a widespread distinctive sequence of tuffs which underlies the disconformity in the eastern part of the basin. The disconformity marks the beginning of widespread faulting and folding in the central part of the Olduvai basin, which is reflected in an increased area of grassland and a decrease in the size of the lake. The disconformity also marks an abrupt faunal change in Bed II: water- and swamp-dwelling forms predominate below the disconformity, and forms favoring open savannah and riverine conditions predominate above it. Deposits of Beds II span the period from 1.70 to about 1.15 m.y.a., and the disconformity has an age of about 1.60 m.y.a.

Bed II is geologically of special interest because of its many lithofacies and environments. The lower part of Bed II, beneath the disconformity, comprises lake deposits, lake-margin deposits, alluvial-fan deposits, and eolian deposits (see Fig. 23, Chap. 5). Lake deposits are principally claystone and occupy a broad area in the central part of the basin (see Fig. 38, Chap. 5). The lake was saline, alkaline, and fluctuated in level and extent. Lake deposits are bordered by lake-margin deposits, which are principally claystone and tuff and contain the bulk of the faunal remains known from the lower part of Bed II. Lake-margin deposits interfinger eastward with alluvial-fan and eolian deposits. The alluvial-fan deposits consist principally of stream-worked tuff which represents a continuation of the Bed I eruptive phase of Olmoti. These deposits are overlain by the eolian deposits, which are principally wind-worked tuffs and constitute most of the Lemuta Member. The lower eolian deposits consist principally of detritus eroded by wind from the surface of the alluvial fan; the upper eolian deposits are principally tephra from an unknown source to the east or northeast.

Above the disconformity are lake deposits, a mixed assemblage of fluvial-lacustrine deposits, and fluvial deposits. The eastern and western fluvial-lacustrine deposits differ in lithology and environment and are classed as separate facies. Lake deposits are principally claystone and have mineralogic features which show that the lake was saline. The lake occupied a much smaller area than the lake which existed prior to the disconformity, and the saline lake further decreased in size upward within Bed II and disappeared shortly before the end of Bed II deposi-

tion. Lake deposits are bordered by the fluvial-lacustrine deposits. The eastern fluvial-lacustrine deposits consist principally of claystone and sandstone and record a complex, shifting pattern of sedimentation in which lacustrine and lake-margin sediments alternate with fluvial and eolian sediments. Fluvial sandstones and conglomerates are concentrated in a main channelway that drained into the east end of the lake (see Fig. 34, Chap. 5). The channelway was fed by tributary streams from both the volcanic highlands to the southeast and basement terrain to the north and northwest. This lithofacies has most of the known faunal remains of Bed II above the disconformity. The western fluvial-lacustrine deposits consist principally of sandstone, and carbonate rocks and claystone form the bulk of the remainder. The proportion of lacustrine sediment is highest in the lower part of the facies, and fluvial sediments are most abundant in the upper part. Faunal remains occur widely but are less common than in the eastern fluvial-lacustrine deposits. A fluvial facies is found only in the eastern part of the basin and is termed the eastern fluvial deposits. These deposits consist of claystone and lesser amounts of sandstone, conglomerate, and mudflow deposits.

Bed II is particularly important from the standpoint of archeology, because it contains numerous sites representing three Paleolithic industries: Oldowan, Developed Oldowan, and Acheulian. Only Oldowan sites are known to occur in the lower part of Bed II, whereas Developed Oldowan and Acheulian sites occur in the middle and upper parts of Bed II. Remains of *Homo habilis* have been found in the lower part of Bed II, and remains of *Homo erectus* have been found near the top of Bed II. *Australopithecus* cf. *boisei* occurs both high and low in Bed II. About two-thirds of the known archeologic sites are in the eastern fluvial-lacustrine deposits, and nearly all of the remainder are either in lake-margin deposits or in the western fluvial-lacustrine deposits (see Table 32, Chap. 12). Developed Oldowan sites were in general situated closer to the lake than were coeval Acheulian sites.

A widespread episode of faulting affected the Olduvai basin about 1.15 m.y.a., causing erosion of Bed II and changing the paleogeography. Beds III and IV were deposited on this changed terrain. Beds III and IV are distinguishable stratigraphic units only in the eastern part of the basin, where Bed III is 4.5 to 11 m thick and Bed IV is 2.4 to 10 m thick. The thickness of Beds III and IV differs on different fault blocks, pointing to contemporaneous fault movements (see Fig. 42, Chap. 6). Bed III is dominantly a reddish-brown deposit, chiefly of volcanic detritus, whereas Bed IV is dominantly a gray or brown deposit containing both metamorphic and volcanic detritus. Beds III and IV can be subdivided and correlated only to a limited extent on the basis of marker beds. Bed III interfingers to the northwest with sediments similar to those of Bed IV, and from the zone of interfingering to the western margin of the basin, Beds III and IV are no longer recognizable units. Here they are combined into a single unit, termed Beds III-IV (und.). Bed III spans the period from about 1.15 to 0.80 m.y.a., and Bed IV spans the period from about 0.80 to 0.60 m.y.a.

Beds III and IV were deposited by streams on an alluvial plain which received metamorphic detritus from the west and northwest and volcanic detritus from the south and east. These deposits are subdivided into three lithofacies: eastern fluvial deposits, western fluvial deposits, and fluvial-lacustrine deposits. Claystone is the principal type of sediment, and sandstone is next in abundance in all three lithofacies. The eastern fluvial deposits are reddish-brown, consist of volcanic detritus, and were laid down on an alluvial surface by braided streams that flowed intermittently. Bed III constitutes the bulk of these deposits. The western fluvial deposits are dominantly gray and brown, consist chiefly of metamorphic detritus, and were deposited on floodplains and in channels of meandering streams that flowed through much of the year. Streams from the volcanic highlands and from the metamorphic terrain to the northwest joined to form a main trunk stream or drainageway which flowed to the east into a semilacustrine drainage sump, which for Bed III lay about 7 km northeast of the sump for the upper part of Bed II (see Fig. 45, Chap. 6). The paleogeography of Beds III and IV differs primarily in the location of the main drainageway, which for Bed IV was about 0.5 km south of the Bed III drainageway. The fluvial-lacustrine facies accumulated in the Bed III drainage sump. These sediments are a mixture of metamorphic and volcanic detritus, and they exhibit fluvial and lacustrine features. The Bed IV sump lay to the southeast of the Bed III sump, and its sediments remain hidden beneath younger deposits (see Fig. 46, Chap. 6). Most of the faunal remains of Beds III and IV are found in the western fluvial deposits, and they are concentrated in sediments of the main drainageway. Numerous archeologic sites are in Beds III and IV, and these represent both Developed Oldowan and Acheulian industries. About three-quarters of the archeologic sites are in sediments of Bed IV age, and about one-quarter are in deposits correlative with Bed III. Sites of both ages are concentrated in sediments of the main drainageway and western alluvial plain (see Figs. 43, 44, Chap. 6; Table 32, Chap. 12).

The Masek Beds are the latest deposits prior to erosion of the gorge. They have a maximum thickness of about 25 m and occur over a slightly larger area than Beds I and II. The thickness of the Masek Beds varies strikingly in relation to faults, with 12 to 15 m common on the downthrown and 1.5 to 4.5 m common on the upthrown side. This pattern is proof of contemporaneous fault movements, and displacements are documented for all faults along which the thickness of the Masek Beds is known (see Fig. 47, Chap. 7). The Masek Beds disconformably overlie Beds III-IV in most places, and in a few places the Masek Beds fill channels 5 to 6 m deep cut into the underlying deposits. Where exposed in the gorge, the Masek Beds consist of about equal amounts of eolian tuff and detrital sediment. The Masek Beds are subdivided into two units, the lower of which is the thicker. The thinner, upper unit is named the Norkilili Member. Two marker tuffs can be recognized widely in the lower unit. The estimated time span of the Masek Beds is from about 0.6 to 0.4 m.y.a.

The Olduvai basin was an alluvial plain during the deposition of the Masek Beds, and the paleogeography resembled that of Beds III and IV except for the location of the main drainageway, which lay to the south of the drainageways for Beds III and IV (see Fig. 51, Chap. 7). The drainage sump lay between the First and Second faults, to the southeast of the sumps for Beds III and IV. The lower unit of the Masek Beds comprises an eastern fluvial facies, a western fluvial facies, and an eolian facies. The two fluvial facies are similar in most respects to the eastern and western fluvial facies of Beds III and IV. The eolian facies consists almost entirely of wind-worked tephra erupted from Kerimasi. Deposits of the eolian and western fluvial facies are interbedded over broad areas, unlike the lithofacies of Bed I through Beds III-IV, which are restricted to different areas. The Norkilili Member consists largely of eolian tuff and is regarded as a single lithofacies. Vertebrate remains are relatively scarce in the Masek Beds and have been found only in the western fluvial deposits. One archeologic site, of the Acheulian industry, is known, and it is in sediments of the main drainageway.

The Ndutu Beds were deposited over a lengthy period of intermittent faulting, erosion, and partial filling of the gorge (see Fig. 57, Chap. 8). The Olbalbal graben had now become the drainage sump, and it has remained the lowest part of the basin up to the present time. The Ndutu Beds are subdivided into two units, the lower of which is represented by relatively small patches of sandstone, conglomerate, and tuff at various levels in the sides of the gorge. The maximum thickness of sediments in any one place is 14 m. The upper unit consists principally of eolian tuffs, which were deposited after the gorge had been partly filled by fluvial sediments of the lower unit to make a relatively wide valley bottom (see Fig. 54, Chap. 8). Tuffs of the upper unit accumulated to a maximum thickness of 24 m over the sides and in the bottom of the gorge. The upper and lower units in the gorge each contain one marker tuff. The Ndutu Beds are represented on the plain by a relatively thin layer of eolian tuff. Oldoinyo Lengai was the principal source of tephra in the Ndutu Beds. Deposition of the Ndutu Beds was terminated by a final, severe phase of faulting which resulted in further subsidence of the Olbalbal graben and in deeper erosion of the gorge. The upper unit spans the period from about 32,000 to 60,000 yr. B.P., and the lower unit presumably spans most of the interval between 60,000 and 400,000 yr. B.P.

Widely scattered vertebrate remains are in both the upper and lower units of the Ndutu Beds. Two archeologic sites with rather nondescript material of M.S.A. (Middle Stone Age)

affinity have been found in the Ndutu Beds, one near the top of the upper unit, and the other near the top of the lower unit.

The Naisiusiu Beds were deposited after the upper unit of the Ndutu Beds had been largely stripped away and the gorge had been eroded to very nearly its present level. The Naisiusiu Beds consist chiefly of eolian tuff, which occurs widely in the sides and bottom of the gorge and over the plain. Sandstones and conglomerates are exposed locally in or near the bottom of the gorge. Most exposures are between 50 cm and 3 m in thickness, and the thickest section is 10 m. One marker bed has been identified. Eruptions of Oldoinyo Lengai supplied the tephra of the Naisiusiu Beds. Radiocarbon dates suggest that the Naisiusiu Beds were deposited from about 22,000 to 15,000 yr. B.P. The one known archeologic site contains a microlithic assemblage and has an age of about 17,000 yr. B.P. A skeleton of *Homo sapiens* (the human skeleton reported by Reck, 1914a) from an intrusive burial in the sides of the gorge yielded a C-14 date of $16,920 \pm 920$ yr. B.P. (Protsch, 1973), indicating that it falls within the age limits of the Naisiusiu Beds.

The Naisiusiu Beds were eroded and the river further entrenched in early Holocene time, and in late Holocene time fluvial sediment accumulated to a thickness of 2 to 5 m in the bottom of the gorge. These fluvial sediments are presently being eroded. An ash deposit termed the Namorod Ash is interbedded near the top of this alluvial sequence. Olbalbal contains an assortment of fluvial, eolian, and semilacustrine sediments of Holocene age. Holocene sediments of the plain comprise a laminated calcrete overlain by an unconsolidated deposit of wind-worked Namorod Ash intermixed with older tephra and detrital sediment. To the north of the gorge the unconsolidated sediment is represented by a belt with barchan dunes and their trailing ridges. Both the Namorod Ash and the older tephra were erupted from Oldoinyo Lengai. Fluvial sediments beneath the Namorod Ash in the bottom of the gorge have yielded a C-14 date of $1,370 \pm 40$ yr. B.P., and calcretes of the plain have given C-14 dates ranging from $2,190 \pm 105$ to $9,130 \pm 130$ yr. B.P.

28

4

BED I

STRATIGRAPHY AND DISTRIBUTION

Bed I is the lowermost mappable unit of the Olduvai Beds, and Reck (1951) defined it as the sequence of tuffs between lava flows in the bottom of gorge and "lacustrine marls" of Bed II. This description seems to fit best the relationships at sites HWK (loc. 42) and FLK (loc. 45), where a widespread tuff (Tuff IF of this report) would be the uppermost stratum of Bed I, as L. S. B. Leakey has so considered it (1965). Bed I by this definition forms only the upper part of a conformable sequence of beds to the west of the Fifth Fault, west of the limit of the lava flows where the base of Bed I as defined by Reck cannot be recognized. In order to make Bed I a mappable unit where the lavas are absent, it was redefined (Hay, 1963a) to include the entire sequence of tuffs and clays between the Naabi Ignimbrite and Bed II (Figs. 8, 9). Where exposed fully in the western part of the gorge, its maximum thickness is approximately 43 m.

Bed I in the eastern part of the gorge comprises a series of lavas overlain by as much as 40 m of sedimentary deposits — mostly reworked tephra. The lavas total 21 m in the one place where their full thickness is exposed, near the Third Fault (loc. 9a). Here they are underlain by 2.4 m of tuff and claystone exposed in an excavation. Beneath the tuff and claystone is a quartz trachyte ignimbrite. Earlier I designated the tuffs and claystones beneath the lavas as the lower Member of Bed I (Hay, 1967a). The lavas were termed the Basalt Member, and the beds above were named the Upper Member. These terms have not proved useful and will no longer be used (Fig. 5, Chap. 1).

The ignimbrite beneath Bed I near the Third Fault is thought to be from Ngorongoro in view of its quartz trachyte composition and K-Ar date of 2.0 m.y.a. It is much closer in its mineral composition to the pale red welded tuff of Ngorongoro than it is to the Naabi Ignimbrite, and very probably it interfingers westward into the lower part of Bed I.

Bed I occupies most of the Olduvai basin, and it may extend northeast into the southern part of the basin of Lake Natron (Fig. 1, Chap. 1). The southern and western margins of Bed I are exposed in only a few places. Bed I pinches out against basement rocks and the Naabi Ignimbrite in the western part of the gorge. To the southwest, near the Kelogi inselbergs (locs. 99 to 101), it pinches out over the Laetolil Beds. Tephra deposits of Bed I also pinch out between the mouth of the gorge and locality 157, 2 km to the south, where tuffs representing the lower part of Bed II overlie a mafic trachyandesite lava flow mineralogically similar to the lowermost flows of Bed I at the Second and Third faults (Fig. 10). If, as seems likely, this is a Bed I flow, then the lack of Bed I tephra implies offset by faulting during or shortly after deposition of Bed I. Bed I extends at least 15 km to the northeast of the gorge. Bed I lavas 45 to 60 m thick are exposed in fault scarps 5 to 12 km north of the gorge, and tuffs and claystones crop out along the east-west fault that parallels the gorge 5 km to the north (locs. 201, 202). The most distant outcrops observed are 15 km to the northeast (loc. 200), where the uppermost 7 m of Bed I are exposed. Hills of basement rock 12 km north of the gorge very likely form the northern boundary of the Bed I sedimentary basin, although deposits of the same age may possibly underlie the plains area separating these hills from the highlands 3 to 5 km farther north.

Tuffs provide the principal basis for correlations within Bed I. The six tuffs most useful in correlating have been designated Tuff IA, IB, IC, ID, IE, and IF (Hay, 1971). Tuffs IA, IC, ID, and IF are air-fall tuffs. Tuff IB comprises an ash-flow tuff, both primary and reworked

BED I

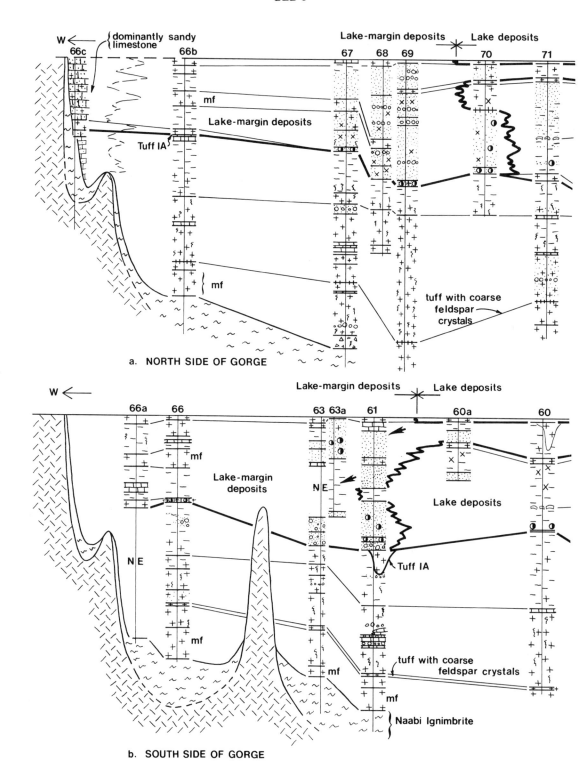

a. NORTH SIDE OF GORGE

b. SOUTH SIDE OF GORGE

30

BED I

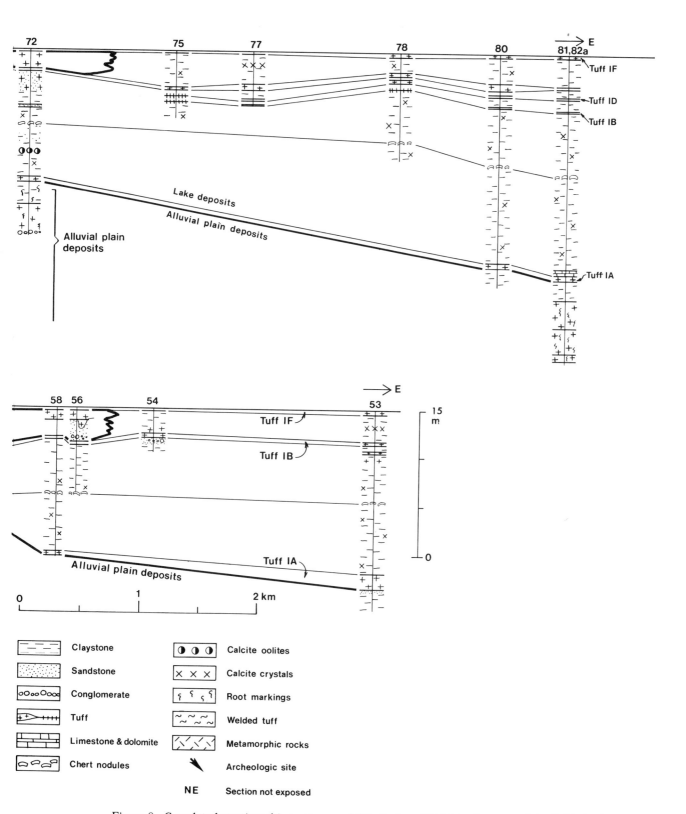

Figure 8. Correlated stratigraphic sections and lithofacies of Bed I to the west of the Fifth Fault. Top of Tuff IF is taken as a horizontal datum.

31

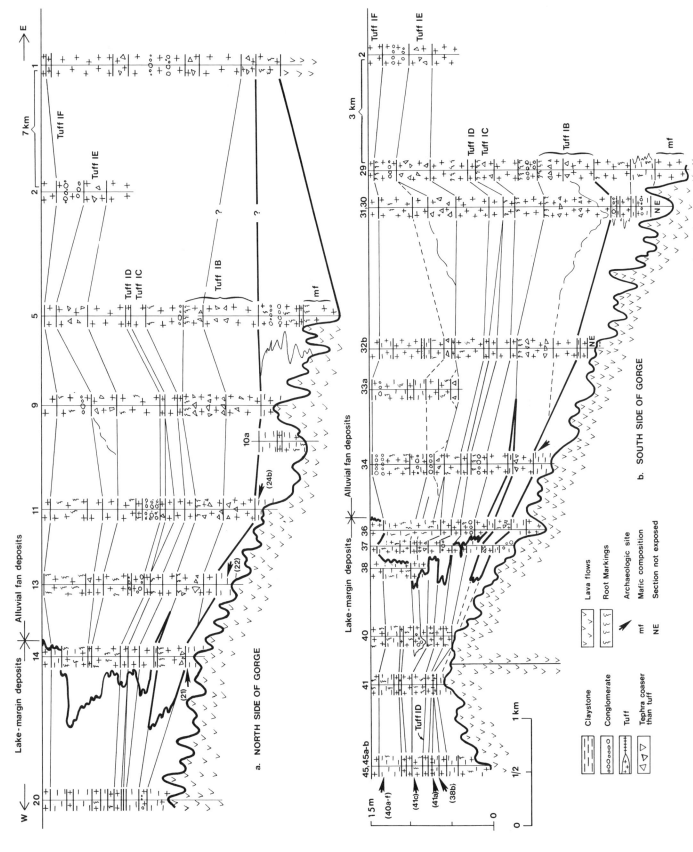

Figure 9. Correlated stratigraphic sections of Bed I in the eastern part of the Main Gorge. Top of Tuff IF is taken as a horizontal datum. Bracketed numbers represent archeologic sites of M. D. Leakey (1971a).

32

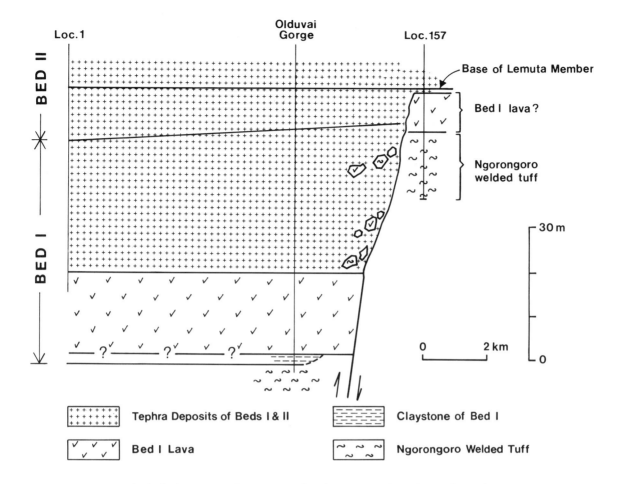

Loc.1   Olduvai Gorge   Loc.157

Base of Lemuta Member

Bed I lava?

Ngorongoro welded tuff

30 m

0   2 km

Tephra Deposits of Beds I & II   Claystone of Bed I

Bed I Lava   Ngorongoro Welded Tuff

Figure 10. Diagram in which faulting is used to account for the stratigraphic relationships of Bed I and the Lower part of Bed II along the western margin of Olbalbal. The base of the Lemuta Member is taken as a horizontal datum. The Olduvai Gorge section is a composite, based on exposures at localities 2, 5, 9, and 9a. In this explanation, a scarp was maintained by fault displacements during eruption of the Bed I tephra deposits.

(Plate 1), and a widespread air-fall tuff. Tuff IE is an ash-flow tuff. The oldest of the artifacts and hominid remains lie a short distance beneath Tuff IB; the others lie at varying levels between Tuffs IB and IF. The lithology and distribution of the marker tuffs are summarized in the following paragraphs, and salient mineralogic data are given in Table 8. Color of the marker tuffs

Table 8. *Mineralogic Data Regarding Marker Tuffs of Bed I*

Tuff	Distribution	Composition and properties of feldspar[a]	Mafic minerals	Refractive index of glass
Tuff IA	West of Fifth Fault (e.g., loc. 80)	Dominantly andesine/labradorite	Rare	Dominantly mafic glass; $n \sim 1.605$
Tuff IB	a. Basal 2 cm (ash-fall tuff) in vicinity of Second Fault	Anorthoclase commonly 3-4 mm long; $n_X = 1.524$, $n_Z = 1.532$	Augite; $n_X = 1.710$	Glass is altered
	b. Ignimbrite, eastern part of Main Gorge	Anorthoclase 2-3 mm long; $n_X = 1.526$, $n_Z = 1.534$; 3.2-3.5% K; $Or_{23}Ab_{74}An_3$ (average of 2 analyses)	Augite; $n_X = 1.707$	$n = 1.528$
	c. Ash-fall tuff, vicinity of FLK	Anorthoclase; $n_X = 1.526$, $n_Z = 1.534$	Augite; $n_X = 1.707$	Glass is altered

33

Table 8. *Mineralogic Data Regarding Marker Tuffs of Bed I (Continued)*

Tuff	Distribution	Composition and properties of feldspar[a]	Mafic minerals	Refractive index of glass
	d. Basal part of ash-fall tuff, west of Fifth Fault (e.g., loc. 80)	Anorthoclase; $n_X = 1.524$; $n_Z = 1.532$	Augite; $n_X = 1.710$	Glass is altered
	e. Upper part ash-fall tuff, (loc. of d)	Anorthoclase; $n_X = 1.526$; $n_Z = 1.534$	Augite; $n_X = 1.707$	Glass is altered
Tuff IC	FLK (loc. 45) and area to east; also loc. 49	Anorthoclase; $n_Z = 1.535$; 2.0-2.2% K	Augite; n_X n.d.	$n = 1.530$
Tuff ID	a. FLK (loc. 45) and area to east	Dominantly oligoclase 2-4 mm long; $n_X = 1.538$-1.540; $n_Z = 1.544$-1.550; 0.8-0.9% K	Augite; $n_X = 1.700$	$n = 1.524$
	b. West of Fifth Fault (e.g., locs. 49, 80)	Dominantly oligoclase; 0.7-0.8% K	Augite dominant; little brown hornblende and olivine	Glass is altered
Tuff IE	Eastern part of Main Gorge	Anorthoclase 1-3 mm long; $n_X = 1.526$, $n_Z = 1.536$; 2.27% K	Augite dominant; little brown hornblende and olivine	$n = 1.525$
Tuff IF	a. FLK (loc. 45) and area to east	Anorthoclase 1-2 mm long; $n_X = 1.527$; $n_Z = 1.539$; 2.74% K	Augite dominant; $n_X = 1.698$; brown hornblende common	1.527
	b. Kelogi (loc. 99)	Anorthoclase; $n_X = 1.526$; $n_Z = 1.534$; 3.15% K	n.d.	1.525
	c. Near Fifth Fault (e.g., locs. 49, 80)	Anorthoclase 0.5-1.0 mm long; $n_X = 1.526$-1.530; $n_Z = 1.536$-1.540; 2.48% K	Augite dominant; $n_X = 1.695$-1.700; brown hornblende common	Glass is altered
	d. West of Fifth Fault (locs. 58-66)	Composition varies at different localities; e.g., dom. andesine (An_{40}) at loc. 58; oligoclase (An_{10-20}) at loc. 66; anorthoclase ($n_X = 1.525$; $n_Z = 1.533$) at locs. 61, 64.	Augite; $n_X = 1.705$ at loc. 66; brown hornblende present locally	1.525-1.526 (locs. 61, 64, 66)

[a]Refractive indices are considered accurate ± 0.002. Potassium values were obtained with flame photometer in the course of K-Ar dating. The composition of the Tuff IB ignimbrite was determined by flame photometer and x-ray fluorescence.

depends largely on the nature and extent of alteration and thus is of little value in correlating. The tuffs are pale gray to light olive gray where unaltered. They are yellow where altered to zeolites in the eastern part of the Main Gorge, and they can be white, yellow, or orange to the west of the Fifth Fault where altered to zeolites or to K-feldspar.

Tuff IA

This is a fine-to medium-grained vitric tuff found only to the west of the Fifth Fault. It is 60 to 150 cm thick and is laminated or ripple-

marked in the eastern exposures (for example, locs. 53, 80, 81) and generally massive and root-marked to the west (locs. 58, 71, 72). The tuff is at least 90 percent vitric shards, which are of both basalt (or trachyandesite) and trachyte composition. Plagioclase forms most of the remainder. Glass is nearly everywhere altered to zeolites, with or without montmorillonite.

Tuff IB

This is vitric trachyte tuff and lapilli tuff found over a distance of 13 km westward from the Second Fault. Between the Second Fault and

DK (loc. 13) it is a compound deposit 4.5 to 7 m thick and comprising a thin (5 to 8 cm) ash-fall stratum overlain by an ash-flow deposit as much as 4 m thick and containing abundant pumice bombs as much as 10 cm in diameter. A deposit of stream-worked ash-flow materials overlies the primary deposit and extends westward. A paleosol is developed over the top of these deposits. Farther west, in the vicinity of localities 20, 41, and 45, Tuff IB is an ash-fall tuff, both primary and reworked, that is 50 cm to 1 m thick. Locally it is graded from a basal crystal-rich layer up into finer-grained vitric tuff. To the west of the Fifth Fault it is about 15 cm thick and graded. Tuff IB averages 80 to 95 percent glass, and most of the remainder is anorthoclase. Both the minerals and a chemical analysis indicate a sodic trachyte composition (see Hay, 1963b). The lowermost ash-flow deposit in locality 1, to the north of the gorge, may well represent Tuff IB. Correlation is strongly suggested both by lithologic similarity and by similar refractive indices of glass and anorthoclase.

Tuff IC

This is a medium- to coarse-grained vitric tuff of trachyte composition. It can be traced from the Second Fault west to FLK (locs. 45, 45b) and was identified on the basis of mineralogy much farther to the west (loc. 49). It is generally stream-worked and 1.5 to 2 m thick in its eastern exposures, but to the west (locs. 20, 45), it is an ash-fall tuff about 30 cm thick that has been only slightly reworked. Pumice bombs and lapilli are common in the stream-worked parts of the deposit but are absent in the ash-fall tuff.

Tuff ID

This is a coarse-grained crystal-vitric tuff and lapilli tuff which extends continuously between the Second Fault and FLK (loc. 45). It is a primary ash-fall deposit about a meter thick which commonly has a basal zone of unconsolidated ash and lapilli that becomes finer grained upward. Locally it is reworked, and in a few places it is conglomeratic. West of the Fifth Fault, it is a coarse vitric tuff 10 cm thick that lies about 1 m above Tuff IB (for example, locs. 49, 80). Sodic plagioclase is the dominant feldspar, suggesting a sodic trachyandesite composition for the tuff.

Tuff IE

This is a trachytic lapilli tuff of ash-flow origin. It is a massive, unsorted deposit 1.2 to 4.1 m thick that extends about 6 km westward from the mouth of the gorge (loc. 2). Its upper surface is commonly channeled, and it is eroded away in many places. Pumice bombs are as much as 25 cm long, and there are many angular fragments of lava as much as 2.5 cm long. The tuff matrix is about 85 percent vitroclasts, 5 percent rock fragments, and 10 percent crystals, most of which are anorthoclase.

Tuff IF

This is the most widespread of the Bed I tuffs, extending throughout the length of the Side Gorge and cropping out as far south as Kelogi (loc. 99). It is present 5 km north of the Main Gorge (locs. 201, 202) and very likely extends farther northeast (loc. 200). It is laminated and presumably lacustrine over a wide area. Moreover, over a sizable area (for example, locs. 36-45, 88a 99), it contains a single lapilli layer 8 to 10 cm thick within otherwise medium- to coarse-grained tuff. It is at most 2.6 m thick, and it thins overall from east to west. There is, however, considerable variation in the eastern part of the Main Gorge, and the tuff is missing in a few places (locs. 7, 13). To the east of DK (loc. 13) and WK (loc. 36), it is difficult to distinguish from several other tuffs and was earlier miscorrelated with Tuff IC (see Hay, 1963a). It thins westward from 45 to 30 cm within lake deposits to the west of the Fifth Fault. Farther west, in lake-margin deposits, it is lenticular, as much as a meter thick, and is locally difficult to distinguish from the overlying reworked tuffs of Bed II. Diameter of pumice fragments decreases overall westward, but rounded pumice cobbles occur in locality 64.

Tuff IF averages about 75 percent vitroclasts, 20 percent rock fragments, and 5 percent crystals. The rock fragments are chiefly trachyte and include a distinctive aegirine-bearing type at least as far west as MLK (loc. 53). Farther west, the tuff is variable and locally consists chiefly of rock fragments and crystals of probable trachyandesite composition. Tuff IF is a product of at least several eruptions. This is indicated not only by its variable composition in different places but by rootmarked horizons within the tuff (for example, locs. 40, 45b).

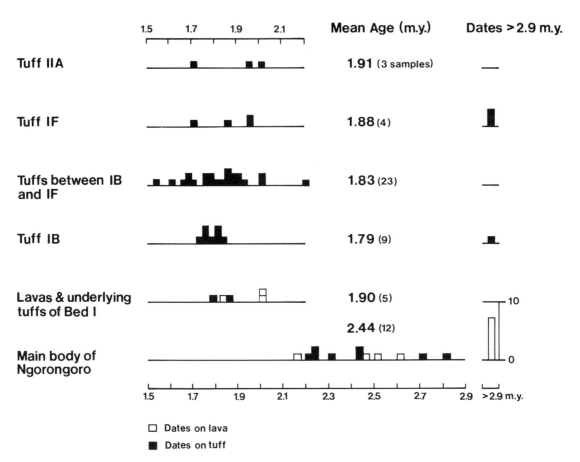

Figure 11. Histogram of K-Ar dates from Bed I and the caldera walls of Ngorongoro.
Data are taken from Curtis and Hay (1972).

AGE OF BED I

Three geophysical dating techniques have been applied to Bed I, giving for at least the upper part one of the firmest dates of any fossiliferous lower Pleistocene stratigraphic unit. Fifty-seven K-Ar dates have been obtained from Bed I, clearly indicating an age on the order of 1.7 to 1.8 m.y.a. for the fossiliferous deposits (Fig. 11; Curtis and Hay, 1972). The high precision of dates for Tuff IB (1.79 ± 03) probably reflects the fact that six of the ten dates are from crystal concentrates obtained from unaltered pumice in an ash-flow tuff. These were particularly easy to clean of air argon, and the precision of the dates is comparable to that obtained from good standards. The greater scatter of dates from the overlying tuffs is partly attributable to a higher percentage of air argon.

The dating of Tuff IF deserves comment as most of the seven dates suggest contamination, yet no basement detritus was detected microscopically. The dates range from 1.71 to 8.52 million years. Curiously, the youngest date is from crossbedded reworked tuff (KA 1787), whereas the oldest is from tuff that was deposited in quiet water and shows no evidence of reworking (KA 1758). If contamination is the cause of the older dates, the contaminants must have been erupted with the ash, not introduced in reworking by water. The dates from Tuff IF are comparable with those of porphyritic lava flows of Ngorongoro, where feldspar phenocrysts evidently acquired radiogenic argon in the magma chamber and subsequently lost only part of it. Argon contamination is strikingly illustrated by feldspar crystals in a tuff near the base of Bed I in the western part of the gorge (KA 1654). Hand-picked coarse euhedral crystals of anorthoclase gave a date of 47.8 m.y.a.!

The lavas have not yielded precise dates,

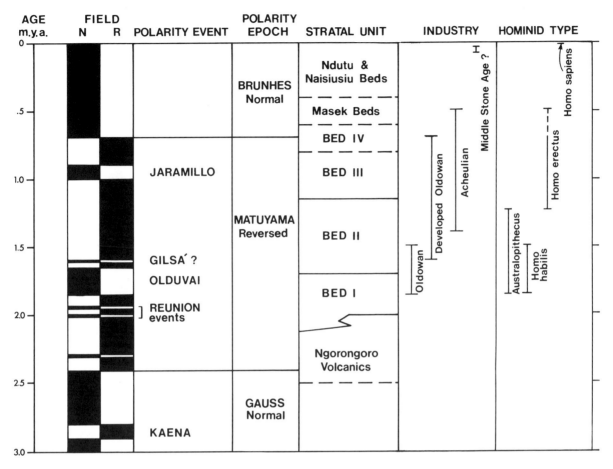

Figure 12. Stratigraphic and archeologic units and hominid types of the Olduvai region matched against the magnetic polarity scale, which is modified after Cox (1969).

despite repeated measurements (L. S. B. Leakey et al., 1962; Evernden and Curtis, 1965; Curtis and Hay, 1972). The last three and most satisfactory runs by Curtis and Evernden gave dates of 1.83 to 2.04, averaging 1.96 ± 0.09 m.y.a. Two dates from reworked tuff beneath lavas have a mean age of 1.82 m.y.a., suggesting that the dates on the lavas are slightly too old. The lowermost lava was more recently dated by F. J. Fitch and J. A. Miller using the conventional K-Ar method. Four runs yielded 1.86 to 1.87 ± 1.9 m.y.a. (1971; Geochronological Report no. FMK/784 by Fitch-Miller rock-dating group). This large standard deviation emphasizes the difficulty in obtaining reliable K-Ar dates from these lavas.

A fission-track date of 2.03 ± 0.28 m.y.a. was obtained from two groups of glass fragments from Tuff IB (Fleischer et al., 1965). At the time, this result was highly significant as a con-

firmation of the K-Ar dating of Bed I, the reliability of which had been questioned (Koenigswald et al., 1961; Straus and Hunt, 1962).

More recently, the magnetic stratigraphy has been worked out (at least roughly) for the Olduvai Beds (Fig. 12). These results, compared with the scale derived from worldwide measurements, aid in estimating the time span of Bed I. The earlier data from Olduvai Gorge showed that a normal polarity event, the Olduvai event (the Gilsá of some authors), is recorded by lavas and ignimbrites of Bed I and the lower part of Bed II (Grommé and Hay, 1971).

Bed I was subsequently sampled to the west of the Fifth Fault (Table 9), both to see whether reversed polarity could be detected in the stratigraphic interval correlative with the lavas and overlying tuffs, and to work out the magnetic stratigraphy of Bed I below the level of the lavas. The samples were either ferruginous zeoli-

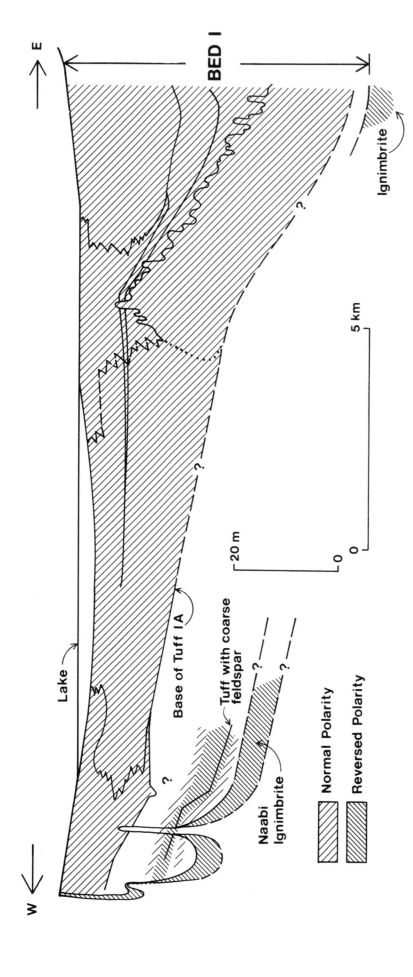

BED I

E

W

Lake

Base of Tuff IA

Tuff with coarse
feldspar

Naabi
Ignimbrite

Ignimbrite

Normal Polarity

Reversed Polarity

20 m

0

0   5 km

Figure 13. Geomagnetic polarity of Bed I and underlying ignimbrites. Polarity is superimposed on the cross-section based on exposures on the south side of the Main Gorge. The stratigraphy of Bed I is shown more fully in Figure 14.

Table 9. *Polarity Measurements from Bed I to the West of the Fifth Fault*[a]

Field/Lab. No.	Stratigraphic Position, Lithology	Polarity
FB 107	Zeolitic tuff of Tuff IF, loc. 80	N
FB 108	Same as above	N
FB 102	Zeolitic tuff of Tuff ID, loc. 80	N
FB 104	Same as above	N
RH 17	Sandy limestone 3 m above Tuff IA, loc. 66c	N
RH 18	Sandy limestone 30 cm above Tuff IA, loc. 66c	N
FB 100	Zeolitic tuff of Tuff IA, loc. 80	N
FB 105	Same as above	N
75-1-4F$_3$	Calcareous concretion in Tuff IA, loc. 61	N
73-12-26F	Sandy limestone 9.1 m below top of Tuff IA, loc. 61	N
74-1-4F$_2$	Sandy, tuffaceous limestone 9.8 m below top of Tuff IA, loc. 61	N
74-1-4F$_1$	Sandy limestone 11.2 m below top of Tuff IA, loc. 61	R
73-12-29H	Clayey, tuffaceous limestone approximately 10 m below Tuff IA, loc. 66b	R
73-12-30J	Sandy limestone from loc. 67, within stratigraphic limits of samples 9.1-11.2 m below Tuff IA at loc. 61	R
73-12-30H	Clayey limestone stratigraphically below 73-12-30J and 2.7 m above Naabi Ignimbrite, loc. 66	N

[a]Samples were collected by F. H. Brown and R. L. Hay; magnetic measurements are by A. Brock.

tic tuff, of untested reliability, or brown sandy and clayey limestone, of the type which generally gave reliable results in Bed II. The nine samples from Tuff IA to the top of Bed I gave normal polarity, which fits with results in the eastern part of the gorge (Fig. 13).

Four of seven samples from the lowermost levels of Bed I gave reversed polarity, and the other three were normal. These seven measurements, if treated as reliable, suggest a reversed interval 3 to 6 m thick in the lower part of Bed I and bounded above and below by rocks with normal polarity. If the three samples with normal polarity acquired their magnetization in a later period of normal polarity, the reversed interval could be much thicker. No polarity information is available for the 9-meter thickness of sediments between Tuff 1A and the uppermost of the seven samples.

These data strongly suggest that the base of the Olduvai event lies below Tuff IA, and they are compatible with interfingering of reversed welded tuffs having an age of 2.0 to 2.1 m.y.a. into the lower part of Bed I. The upper limit of the Olduvai event is at the base or within the lower 75 cm of the Lemuta Member, as discussed more fully in a subsequent chapter.

The Olduvai event lasted 150,000 to 200,000 years, based on age estimates on deep-sea cores (Opdyke, 1972). Dates ranging from 1.79 to 1.87 m.y.a. have been placed on the lower limit

and 1.60 to 1.71 m.y.a. on the upper limit of the Olduvai event (Grommé and Hay, 1971; Shuey et al., in press). Age limits of 1.65 to 1.85 m.y.a. are taken here (Fig. 13) as most compatible with modern worldwide data (for example, Dalrymple, 1972; Opdyke, 1972) and with the position of Tuff IB (1.79 ± 0.03 m.y.) well above the base of the stratigraphic section recording the Olduvai event. About three-quarters of the Olduvai event, or about 150,000 years, fall within Bed I on the basis of the relative stratal thickness of sediments in those parts of Beds I and II spanned by the Olduvai event. On this basis, the top of Bed I has a probable age of 1.70 m.y.a. If Tuff 1A lies at or near the base of the section representing the Olduvai event, the upper, fossiliferous part of Bed I probably represents about half of the 150,000 years tentatively assigned to Bed I. A figure of 75,000 years accords with geologic features such as the weak development of paleosols and the generally slight degree to which the uppermost Bed I lava is eroded.

LITHOFACIES AND ENVIRONMENTS

Bed I is subdivided into five lithofacies: lake deposits, lake-margin deposits, alluvial-fan deposits, alluvial-plain deposits, and lava flows. Facies relationships are shown in Figure 14, which is a reconstructed east-west cross-section

W ←

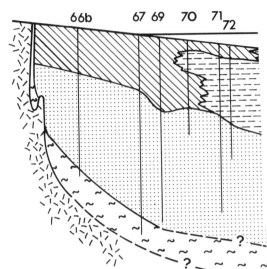

a. NORTH SIDE OF GORGE

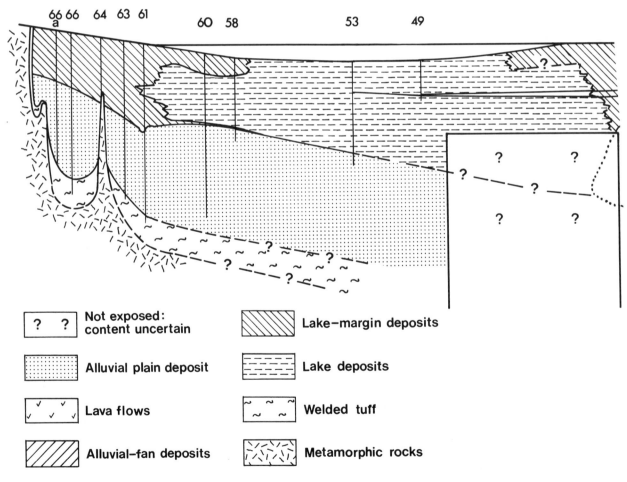

? ?	Not exposed: content uncertain	
(dotted)	Alluvial plain deposit	
v v v	Lava flows	
(hatched)	Alluvial-fan deposits	
(hatched)	Lake-margin deposits	
(dashed)	Lake deposits	
~ ~	Welded tuff	
(metamorphic pattern)	Metamorphic rocks	

40

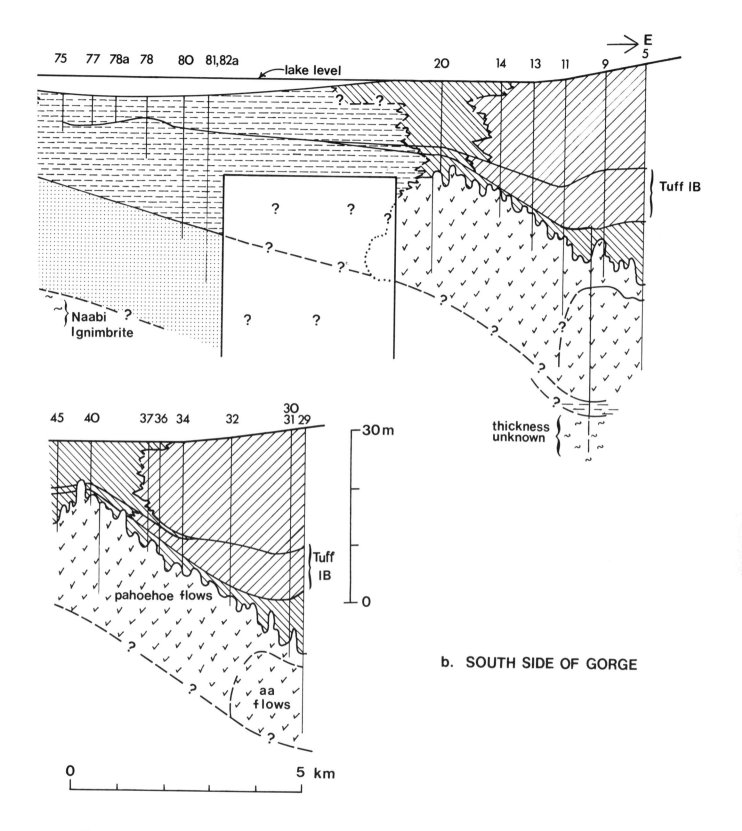

b. SOUTH SIDE OF GORGE

Figure 14. Longitudinal sections showing lithofacies of Bed I as exposed along the Main Gorge. Sections are reconstructed with the top of Bed I approximately as it would have appeared when Tuff IF was deposited. Vertical lines represent measured stratigraphic sections.

Table 10. *Composition of Lithofacies in Bed I*

Lithofacies	Percent claystone	Percent sandstone	Percent conglomerate	Percent tuff	Limestone and dolomite	Other
Lake deposits	80	2	tr.	15	3	Tr. chert
Lake-margin deposits (eastern)	49	tr.	tr.	49	2	Tr. siliceous earth
Lake-margin deposits (western)	40	42	0.3	14	4	
Alluvial-fan deposits	2	0	2	96[a]	0	
Alluvial-plain deposits	28	4	3	56	9	
Lava flows	0	0	0	tr.	0	>99% lava flows

[a]Includes about 10 percent of pyroclastic deposits coarser than tuffs.

through Bed 1. The lake deposits accumulated in a perennial lake in the lowest part of the basin. Lake-margin sediments were laid down on low-lying terrain intermittently flooded by the lake. Lava flows and alluvial-fan deposits are along the eastern margin of the basin, and they interfinger westward with lake-margin deposits. Alluvial-plain deposits are exposed only in the western half of the basin, where they underlie lake and lake-margin deposits. Lake-margin deposits of the eastern and western parts of the basin differ in lithology as well as fossils and archeologic content, and the two assemblages of lake-margin sediments will be described separately under the headings *eastern lake-margin deposits* and *western lake-margin deposits.* The lake-margin sediments to the north and south of the gorge are included with the eastern lake-margin deposits.

The following sections describe the lithofacies of Bed I and interpret them in terms of depositional environment. The lake deposits are described first, as the nature of the lake is basic to an understanding of the lake-margin deposits. Facies descriptions will then be given in the following sequence: western lake margin, eastern lake margin, alluvial fan, lava flows, and alluvial plain.

Lake Deposits

The lake deposits are exposed westward about 6.5 km from the Fifth Fault and extend an undetermined distance to the east between the Fifth Fault and the fault near FLK (loc. 45) beneath younger sediments. They have a maximum thickness of 26 m near the Fifth Fault (for

example, loc. 80), and thin to about 10 m near the western margin. Lake deposits interfinger and intergrade westward with lake-margin deposits over a zone about 2 km wide. Tuff IA is taken as the basal bed of the lacustrine facies. Where the tuff lies slightly above the base of the lake deposits (locs. 80, 81), the underlying lake deposits are placed in the alluvial-plain deposits. The lake deposits are chiefly claystone but include tuff, sandstone, limestone, dolomite, chert, and conglomerate (Table 10).

Claystones are generally waxlike, massive to laminated, and greenish in color. Authigenic minerals, that is, minerals formed in place, constitute a substantial amount of the claystones (Hay, 1970a). Widespread and abundant are calcite, dolomite, K-feldspar (= potassium feldspar), and altered pyrite; relatively rare are calcite replacements of gaylussite and trona(?). Calcite crystals constitute, on average, between 5 and 10 percent of the claystones. They are unusual in their large size (generally 0.2 to 2 mm long), euhedral shapes, and varied crystal habits at different levels (Plate 9). Calcareous claystones grade into clayey limestones consisting largely of coarse calcite crystals. Limestones are widespread, and dolomites are found in the eastern half of the facies. Most of the limestones are of coarse calcite crystals similar to those of the claystones. Dolomites are characteristically very fine-grained. Chert nodules (Plate 8) are both widespread and abundant in a single horizon near the middle of the facies. Sandstones and pebble conglomerates are present in a few places in the western part of the facies.

Tuff beds are laterally extensive, evenly bedded, and decrease westward in both grain size and thickness. Most of them are vitric tuffs of original trachyte composition in which the glass has been altered to phillipsite and K-feldspar. The proportion of K-feldspar is highest in tuffs of the western part of the facies.

These sediments accumulated in a shallow saline lake of fluctuating level and extent. The greenish pyritic clays are a typical lacustrine sediment, and the laminations in clays and even bedding of tuffs and dolomites are evidence of quiet waters. A fluctuating level is indicated by the variable western margin of the facies and interbeds of pebbly sandstone within waxlike claystones. Sand-size claystone pellets at a few localities reflect erosion of nearby mudflats by wind. Casts of the soluble sodium-carbonate minerals gaylussite and trona(?), although rare, are proof of a highly saline sodium-carbonate lake. Additional evidence is provided by the dolomite and K-feldspar (Hay, 1966, 1970a). The chert nodules are of a type found only in deposits of saline sodium-carbonate lakes (Hay, 1968; Eugster, 1967, 1969).

The large euhedral calcite crystals deserve special comment, as only a single other occurrence of this type has been reported, as far as I am aware (Isaac, 1967). At least some of the crystals grew in lake-bottom muds at depths sufficiently shallow (a few centimeters or less?) for wave action to rework the muds and concentrate the crystals in thin layers. Microprobe analyses by W. Wright of three different types of crystals show that they are nearly pure calcite with a few tenths of a percent of SrO and MgO. They contain surprisingly high uranium concentrations (40 to 50 ppm) as determined by fission-track activation, yet they have relative few spontaneous fission tracks, indicating that the tracks have been healed through time (MacDougall and Price, 1974). Isotope measurements, by J. R. O'Neil, were made in order to see if crystals of different habits differ in their isotopic composition and hence in salinity of the fluid from which they crystallized. Both the O^{18}/O^{16} and C^{13}/C^{12} ratios are very high (Fig. 15) and point to crystallization in a brine. No correlation is evident between crystal form and isotopic composition, and no clear trend in salinity of the lake is indicated by the sequence. The O^{18}/O^{16} ratio in a chert nodule points to saline lake water (O'Neil and Hay, 1973).

Western Lake-Margin Deposits

The western lake-margin deposits are 7 to 12 m thick and extend westward 3 km from the zone of interfingering with the lake deposits. These sediments are chiefly sandstone and claystone but include a substantial amount of tuff and some limestone (Table 10). Sandstones are chiefly quartz and feldspar of basement origin. Oolites are in many of the easternmost sandstones, and claystone pellets are abundant in some of those to the west. Clasts of the Naabi Ignimbrite predominate in most conglomerates, the remainder consisting largely of basement rock. Claystones are grayish brown or pale olive and contain rootmarkings. Some are sandy, and the sandier of the claystones contain claystone pellets. Reworked mafic vitric tuff and tuffaceous sandstone of Tuff IA form the basal bed of the facies, and an evenly laminated mafic vitric tuff 2 m thick lies near the middle of the western half of the facies. Trachyte tuffs are in the upper 4.5 m of the facies, and they correlate with the sequence from Tuff IB to Tuff IF in the lake deposits to the east. They trachytic tuffs are thin bedded in their eastern exposures and massive and rootmarked to the west. Rounded pumice cobbles as much as 15 cm in diameter were noted in Tuff IF at locality 64.

Nodular limestones are widespread and become more common to the west. They vary greatly in their content of sand and locally grade into calcareous sandstone. One of the few samples analyzed mineralogically proved to be dolomite rather than limestone. Laminated algal limestone (tufa) occurs near the base of the facies at locality 63.

The westernmost part of the facies, at locality 66c, comprises Tuff IA overlain by 6 m of yellowish-brown sandy, tuffaceous limestone. Both Tuff IA and the limestones lie with a depositional contact against a scarp eroded in the Naabi Ignimbrite, and they contain talus blocks of the ignimbrite and of gneiss from basement exposures a short distance to the west. Much of the sand-sized detritus in the limestones has pelletoid coatings of clay, pointing to eolian transport. The calcium carbonate of the limestones is probably pedogenic in origin. The limestones and Tuff IA at this locality do not exhibit any lacustrine features and are included within the

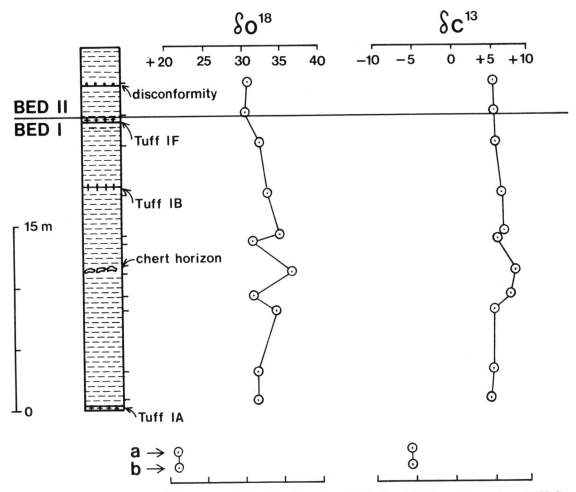

Figure 15. Isotopic composition of calcite crystals in lake deposits of Beds I and II. Samples are from RHC (loc. 80) except for the topmost one, from locality 49. Included for comparison are values of two calcite nodules (a, b) from lake-margin deposits of Bed I. Data were provided by J. R. O'Neil (1972, personal communication).

western lake-margin deposits only because their very localized occurrence does not warrant establishing a separate facies of eolian-colluvial deposits.

The lake-margin deposits contain a few narrow, steep-sided stream channels, filled with tuff or conglomerate. Two channels near the base of the facies are oriented N65°E and N85°E; two others near the top are N50°E and N80°E. The four channels average N70°E. The deepest channel is at locality 61 and has a depth of 3.4 m. As the detritus in this facies is from westerly sources, the channels must have been cut by streams flowing northeast.

Sparse faunal remains occur throughout the facies, and a skull of *Kobus sigmoidalis* (waterbuck) was collected near the base of the facies in locality 63. Rootmarkings represent both grass-

land and marshland vegetation.

This facies has both lacustrine and fluviatile features, indicating that it accumulated on terrain intermittently flooded by the lake. Oolitic sandstone indicates shallow, wave-agitated lake water, and the laminated vitric tuff suggests quiet water. Sand-free claystones were deposited in quiet water, either of a lake or floodplain. Lenticular conglomerates and channeling are fluvial features, and if the streams flowed into the lake perpendicular to its margin, then the shoreline was oriented about N20°W. The stream channel 3.4 m deep was near the margin of the lake, and its depth may be a rough measure of the fluctuation in lake level. Claystone pellets probably reflect erosion of mudflats by wind at a time of low lake level. In view of their large size, the rounded pumice cobbles probably floated westward across the lake to their present position.

The lake water flooding the lake-margin terrain was overall less saline than that in the center of the basin. Vitric tuffs are extensively altered to zeolites only in the eastern part of the facies, and both authigenic K-feldspar and evidence of soluble salts are lacking.

Eastern Lake-Margin Deposits

Lake-margin deposits to the east of the lake beds are exposed almost continuously in the Main Gorge above the lavas over an east-west distance of 5.4 km. These are as much as 17 m thick, and an additional 1.4-meter thickness is present beneath the lavas. Also included in these deposits are 5 m of beds near Kelogi, to the southwest (locs. 99, 100), and 7 m of beds to the north of the gorge (locs. 201, 202).

The vertical sequence of lake-margin deposits above the lavas is subdivided at the base of Tuff IB into lower and upper units. The lower unit extends from the Second Fault west to FLK-NN (loc. 45b), a distance of 5.4 km. The upper unit is restricted by alluvial-fan deposits to the western half of this area. The lower unit, except for its easternmost exposures, is 0 to 6 m thick, the variation reflecting the uneven surface of the lavas. It consists largely of claystone but contains a few trachytic tuffs. Granule- and pebble-size clasts of basement and volcanic rock occur widely on the north side of the gorge. These sediments grade eastward into a section that is dominantly tuff and as much as 11 m thick. These easternmost deposits have some lithological similarity to the overlying alluvial-fan deposits and some mineralogical similarity to the underlying lavas. They are, however, included within the lake-margin deposits in view of their limited exposure and uncertain genetic relationship. The upper unit of lake-margin deposits is about two-thirds tuff and one-third claystone, and it also includes limestone, pumice-pebble conglomerates, and siliceous earth.

Overall, the eastern lake-margin deposits are about 98 percent tuff and claystone, which are in roughly equal proportions. There is perhaps 2 percent of limestone and far less than a percent each of conglomerate and siliceous earth. Paleosols are weakly developed at many hori-

zons over both tuffs and claystones. Where developed on claystones, they are crumbly, brown or brownish-gray, rootmarked zones. Paleosols on tuffs are rootmarked zones as much as 30 cm thick in which much of the finest-grained vitric ash is weathered to montmorillonite.

Claystones are both tuffaceous and nontuffaceous and can be pale olive or various shades of yellow and brown. Rootmarkings are common, particularly above the level of Tuff IC. Waxlike claystones relatively free of sand are much more common below Tuff IC than they are above. Calcite concretions are widespread and abundant, and calcite replacements of gypsum rosettes ("desert roses") have been found beneath Tuffs IB and IF.

Tuffs and lapilli tuffs are dominantly trachytic but include some of basaltic and trachyandesitic composition. Basaltic deposits are confined to the eastern margin of the facies, where they overlie the lava flows. The basal bed of basaltic tuff contains lapilli of basaltic spatter as much as 4 cm long. The other, more widespread tuffs vary considerably in the degree and nature of reworking and provide considerable information about the environment in which they were deposited. A few tuffs represent single showers of ash that fell on the land surface and were generally reworked only slightly if at all (for example, Tuffs IB and IC at loc. 45b). A massive 1.2-meter-thick bed above Tuff ID is rootmarked and contains diatoms and ostracods in addition to pyroclastic materials. It is a product of several eruptions in which the ash was deposited either in marshland or on a land surface intermittently flooded by lake water. A few tuffs are for the most part evenly laminated and are dominantly lacustrine. Tuff IF is a lacustrine deposit, and its thickness ranges from 15 cm to 2.1 m and varies systematically on a regional scale, suggesting that the greater thicknesses accumulated in deeper water (Fig. 16). Tuff IF is also laminated and waterlaid at localities 201 and 202, to the north of the gorge and locality 99 to the southwest.

Limestones occur both as coarsely crystalline concretions cemented together and as fine-grained beds of irregular shape. Both types occur in claystones and were probably precipitated from ground water at shallow depth. Isotopic values from two limestone concretions suggest that the calcite crystallized from relatively fresh

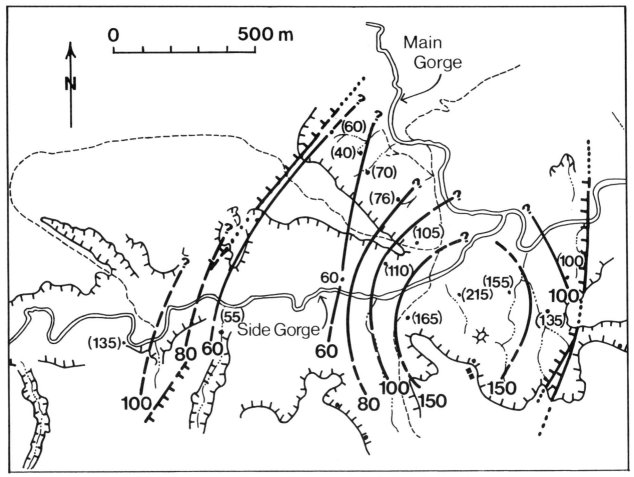

Figure 16. Map with thickness measurements, in centimeters, of Tuff IF and inferred isopachs, or contour lines of equal thickness. Area includes the junction of the Main and Side gorges. Locality numbers are given in Figure 6.

water (Fig. 15). A 60-centimeter thickness of siliceous earth occurs at one place in the widespread rootmarked massive tuff that overlies Tuff ID. In a few places elsewhere the tuff is siliceous and earthy. The topmost 3 cm of claystone beneath Tuff IF at locality 46a has laminae of siliceous earth consisting chiefly of silicified plant remains.

A variety of basement and volcanic detritus, commonly accompanied by fragments of bone, makes up the assemblage of granule- (2 to 4 mm) and pebble-size (4 to 64 mm) clasts in the lower unit between localities 10 and 20. Clasts may be either isolated or concentrated in thin, lenticular beds. The basement clasts are varied and include materials (for example, pink pegmatitic feldspar) from an unknown source, probably to the north or northwest. Clasts of volcanic rock are chiefly welded tuff, both from Ngorongoro and the Naabi Ignimbrite, but in-

clude varied lavas, some of which is from Engelosin. Conglomerates elsewhere in the lower unit are chiefly of lava and welded tuff from Ngorongoro. Pumice-pebble conglomerates predominate above the level of Tuff IB.

Fossilized leaves are rare, and little is yet known of the pollen, but swamp and shore vegetation is indicated by abundant coarse, generally unbranching vertical root channels and casts. These are commonly 10 to 30 cm long, 2 to 6 mm wide, and could have been made by *Typha,* and some kinds of shore grasses and reeds. Fossil rhizomes of papyrus (probably *Cyperus papyrus*) point to marshland or shallow water. Silicified remains of a water plant (cf. *Potamageton*) were identified by Howard Schorn (1971, personal communication). Diatoms and encysting cases of chrysophyte algae are in the siliceous earth and siliceous, earthy tuff that overlies Tuff ID (for instance, at

46

loc. 85); they point to fluctuating lake level and periodically saline, alkaline conditions (Hay, 1973). Ostracods are in a few samples of the siliceous tuff unit above Tuff ID, and freshwater snails were collected from clays of the "Zinjanthropus" level, which underlie Tuff IC (M. D. Leakey, 1971a). Fossil remains of urocyclid slugs have been found at two levels below Tuff IC. These suggest damp conditions in evergreen forests "where the rainfall exceeds 35 inches per year or where damp conditions are maintained by regular mists" (Verdcourt, 1963).

Vertebrate remains are both varied and abundant, and the various forms represent a wide variety of habitats (see L. S. B. Leakey, 1965; M. D. Leakey, 1971a). At many levels are remains of large mammals: proboscideans (principally elephants), rhinoceros, equids, suids, hippos, bovids, and giraffids. M. D. Leakey gives the number and proportions of specimens of identifiable larger mammals from excavated sites (1971a). In this tabulation the bovid remains, principally of gazelles and other antelopes, are much more abundant between the "Zinjanthropus" level and Tuff IF than they are in the underlying deposits. This difference reflects either a real difference in the fauna or the increasing importance of bovids in the hominid diet. There is, moreover, a significant difference in bovid types between the lower and upper horizons. Alan Gentry has kindly provided the following summary of detailed studies by himself and Anthea Gentry (1973, personal communication):

The commonest bovid remains at DK and FLK-NN [the lower levels] are of Reduncini, the tribe containing living reedbuck, kob, lechwe and waterbuck. Alcelaphini, the tribe containing living wildebeest, hartebeest and topi, are represented but not very common. Reduncine antelopes of the present day always live close to water, while alcelaphines are more characteristic of drier open country. Both tribes graze rather than browse, but the alcelaphines have higher-crowned cheek teeth and limb bones with better cursorial adaptations. FLK [the "Zinjanthropus" level] has fairly diversified bovids. At FLK-N [the uppermost fossiliferous levels in Bed I] the Reduncini are very rare and there is an overwhelming preponderance of Alcelaphini, especially the extinct *Parmularius altidens*. There is also a lot of *Antidorcas recki*, an extinct relative of the living springbok, which is well known as an inhabitant of dry open country.

Small mammals (rodents, gerbils, carnivores) are abundant in some levels of the excavations at FLK and FLK-N, and the rodents are of particu-

lar ecological interest. Recent studies by J. J. Jaeger (in press) on the rodents add considerable information to the earlier conclusions of Lavocat (pp. 17-19 in L. S. B. Leakey, 1965). Based on a comparison with living species, the Murid fauna of Bed I reflects an environment more humid than that of Olduvai at the present time. Several of the Murid types point to marshland and humid prairie. The rodent fauna changes significantly from the lower levels at FLK and FLK-NN between Tuffs IB and ID, and the uppermost levels of Bed I at FLK-N. This change is marked by an increasing proportion of gerbils, the appearance of Bathyergids and Sciurids (ground squirrels) and the reduction in proportion of Murids. Quantitative analysis of the Murid fauna shows that two forms at the lower levels (*Denomys* and *Grammomys*) are absent in the upper levels. These two genera could indicate the presence at this lower level of thickets (Jaeger, 1973, personal communication). The arboreal rodent *Thallomys* lives on *Acacia* of small size in wooded grassland.

The fossil avifauna of Beds I and II is one of the largest heretofore known. P. Brodkorb has summarized the general nature of the avifauna as follows (1973, personal communication):

Abundant remains of aquatic birds indicate that water was much more plentiful throughout the time of deposition of the earlier strata that form Beds I and II than is the case under the present semidesert conditions. The water birds included swimmers and divers such as grebes, cormorants, pelicans, and many ducks. Larine birds were represented by gulls, terns, and skimmers. Waders were abundant and included flamingoes, herons, storks, rails, jacanas, plovers, sandpipers, and stilts. The presence of flamingoes at several sites in both Bed I and Bed II indicates the proximity of brackish water. Seed-eaters include francolins, quail, hemipodes, and several species of doves. They demonstrate the presence of grassland.

Aquatic birds are most common below Tuff IC, and land birds (perching birds, quail, parrots, and so forth) are most common in the uppermost levels at FLK-N.

Crocodile remains are widespread below Tuff IB and are especially common at DK (loc. 13). Remains of fish were collected from the "Zinjanthropus" level at FLK (loc. 45) and from the uppermost part of Bed I at FLK-N. Regarding the fish, Greenwood and Todd state:

Site FLK-NI is of particular interest because it has yielded only the remains of Cichlidae. Samples from this site are quite small, but, by analogy with other sites, at least a few clariid remains would be expected. Clariids are some of the most ubiquitous fishes in present day African fresh waters. Their absence, when cichlids are present, might indicate strongly saline (especially alkaline) conditions, as in Lake Magadi today. Since earlier and later fish-bearing deposits contain both clariids and cichlids, there does seem to be a possibility that the lake was more saline during the deposition of the FLK-NI bed than at other periods [1970, p. 241].

To summarize, a wealth of paleontologic data points to widespread marshland and the frequent occurrence of standing water up through deposition of the siliceous earthy tuff above Tuff ID. Drier conditions are indicated by the fauna from excavated levels in the uppermost 1.2 m of strata below Tuff IF at FLK-N.

The eastern lake-margin deposits accumulated on relatively flat terrain that was intermittently flooded and dried in response to changes in level of the lake (Hay, 1973). Initially, the lava surface had a local relief of at least 6 m, which was reduced to about 1 m as the lavas were buried by sediments. Claystones and widespread water-laid tuffs were deposited at times of high water level, whereas paleosols, hominid occupation sites, eroded surfaces, and extensive land-laid tuffs represent periods of exposure. Prior to Tuff IB, the zone of marginal terrain along the eastern margin of the lake was at least 5.4 km wide; after Tuff IB, the zone was narrowed to about 2.1 km by westward growth of an alluvial fan. Most of the fluctuations may have been relatively short — either seasonal or involving no more than a few tens of years. A longer-term paleogeographic or climatic fluctuation is probably indicated by the fauna collected at several horizons through 1.2 m of claystone beneath Tuff IF at FLK-N. This fauna points to substantially drier conditions than prevailed at lower fossiliferous horizons.

The pattern of zeolitic alteration in lacustrine tuffs represents a salinity gradient at times of flooding, with the greater degree of zeolitic alteration and therefore the highest salinities on the northwest (toward the lake center) and the freshest water on the southeast, nearest the stream inlets. This gradient is shown most clearly in the lacustrine part of Tuff IF, which is zeolitic in localities 45 and 85 and unaltered to the southeast (for example, locs. 38, 40).

Prior to eruption of Tuff IB, streams from Ngorongoro carried pebbles of lava and welded tuff westward over the mudflats. It is not clear, however, as to how the granules and pebbles of basement rock (quartz, pink feldspar, and gneiss) and Engelosin phonolite were transported southward and southeastward for a considerable distance over the same mudflats. The clasts are in claystones rather than in beach deposits, and sheetwash associated with torrential downpours to the north seems the most likely way of transporting coarse sediment southward over mudflats. Kay Behrensmeyer has suggested that some of these clasts may have been caught and transported in the hooves of equids and the larger bovids as in the Amboseli area of Kenya at the present time (1974, personal communication).

Alluvial-Fan Deposits

Alluvial-fan deposits lie along the east side of the basin and interfinger westward with lake-margin deposits. They are exposed in the gorge over an east-west distance of 7.5 km, and they crop out intermittently for a distance of 7 km to the northeast along the west side of Olbalbal. The uppermost part of the facies is exposed 15 km northeast of the gorge (loc. 200).

The alluvial-fan deposits are almost entirely of eruptive origin, and they range from 21 to 28 m in thickness. Tuff IB is the basal bed of these deposits. Approximately 15 percent of the facies consists of ash-fall and ash-flow deposits that have not been reworked. Ash-flow deposits exposed in the gorge comprise Tuff IE and the massive part of Tuff IB that overlies a basal ash-fall stratum. Their magnetic polarity shows that they were emplaced at high temperature (Grommé and Hay, 1971). The reworked pyroclastics were deposited chiefly by streams but include some layers emplaced as mud flows. A few of the fine-grained tuffs are thinly laminated. Overall, the alluvial-fan deposits are about 90 percent glass and 10 percent crystals and rock fragments. Nearly all are of sodic trachyte composition, and a very few are probably trachyandesite.

Brown, rootmarked paleosols 15 to 60 cm thick are widely developed on Tuffs IB and IC and on tuffs that underlie Tuffs IC and IF. More localized and poorly defined paleosols are developed at a few other horizons above Tuff IC in the vicinity of the Second and Third faults, and rootmarkings are through most of the section near the western margin of the facies. Even where developed on thinly bedded tuffs, paleosols are massive, thus showing the churning effect of roots and burrowing animals, but they are immature, as only a small amount of clay has been formed by weathering.

Conglomerates form lenticular beds that are coarsest and most abundant in the eastern part of the facies. Clasts are chiefly of cobble and pebble size, but boulders as much as 60 cm in diameter are not rare. The clasts are of pumice, obsidian, lava, welded tuff, and zeolitic lapilli tuff. Most of these are from Olmoti, and the remainder are from Ngorongoro.

Stream-channel alignments show that the streams flowed to the west-northwest. Nearly all of the seventeen measured channels are 1 to 1.6 m deep and filled with conglomerate. Eight of these channels are cut into the ash-flow deposit of Tuff IB; the others lie at various horizons above. Their orientations range from about east-west to N35°W, averaging N73°W. Eleven of these, including all of the deeper and better-defined channels, are between N60°W and N85°W.

These deposits accumulated on the lower part of an alluvial fan sloping northwestward from Olmoti and Ngorongoro to the lake-margin terrain. The relatively consistent stream-channel alignments and coarse size of much debris point to significant slopes, although some of the laminated sediments indicate local areas of ponded water. An average slope of 1:300 (3.3 m/km) is used in Figure 14. The fan was sparsely vegetated by comparison with the lake-margin terrain, for rootmarkings are relatively rare except in the few paleosols. Remains of few vertebrates have been found in these deposits (see Hay, 1973), and the scarcity of remains may reflect either low average concentration of animals, highly seasonal concentrations of animals, or poor conditions for preserving remains.

Olmoti was the source of pyroclastic materials in the fan. Average clast size is coarsest and ash-flow tuffs are most numerous in locality 1, the exposure closest to Olmoti. The average stream-channel alignment (N73°W), when extrapolated upslope to the southeast, touches the volcanic highlands near the junction of Ngorongoro and Olmoti. As the central crater of Ngorongoro had been extinct long before this time, Olmoti is indicated as the source. Most of the lava and tuff clasts in conglomerates are of types found on Olmoti.

Lava Flows

Lava flows are exposed almost continuously in the gorge from the Second Fault west to FLK-NN, a distance of 4.5 km. They have a total thickness of 21 m where they are fully exposed upstream from the Third Fault (loc. 9a). Lavas of Bed I also crop out along the western margin of Olbalbal (locs. 1, 157), and 45 to 60 m of flows are exposed in fault scarps 5 to 12 km north of the gorge. The lava sequence in the gorge near the Second and Third faults comprises a basal thick flow of trachyandesite overlain by several thin flows of olivine basalt with ropy or pahoehoe surfaces. Olivine basalt is the only type found to the west of the waterfall upstream from the Third Fault. These basalts have a maximum exposed thickness of 15 m in the horst between KK (loc. 41) and MCK (loc. 40). Numerous thin, tonguelike sheets 30 cm to 1 m thick may be piled on one another, as near HWK (loc. 42), and there are many elongate to round lava mounds, or tumuli, between 1 and 4.5 m high.

Perfect preservation of pahoehoe crusts on the basalts shows that they spilled out in rapid succession, almost certainly from a vent nearby rather than from a summit crater of one of the large volcanoes. Directional features in the olivine basalts vary widely in orientation but suggest a vent to the south of the gorge. A southern source is consistent with the lack of similar flows in exposures to the north. Basaltic tuff and lapilli tuff overlying the basalt flows at the Second Fault may well have originated from the same nearby vent.

Trachyandesites to the north and southeast of the gorge are mineralogically similar to those beneath the olivine basalts in the gorge. The only directional feature noted on a trachyandesite flow was smooth, parallel grooving with a bearing of N55°W on the bottom of the flow exposed at the waterfall upstream from the Third Fault. The grooving fits with an origin from either Olmoti or Ngorongoro, but K-Ar dates and paleomagnetic measurements place the flows within the eruptive span of Olmoti. Moreover, the trachyandesites are thickest to the

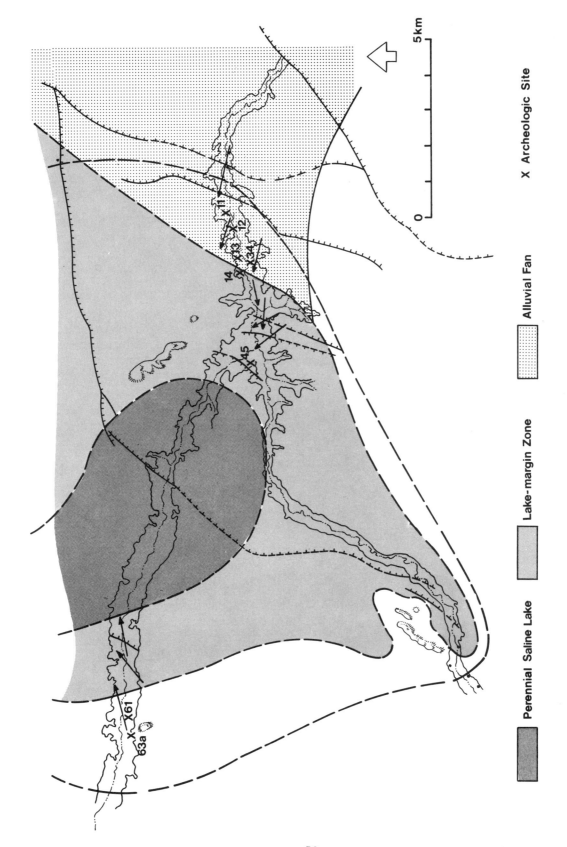

Figure 17. Paleogeography of Bed I for the interval between Tuff ID and the top of Bed I. Limits of lake-margin zone and alluvial fan are based on the paleogeography of Tuff IF. Arrows indicate direction of stream flow as based on channel measurements.

Perennial Saline Lake

Lake-margin Zone

Alluvial Fan

X Archeologic Site

5 km

50

north of the gorge, directly opposite Olmoti, and similar lavas are absent on the outer slopes of Ngorongoro, above the youngest ignimbrites.

Alluvial-Plain Deposits

Alluvial-plain deposits constitute the lower part of Bed I to the west of the Fifth Fault, where they crop out discontinuously over an east-west distance of 9 km. They range in thickness from 12.4 to 21 m except where they thin and pinch out over the top of a buried inselberg (loc. 64). These deposits are chiefly tuff and claystone but include limestone, conglomerate, and sandstone. Tuffs form most of the lower two-thirds of the facies, and nearly all of the claystones are in the upper one-third.

Tuffs are characteristically massive or crudely bedded, clayey, and yellowish-brown. Root-markings are abundant, and most of the tuffs appear to have been homogenized by pedogenic processes. Tuffs are dominantly crystal-vitric, and the crystal content indicates a quartz trachyte composition. Detrital sand forms a substantial proportion of many of the reworked tuffs. Much of the glass is weathered to montmorillonite clay. Claystone pellets are common in some samples, as are clay-coated, pelletoid grains, especially of pumice. At or near the base of the tuffaceous sequence is a dark gray, reworked basaltic vitric tuff. Near the middle of the tuff sequence is a widespread quartz trachyte tuff, generally about 30 cm thick, which is characterized by coarse anorthoclase crystals and by numerous rock fragments as much as 2.5 cm in length. One of the fragments is nepheline phonolite of Sadiman type, and it suggests that Sadiman was in existence by this time. This tuff is the coarsest of those in the alluvial-plain deposits. Ngorongoro is the most likely source for the trachytic tuffs in view of their mineralogical similarity to the Ngorongoro ignimbrites.

Claystones form massive beds that have finely textured rootmarkings and contain detrital sand. Sandstones and conglomerates consist largely of detritus eroded from the metamorphic basement and Naabi Ignimbrite. A few of the conglomerates fill narrow, steep-walled channels 0.5 to 1 m deep, and the two measured channels are oriented N70°E and N75°E.

Limestones form massive beds generally 30 cm to a meter in thickness. Most of them are highly tuffaceous and represent reworked tuffs cemented and replaced by calcite. A few beds appear to represent caliche horizons, or pedogenic concentrations of calcium carbonate. Limestone is widespread at a horizon near the middle of the facies, a few meters above the tuff with coarse anorthoclase crystals.

The only faunal remains known from these deposits are suid and elephant teeth, noted in claystone near the top of the facies in locality 71.

The clayey tuffs in the lower part of the facies probably accumulted in a steppe environment where the vegetation was at least seasonally insufficient to prevent considerable reworking of sediment by wind. The degree of weathering and clay formation suggests a rather slow rate of accumulation. Rootmarkings are similar to those produced by grass, and claystone pellets and clay-coated grains indicate eolian transport. Conglomerates and sandstones were deposited by streams flowing northeast, and the rootmarked sandy claystones probably accumulated on floodplains adjacent to the streams. These alluvial-plain deposits grade upward into lake- and lake-margin deposits.

ENVIRONMENTAL SYNTHESIS AND GEOLOGIC HISTORY

Bed I was deposited in a lake basin at the western foot of the volcanic highlands. The lake did not have an outlet and consequently fluctuated in level, alternately flooding and exposing a broad marginal zone (Figs. 17, 18). The initial paleogeography of the lake basin remains obscure because of the very limited exposure below the lavas of Bed I in the eastern part of the gorge (Fig. 14). From stratigraphic and topographic relationships (Fig. 14), it seems likely that the lavas were erupted at about the time that Tuff IA was deposited. Presumably the lavas displaced the eastern shoreline and the overall position of the lake a substantial distance to the west, where it remained during the deposition of Bed I.

The basin was structurally modified during the deposition of Bed I. Faulting is suggested by the pinchout of tephra deposits less than 2 km

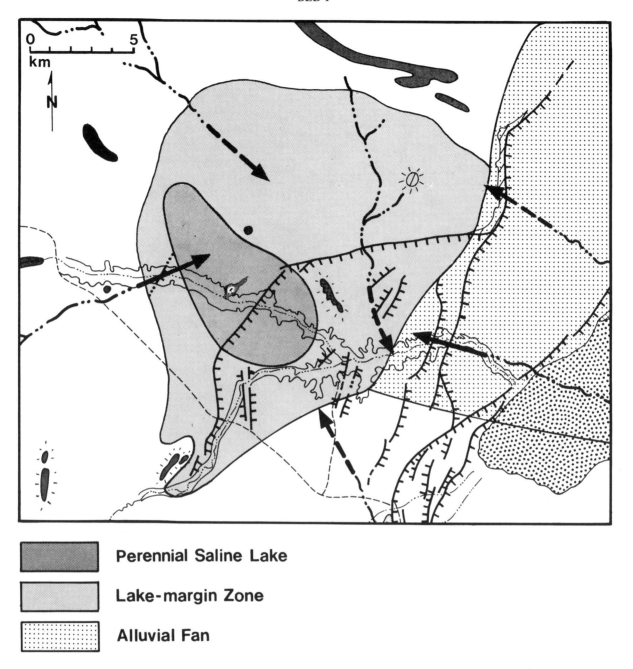

Perennial Saline Lake

Lake-margin Zone

Alluvial Fan

Figure 18. Paleogeography of the Olduvai region for the interval between the base of Tuff IB and the top of Bed I. The extent of the lake-margin zone and the alluvial fan are based on the paleogeography of Tuff IF. Solid arrows indicate flow direction of streams as based on channel measurements, and dashed arrows show flow direction inferred from clast composition and regional paleogeography.

south of the mouth of the gorge (see Fig. 10), and eastward tilting of the basin as a whole is indicated by anomalous paleogeographic relationships of Tuff IB compared to the top of Bed I (Fig. 14). The top of Bed I, represented by Tuff IF, must have been very nearly horizontal in the eastern lake-margin terrain, but a reconstructed cross-section of Bed I drawn on this basis places lacustrine clay above Tuff IB at FLK (loc. 45), topographically higher than stratigraphically equivalent stream-laid sediment to the east. Moreover, if any reasonable water depth is postulated for deposition of Tuff IF in the center of the lake, then Tuff IB, here deposited in lake water, lies either at the same or at a higher elevation than it does at the eastern margin of the lake, where it is a subaerial ash-fall deposit. The basin appears to have been tilted slightly downward to the east, and the area occupied by the lavas may in addition have been warped into a shallow syncline.

The average diameter of the lake fluctuated between limits of about 7 and 25 km during all but the most extreme periods of desiccation. The lake was widest prior to the ash flow of Tuff IB, which spread eastward and narrowed the area subject to flooding. It appears to have been smallest, on average, between deposition of Tuffs ID and IF. The lake expanded rather abruptly just before Tuff IF was deposited.

The perennial part of the lake was comparatively shallow at times of low level as indicated by evidence of wave or current action at several horizons, and by desiccation cracks at one level in the central part of the basin. The wide marginal zone of intermittent flooding and drying also fits with a shallow lake. Water-level fluctuations on the order of 1.5 to 3.4 m can account for most or all of the features noted in the lake-margin deposits. The most significant measurement in this regard is a stream channel 3.4 m deep near the western margin of the lake between Tuffs ID and IF.

Water of the perennial part of the lake was saline, alkaline, and rich in dissolved sodium carbonate-bicarbonate. Water flooding the lake-margin terrain varied widely in salinity, with fresh water near stream inlets along the eastern margin of the lake. Frequent fluctuations in level and high salinity of the lake suggest that its level was controlled by the balance of inflow and evaporation.

The climate was probably semiarid though wetter than that in the Olduvai region today. High salinity and fluctuations in lake level are the best evidence that the climate was relatively dry over a long period. Urocyclid slugs, found below the level of Tuff ID, appear to be the best evidence clearly indicating a climate appreciably moister than that of today in the same area. The slugs are commonest in sediments beneath Tuff IB (for example, locs. 13, 20), but they were also found in the "Zinjanthropus" level, beneath Tuff IC. Both the fauna and sediments

immediately beneath Tuff IF at FLK-N (loc. 45a) show that the climate at that time was appreciably drier than that prevailing between eruption of the Bed I lavas and Tuff ID.

The geologic history of Bed I can be briefly summarized as follows:

1. The oldest deposits of Bed I are tuffs overlying the Naabi Ignimbrite in the western part of the gorge. They accumulated on a land surface sloping northeast, and presumably they interfinger with lacustrine sediments in an area not now exposed. Most of the tuffs were probably erupted from Ngorongoro, and the lowermost are very likely 2.0 to 2.1 million years.

2. Approximately 1.85 m.y.a., a large volume of lavas flooded the eastern part of the basin, displacing the lake westward to a position which it occupied for the remainder of Bed I. The bulk of the lavas are trachyandesites from Olmoti; the remainder are olivine basalts erupted from a vent which lay to the south of the gorge.

3. The lake was shallow and fluctuated in depth and areal extent, largely in response to changes in the balance of inflow and evaporation. Lacustrine clays accumulated to a thickness of about 26 m in the center of the basin and were deposited widely to the east over the top of the lavas up to the deposition of Tuff IB.

4. Tuff IB, erupted 1.79 ± 0.03 m.y.a., was the first of a series of deposits produced in a major explosive phase of Olmoti which continued into the lower part of Bed II. These deposits were reworked to form an alluvial fan which displaced the margin of the lake westward from its eastermost limit above the lavas. The fan deposits may well span a period of 50,000 to 75,000 years, as based on the several paleosols and thickness of lacustrine clays between Tuffs IB and IF in the center of the basin.

5. The climate was probably semiarid but somewhat moister than that in the same area today. A period of relative desiccation seems to be indicated for at least the upper part of the interval between Tuffs ID and IF, roughly 1.70 m.y.a.

HOMINID REMAINS AND ARCHEOLOGIC MATERIALS

Evidence of hominid activities in Bed I is largely restricted to the eastern lake-margin deposits. These sediments have thus far yielded fourteen hominid fossils, either *in situ* (nos. 4, 5, 7, 8, 10, 35, 39, 43) or from the surface where assignment to Bed I can reliably be inferred (nos. 6, 24, 33, 44, 45, 46). In addition, three more surface finds may either be from Bed I or Bed II (nos. 27, 31, 42). Eighteen archeologic sites have been found in these deposits (Appendix B). These hominid materials occur at many levels between the lavas and Tuff IF. The geographic distribution of these materials differs considerably with reference to Tuff IB. Below the tuff, they are confined to an eastern area including localities 10 to 14 and 34. Most of these come from the uppermost 2 m of sediment below the tuff. Above Tuff IB, hominid remains and artifacts are found only to the west, in the vicinity of FLK (locs. 45, 45a, 45b, 45c) and HWK (loc. 42). The two groupings are, however, more or less comparable with regard to the former edge of the lake. At DK (loc. 13), fossil papyrus rhizomes and large quantities of crocodile remains testify to permanent water nearby. The stone circle and occupation site at DK overlie a paleosol developed partly on the eroded surface of a tuff and partly on the basalt where it rose above the level of the tuff. The surface of the tuff showed a number of narrow, steep-sided channels 45 to 60 cm deep that resemble game trails leading to the edge of the lake in Ngorongoro (M. D. Leakey, 1971a, p. 22-23).

The occupation site at FLK-NN (loc. 45b, site 38b, level 3) is on the weathered surface of a thin claystone above Tuff IB. "The many rootlet holes and reed casts in Tuff IB and occurrence of numerous fish and amphibian remains, together with bones of waterfowl, indicate that the site was situated near the shores of a lake or by a swamp" (M. D. Leakey, 1971a, p. 42). The

occupation site at FLK (loc. 45), termed the "Zinjanthropus" floor, lies on a lacustrine claystone beneath Tuff IC. The top of the claystone is a weakly developed paleosol, slightly uneven, and with a small channel and oblong depression (M. D. Leakey, 1971a, p. 49). Tuff IC contains abundant rootcasts of marshland vegetation. At FLK-N (loc. 45a), the topmost 1.5 m of claystone beneath Tuff IF has yielded five implement-bearing levels. The lowest level (no. 6) was a butchering site where artifacts were associated with the skeleton of an elephant. The highest level (nos. 1 and 2) was an occupation site comparable in many respects to the "Zinjanthropus floor." Sites at the intermediate levels have no marked concentrations of artifacts. These claystones were exposed at the surface too briefly to develop paleosols. The shoreline may well have been a kilometer or so to the west of these sites for much of the time represented by these levels at FLK-N. Scattered artifacts have been encountered in both excavations and natural exposures at other levels between Tuff IC and level 6 at FLK-N. They seem to be most widespread in the massive tuff overlying Tuff ID, which contains diatoms, ostracods, and rootmarkings of marshland vegetation (sites 41e, f).

Artifacts have been found in two places within the western lake-margin facies (locs. 61, 62a; Appendix B). Neither site has been excavated. The lowest of these sites, at locality 62a, lies stratigraphically below the lowest of the excavated archeological sites to the east. The scarcity of archeological sites in the western lake-margin deposits may reflect a lack of game or appropriate vegetation, or of both. Infrequent flow of the streams along the western side of the lake is another possible contributing factor.

5

BED II

STRATIGRAPHY AND DISTRIBUTION

Bed II is an extremely varied sequence generally 20 to 30 m thick. It is of special geological interest because of the many lithofacies and environments it represents. It is particularly important from the standpoint of archeology because it records the cultural change from Oldowan to Developed Oldowan and Acheulian. Its stratigraphy is complex, and correlation is difficult in many places. Following Reck's usage, Bed II is taken to include the deposits between Tuff IF and reddish-brown sediments of Bed III. Site JK (loc. 14) can serve as a type section, inasmuch as the contacts with Beds I and III are clearly exhibited (Plate 2). Fortunately, Tuff IF is widespread and establishes a uniform base for Bed II through most of the Olduvai region. In the few places where the tuff is missing, the contact can be difficult to locate. The contact between Beds II and III is generally sharp and disconformable in the eastern part of the Main Gorge and in the Side Gorge. Farther to the west, there is commonly some uncertainty about the contact between Bed II and Beds III-IV (und.).

Bed II is found over a somewhat larger area than Bed I. Within the Main Gorge, Bed II pinches out against basement rocks and the Naabi Ignimbrite, only slightly west of the pinchout of Bed I. Bed II overlies the Laetolil Beds near Kelogi (loc. 101), beyond the southernmost extremity of Bed I, and it crops out in the faulted area near the southwestern margin of Olbalbal as much as 6 km southwest of the mouth of the gorge. The lower part of Bed II is exposed along the fault scarp 5 km north of the gorge (locs. 201, 202) and adjacent to another fault scarp 15 km to the northeast (loc. 200). Finally, lapilli tuffs probably representing the lowermost part of Bed II crop out above lavas at the foot of Olmoti 8 km east of the mouth of the gorge.

A widespread disconformity serves to subdivide Bed II into two major units of differing lithology and paleogeographic setting. Deposits below the disconformity have many features in common with Bed I, whereas the overlying deposits have some features in common with Beds III and IV. The different areas of Bed II will now be considered from the standpoint of subdivision and correlation.

Bed II in the Side Gorge and Eastern Part of the Main Gorge

Bed II can be subdivided on a rather fine scale in parts of the Main and Side gorges (Figs. 19, 20), and many individual beds, particularly tuffs, can be used for correlation. The Lemuta Member is a widespread, distinctive sequence of tuffs which underlies the disconformity in the eastern part of the Main Gorge. It consists mostly of eolian tuffs and was earlier named the Eolian Tuff Member (Hay, 1967a). It is widely exposed throughout the eastern 9 km of the Main Gorge and crops out at intervals for a distance of 15 km northeast from the gorge and 6 km to the south. The name Lemuta is taken from Lemuta Hill, to the northwest of the gorge, which is shown on the Ngorongoro special sheet (1 × 250,000) a map published in 1962 by the Survey Division, Ministry of Lands, Forest and Wildlife, Tanganyika. The Lemuta Member has a maximum known thickness of 14.5 m (loc. 1a), but is generally between 3 and 8 m thick. In most places it can be subdivided into lower and upper units of differing lithology and composition. The lower unit differs from the upper by a lighter color, higher proportion of interbedded conglomerate, and the dominance of trachytic over mafic tephra. Limestone is very rare in the lower unit but common in easterly exposures of the upper unit. The upper unit is missing in a

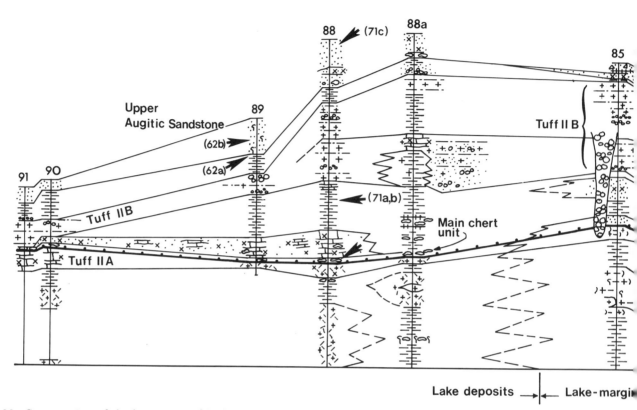

Figure 20. Cross-section of the lower part of Bed II between the base of Bed II and the upper augitic sandstone from Long K (loc. 38) to SHK-west (loc. 91). Base of Bed II is taken as horizontal. Lithologic symbols are the same as in Figure 19. Archeologic sites described by M. D. Leakey (1971a) are indicated by numbers in brackets.

few places (locs. 32a-33b), but it extends westward beyond the lower unit and bifurcates into two beds, the lower of which is Tuff IIA. The lithology of the Lemuta Member is described in more detail as the *eolian facies,* and Tuff IIA is described below together with the other key marker tuffs.

The four principal marker tuffs have been designated Tuffs IIA, IIB, IIC, and IID (Hay, 1971). These tuffs are typically reworked and discontinuous, and some are not readily recognizable as tuffaceous, either because of reworking or contamination by older detritus. The reader should note that the terms Tuff IIA, IIB, and so forth, are names for stratigraphic units, usually single beds, which vary in their content of tephra and in places are not actually tuff. Tuff IIA, for example, is locally a siliceous

earthy claystone with only about 10 percent of tephra. Tuff IID is the most widespread of the marker beds. Several additional tuffaceous units are used here to supplement the principal marker tuffs. These are most numerous and readily identifiable in the Main Gorge between localities 34 and 85. Of these the *bird-print tuff,* between Tuffs IIB and IIC, is especially significant in correlating between eastern and western parts of the Main Gorge.

A thin stratum rich in chert nodules widely overlies Tuff IIA in the Side Gorge. It is termed the *main chert unit* and is correlated on its stratigraphic position and mineral content with the lower part of the lower augitic sandstone, which disconformably overlies the Lemuta Member to the east. This correlation gives the basis for establishing the paleogeography at the time

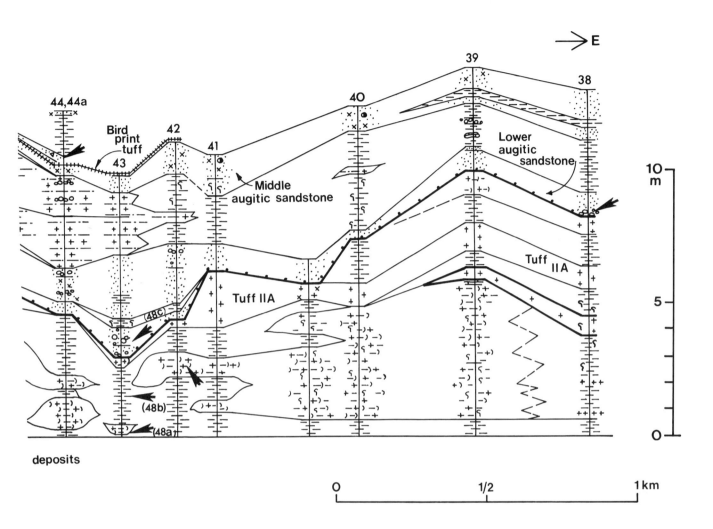

deposits

when chert was used extensively as a raw material for tools (Stiles et al., 1974).

The contact of Beds II and III can be readily demarcated in areas where Beds II and III are lithologically different, but careful examination is needed toward the east, where the upper part of Bed II is reddish-brown and resembles Bed III. Here no single criterion can be used. In some places the contact is a surface of channeling, and in some other places the lowermost stratum of Bed III is a distinctive trachyte tuff. Elephant Karonga (loc. 32) is a key locality in showing that the base of the tuff represents the same horizon as the channeling.

Key tuffaceous beds for correlating in the Side Gorge and eastern part of the Main Gorge are described in the following paragraphs, and mineral data are given in Table 11. The beds are depicted in Figures 19 through 22.

1. *Tuff IIA.* Tuff IIA is a bed of eolian tuff that extends nearly continuously from localities 33*b* and 12*b* west to FLK-NN (loc. 45*b*) in the Main Gorge and to SHK (loc. 91) in the Side Gorge. Tuff IIA in the Main Gorge is 30 cm to 2 m thick, and it grades westward from yellowish-brown to medium gray eolian tuff to cream-colored siliceous earthy, tuffaceous claystone in the vicinity of HWK (loc. 42). In the Side Gorge, Tuff IIA is represented principally by a yellowish bed of calcareous zeolitite and tuffaceous zeolitic limestone 8 cm to 1 m thick. Relatively pure vitric tuff was found in the Side Gorge only at PEK (loc. 88*a*).

Tephra is chiefly of mafic nephelinite composition and consists largely of rock fragments, vesicular vitroclasts, and augite crystals. The vitroclasts are generally 0.2 to 0.7 mm long and contain microphenocrysts of augite and less commonly nepheline, melilite, ilmenite, and perovskite. Biotite crystals, curiously, are common

57

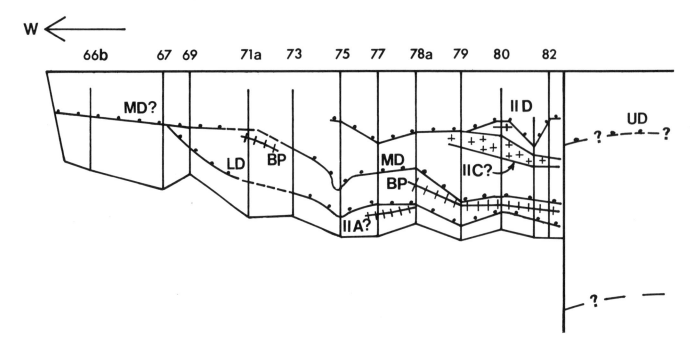

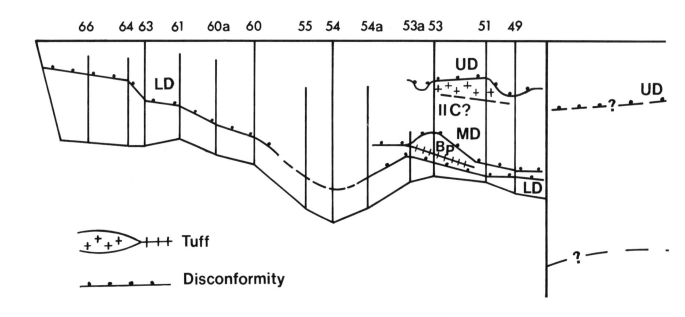

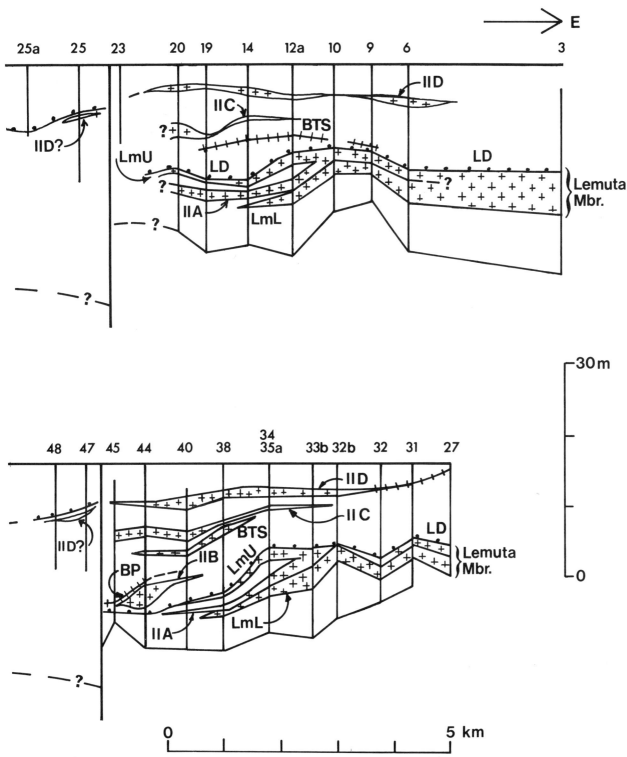

Figure 21. Cross-section showing marker tuffs and disconformities of Bed II as exposed along the Main Gorge. Section is reconstructed with the inferred top of Bed II taken as horizontal. Vertical lines represent measured stratigraphic sections. Abbreviations are as follows: IIA, Tuff IIA; IIB, Tuff IIB; IIC, Tuff IIC; IID, Tuff IID; LmL, lower unit of Lemuta Member; LmU, upper unit of Lemuta Member; BP, bird-print tuff; BTS, brown tuffaceous siltstone; LD, lower disconformity; MD, middle disconformity; and UD, upper disconformity.

Table 11. *Mineral Content of Tephra in Tuffaceous Marker Beds of Bed II*

	Tuff or tuffaceous unit	Olivine	Augite	Sodic augite	Brown Hornblende	Biotite	Ilmenite	Melanite	Perovskite	Sphene	Plagioclase	Anorthoclase	Nepheline	Melilite
Side Gorge and eastern part of Main Gorge	Mafic tuffs above Tuff IID	–	xx	–	+	–	x	–	–	–	+	–	+	–
	Tuff IID	–	x	xx	xx	–	+	–	–	+	–	x	x	–
	Tuff IIC	–	xx	–	–	–	–	–	–	–	+	–	+	–
	Upper augitic sandstone	–	xxx	xx	xx	–	xx	+	x	–	+	–	+	–
	Bird-print tuff	–	x	–	–	–	+	–	–	–	xx	–	–	–
	Middle augitic sandstone	–	xxx	xx	xx	+	xx	–	x	–	+	–	+	–
	Tuff IIB	–	xx	x	+	xx	xx	–	x	–	–	–	x	+?
	Lower augitic sandstone	–	xxx	xx	xx	+	xx	–	x	–	–	–	+	–
	Tuff IIA	+	xxx	–	+	– to x	xx	x	x	–	–	–	x	– to x
	Mafic tuffs below Tuff IIA	+	xx	+	–	–	+	+	– to +	–	–	–	–	–
West of Fifth Fault	Mafic tuffs above Tuff IID	–	xx	–	–	–	x	–	–	–	+	–	+	–
	Tuff IID	–	x	xx	xx	–	x	–	–	+	–	x	x	–
	Mafic tuffs between bird-print tuff and Tuff IID	–	xxx	–	–	–	x	–	–	–	+	–	+	–
	Bird-print tuff	–	x	–	–	–	+	–	–	–	xx	–	–	–
	Mafic tuffs and augitic sandstones	–	xxx	x	xx	+	xx	+	x	–	–	–	+	–
	Mafic tuffs of Lemuta Member(?)	–	xxx	–	x	+	xx	+	x	–	–	–	+	–
	Nephelinite tuff near base of Bed II (locs. 53, 75-78a)	–	–	xx	–	x	+	x	+	x	–	–	xx?	–

NOTE: Frequency symbols are: xxx = abundant, xx = common, x = between rare and common, + = rare, and – = absent.

in the eastern part of the area, but are extremely rare to the west (for example, locs. 45b, 88, 91). The tephra of Tuff IIA is mineralogically indistinguishable from that of the uppermost tongue of the Lemuta Member (locs. 33a-38, 11a-14).

2. *Lower augitic sandstone.* This deposit, chiefly of augite-rich sandstone, disconformably overlies the Lemuta Member on the south side of the Main Gorge between FLK (loc. 45) and WK (loc. 36). It is highly lenticular and as much as 4 m thick. A basal layer, found in the vicinity of HWK (locs. 42-44) and VEK (loc. 85), is a bed of conglomeratic sandstone rich in chert artifacts, the *sandy conglomerate* of M. D. Leakey (1971a, p. 96).

The sandstone is typically 50 to 75 percent augite and 10 to 30 percent volcanic rock fragments, mostly of contemporaneous pyroclastic origin. The mineral composition (Table 11) suggests a mafic nephelinite composition rather similar to the tephra of the Lemuta Member. A scarcity of vitroclasts and zoning of about 10 percent of the augite crystals from colorless to green are petrographic features which serve to distinguish the

tephra of this deposit from that of the Lemuta Member.

3. *Tuff IIB.* This deposit comprises tuffs and tuffaceous sediments of highly varied lithology and mineral composition. It is recognizable in most places between MCK (loc. 40) and SHK (loc. 91), a distance of 3 km, and it ranges in thickness from 30 cm to 4.3 m. Most of the deposit is an orange-brown tuffaceous siltstone and sandstone, but conglomerates are widespread, and claystones are included at PEK (loc. 88a). Claystone clasts of silt and sand size form a substantial part of Tuff IIB and are particularly abundant in the Side Gorge. In most places the tephra is of varied composition and is extensively intermixed with older detritus. The lower half of the deposit, is, however, generally characterized by biotite (Fig. 20). Relatively pure mafic nephelinite tuff with biotite forms the lower 2 meters of Tuff IIB at PEK (loc. 88a).

4. *Middle augitic sandstone.* This is a deposit, dominantly of sandstone, that is similar in most respects to the lower augitic sandstone. It is generally 60 cm to 1.5 m thick and extends from locality 35a to VEK (loc. 85), a distance of 2.5 km, on the south side of the Main

Gorge, and from JK (loc. 14) to locality 20, a distance of 1.4 km, on the north. The tephra is mafic nephelinite, generally similar to that of the lower augitic sandstone except for a slightly higher content of vitroclasts.

5. *Bird-print tuff.* This is a yellow laminated vitric tuff 2.5 to 12 cm thick that is found in most places between localities 45b and 42, to the east, and in many places to the west of the Fifth Fault. In a few places where the tuff is eroded away, clasts of the tuff are in slightly younger conglomeratic sandstone (for example, locs. 88, 88a). It is named the bird-print tuff because it contains abundant footprints of shore birds at several localities (locs. 44, 80).

The bird-print tuff is fine- to medium-grained and is at least 95 to 98 percent vitroclasts, mostly shards. Crystals of plagioclase (An_{45-55}) form most of the remainder. Vitroclasts are in most places altered to montmorillonite and zeolites. Unaltered glass has a refractive index of 1.595, and the tuff is probably of trachyandesite or basalt composition.

6. *Upper augitic sandstone.* Augitic sandstone as much as 3 m thick overlies the bird-print tuff at VEK (loc. 85), and it forms a continuous deposit in the Side Gorge as far west as SHK (loc. 91). Tephra is indistinguishable from that of the lower and middle augitic sandstones except for a lack of biotite. This sandstone may be correlative with 1.5 m of sandy tuffaceous limestone above the middle augitic sandstone at locality 20, and with mafic tuffaceous sediments to the east (locs. 14, 19).

7. *Brown tuffaceous siltstone.* A tuffaceous, fine-grained deposit of yellowish-brown, brown, or orange color is widespread on both north and south sides of the Main Gorge. It is massive and ranges in thickness from 8 cm to 1 m. It is dominantly eolian sediment, and in most places the bed consists largely of silt-size clasts of claystone cemented by zeolites. The pyroclastic fraction is chiefly altered fine-grained vitric material of unknown composition, and the remainder comprises coarser vitroclasts of mafic composition and a crystal fraction of feldspar (chiefly anorthoclase?), augite, sodic augite, hornblende, and biotite. The tephra seems to represent eruptions of varied composition.

8. *Tuff IIC.* Tuff IIC is a mafic vitric tuff and tuffaceous sandstone found widely between the Second Fault and FLK (loc. 45). East of MCK (loc. 40) it is a relatively pure vitric tuff 15 to 30 cm thick that can be either fresh and gray or altered and yellow. Westward in the Main Gorge it is a tuffaceous sandstone 15 cm to 2 m thick that is locally siliceous, clayey, and contains silicified rootcasts (for instance, locs. 44, 85). It is probably represented in the Side Gorge by a bed of tuffaceous sandstone which at SHK (loc. 91) is siliceous, earthy, and yellow.

The vitric tuff consists chiefly of mafic vitroclasts with a refractive index of 1.606. Augite forms most of the remainder. Mafic rock fragments and augite form a substantial proportion of the tuffaceous sandstone found to the west of MCK (loc. 40). A tephrite composition is suggested by a very few crystals of plagioclase and nepheline in the vitroclasts.

9. *Tuff IID.* Tuff IID is a trachytic tuff of distinctive composition and lithology that is nearly continuous on the south side of the Main Gorge, between localities 27 and 45, where it is generally 60 cm to 1.5 m thick. The lower half of the tuff is commonly laminated, and the upper half is massive and rootmarked. It is discontinuous and lithologically more variable on the north side of the Main Gorge. It is pale gray where fresh and yellow or reddish-brown where altered to zeolites. In the Side Gorge it is generally represented by 15 to 30 cm of tuffaceous claystone and sandstone, and pure tuff was found only at MNK (loc. 88).

The tuff is fine- to medium-grained and is about 90 percent vitroclasts, principally pumice, which rarely exceed 1 mm in diameter. Most of the glass is trachytic (n = 1.520), but mafic shards are in a few samples in the Main Gorge and may locally equal or exceed the trachytic vitroclasts in the Side Gorge. Anorthoclase, hornblende, augite, and rare nepheline suggest a phonolitic trachyte composition. Rock fragments are dominantly trachyte and phonolite, and they predominate over pumice in the basal part of the tuff in a few places (for example, loc. 40).

10. *Tuffs above Tuff IID.* Many and varied tuffs and tuffaceous deposits lie between Tuff IID and the top of Bed II. Tuffs in the Main Gorge are dominantly trachytic, whereas those in the Side Gorge are dominantly mafic (basaltic?). The most important bed for correlating in the Side Gorge is a sandy, tuffaceous siliceous earth as much as 3.4 m thick. It is cream-colored and contains abundant siliceous woodcasts and rootcasts. Its tephra includes mafic and trachytic vitroclasts. Earlier, this was miscorrelated with Tuff IID, which lies a short distance below (Hay, 1971). On the basis of trachytic pumice with thin, tabular feldspar crystals, this bed is correlated with a trachyte tuff 50 cm to 1.2 m above Tuff IID at localities 19 and 85. It may well be correlative with tuffaceous earthy claystone overlying Tuff IID at JK (loc. 14).

Bed II in the Western Part of the Main Gorge

Bed II in the Main Gorge west of FLK (loc. 45) and PLK (loc. 23) differs greatly in lithology from its exposures to the east. It can be subdivided widely on the basis of tuffs and disconformities (Fig. 22), but these subdivisions can be correlated only to a limited extent with subdivisions based on marker beds to the east (Fig. 21). Moreover, the top of Bed II can be difficult to locate in this western area, as Bed III is not a

61

distinctive, mappable unit characterized by volcanic detritus and a reddish-brown color.

Only the upper 6 to 10 m of Bed II crop out between FLK and the Fifth Fault. Pits and auger holes were put down at localities 23, 25, and 47, making total measured sections of 13.4, 16.8, and 16.8 m, respectively, at these localities. From localities 24 and 47 westward, the known part of Bed II comprises an upper unit, 5.5 to 10 m thick and chiefly sandstone, disconformably overlying greenish claystones with a few interbedded sandstones and mafic tuffs. There was considerable uncertainty earlier (Hay, 1971) about whether to assign the sandstone-rich section to Bed II or to Beds III-IV (und.). Present evidence strongly favors correlation with the uppermost part of Bed II. The most convincing evidence is similarity of mafic tephra to that of Bed II above Tuff IID, particularly in the Side Gorge. In addition, pebbles almost certainly of Tuff IID are common in some of the sandstones. Clasts of similar tuff are widespread and common in the upper part of Bed II to the west of the Fifth Fault, and a few similar clasts were found 3 to 4 m below the top of Bed II at PLK (loc. 23). Additional features which help to identify the sandstones as belonging to Bed II are ostracods and rounded, sand-size crystals of calcite. Similar rounded crystals are common in the uppermost sandstones of Bed II to the west of the Fifth Fault and are extremely rare and generally confined to the lowermost sandstones of Beds III-IV (und.) within the gorge. Ostracods are known from various places above the level of Tuff IID (for instance, locs. 23, 25, 47) but have not been found in Beds III-IV within the gorge.

Clasts of Tuff IID indicate that the uppermost, sandstone-rich unit of Bed II postdates the tuff. Neither the sandstones nor the disconformity at their base can, however, be directly correlated with features in Bed II to the east. The greenish clays beneath the sandstones contain a few sandstone and tuff interbeds, but none of these is obviously correlative with beds in Bed II to the east. The thickest of these tuffs, 30 to 60 cm thick, is an unusual deposit that is widespread less than a meter below the top of the greenish claystones. It is yellow to orange, and contains abundant molds of trona and gaylussite in a matrix of calcite and clinoptilolite, a zeolite. Only mafic vitroclasts are recognizable, but clinoptilolite is not normally an alteration product of mafic glass. It is possible that formerly dominant trachytic vitroclasts have wholly disappeared in altering to form the clinoptilolite. If this is the case, the tuff probably represents Tuff IID.

The full thickness of Bed II is exposed in many places between the Fifth Fault and the western margin of the basin (Fig. 22). Over much of this area it can be subdivided into four major units on the basis of three disconformities. The lowermost disconformity is 70 cm to about 5 m above the base of Bed II, and it generally separates claystones from overlying tuffaceous sandstones. It can be recognized nearly everywhere except for an area in the vicinity of the westernmost fault, where the contact between claystones and sandstones is gradational (locs. 54-58, 71a-73). This lowermost disconformity is almost certainly correlative with that above the Lemuta Member in the eastern part of the Main Gorge.

The next disconformity, 2.5 to 7 m higher, is obviously erosional to the east, where it separates conglomeratic sandstones from greenish, waxlike claystones, which locally contain the bird-print tuff. Clasts of the bird-print tuff are found locally in the conglomeratic zones above the disconformity. Farther west, the disconformity is not well marked, and it is generally taken as the contact between a dominantly lacustrine and an overlying dominantly fluvial sequence. This disconformity may be correlative with the erosional base of the upper augitic sandstone at localities 85, 88, and 88a. Clasts of the bird-print tuff are at the base of this sandstone.

The highest disconformity is an uneven surface that includes steep-walled channels 2 m deep and lies 5 to 11 m below the top of Bed II. It is typically overlain by conglomerate with clasts of Tuff IID, and it is correlated with the disconformity beneath the uppermost sandstone unit to the southeast (locs. 25, 47, 48). This disconformity was earlier (Hay, 1971) thought to separate Bed II from Beds III-IV (und.), and several archeologic and faunal sites of M. D. Leakey (1971a) in the uppermost part of Bed II were incorrectly assigned to Beds III-IV. These localities and sites are as follows: loc. 25 (FK, sites 12a, 12b), loc. 49 (site 35), loc. 52

(MLK-E, site 34), loc 53 (MLK-W, site 33), loc. 77 (Kar K, site 3), loc. 79 (site 4), and loc. 82a (site 6).

Several tuffs and tuffaceous deposits are widespread in Bed II to the west of the Fifth Fault, and they provide a means of correlating with subdivisions of Bed II to the east and south (Figs. 21, 22). A thin tuff possibly representing the Lemuta Member is found 3 m above the base of Bed II (locs. 77, 78a, 77b). The bird-print tuff lies 3 to 6 m above the base of Bed II and constitutes one of the firmest ties between the eastern and western parts of Bed II in the Main Gorge. Sandstones rich in augitic tephra form most of the interval between the lowermost disconformity and the bird-print tuff. This sandstone sequence is correlated with the combined lower and middle augitic sandstones. Reworked mafic tuffs and tuffaceous sandstones lie at various levels between the bird-print tuff and the third disconformity. The thickest and most continuous of these tuffaceous units, near the middle of Bed II (locs. 79-82), may represent Tuff IIC (see Table 11). A part of Tuff IID was found in place undisturbed beneath the uppermost disconformity at RHC (loc. 80; Plate 3). Elsewhere to the west of the Fifth Fault it is represented only by tuff clasts above the disconformity (Plate 4). Finally, mafic tuffs and tuffaceous sandstones in the uppermost part of Bed II are mineralogically similar to and probably correlative with mafic tuffs above Tuff IID to the east and south, particularly in the Side Gorge (esp. locs. 93, 94). Key tuffs and tuffaceous deposits in Bed II west of the Fifth Fault are described in the succeeding paragraphs and mineralogic data are given in Table 11.

1. *Tuff of the Lemuta Member*. A bed of orange altered vitric tuff is locally (for example, locs. 77-78a) about 3 m above the base of Bed II. The tuff is 2.5 to 15 cm thick and is mineralogically similar to nephelinite tuffs of the Lemuta Member.

2. *Augitic sandstone*. Augite-bearing sandstone constitutes a tuffaceous deposit of lenticular shape as much as 9.4 m thick which lies between the lower and middle disconformities of Bed II. Most of the augite is of pyroclastic origin, and a few mafic tuffs are interbedded in the sequence. In many places the sandstones can be subdivided into two units of differing lithology, which are locally separated by claystone with chert nodules. The lower unit is 15 cm to 1 m thick and is dominantly coarse-grained. It contains abundant coarse euhedral cal-

cite crystals and a relatively small amount of augite and associated tephra. The upper unit thins eastward from a maximum of 8.5 m (loc. 54) to a minimum of 7.5 cm (loc. 79). It commonly has an erosional base and cuts out the lower unit in several places. The upper unit is dominantly medium-grained and contains a substantial proportion of augite and related tephra. On the basis of tephra composition and stratigraphic position, the sandstones and interbedded tuffs are correlated with Tuff IIB and the lower, middle, and possibly the upper augitic sandstones of the Main Gorge. The lower unit of the sequence may possibly be correlative with the lower augitic sandstone.

3. *Bird-print tuff.* The bird-print tuff is 1 to 12 cm thick and indistinguishable in texture and content of pyroclastic minerals from equivalent tuff to the southeast. It lies above the eastern, thinner part of the lenticular unit of augitic sandstones; farther west (loc. 71a), in one of the thicker sections of augitic sandstone, it lies 1.2 m below the top.

4. *Mafic tuffs and tuffaceous sandstones.* Reworked tuffs and tuffaceous sandstones are widespread as lenticular beds between the bird-print tuff and conglomerates with clasts of Tuff IID (Plates 3, 4). The largest concentration of tephra is in 3.7 to 5 m of beds in the upper part of this interval. Tuffs are vitric and consist largely of shards but contain droplets and fragments of ropy spatter. Augite forms 1 to 5 percent of most tuffs and 5 to 15 percent of associated tuffaceous sandstones. The mineral content suggests a tephrite composition. These may be correlative, at least in part, with Tuff IIC.

5. *Tuff IID.* A part of Tuff IID was found undisturbed only at RHC (Loc. 80). This undisturbed portion, 2 m long and 45 to 60 cm thick, lies 1.2 m above cliff-forming tuffaceous sandstones of Tuff IIC(?) (Plate 3). Elsewhere Tuff IID has been stripped away in erosion of the uppermost disconformity, and clasts of the tuff are widespread above the disconformity. Tuff clasts are largest and most angular at RHC, where they have been transported only a short distance (Plate 4).

6. *Tuffs and tuffaceous sandstones above Tuff IID.* Augite-bearing mafic tuffs and tuffaceous sandstones form about a tenth of the section above the level of conglomerates with clasts of Tuff IID. Relatively pure tuff was found in localities 70 and 72; elsewhere, it is contaminated with detrital sand. Vitroclasts are the dominant form of tephra, and the mineral composition suggest a tephrite composition, similar to most of the tuffs above the level of Tuff IID in the Side Gorge.

Bed II near Kelogi and Outside of the Gorge

A 6-meter thickness of Bed II overlies Bed I and the Laetolil Beds near Kelogi (locs. 99-101; see

Fig. 41, Chap. 6). This section consists largely of reworked tuffs and brown claystone which represent the lower part of Bed II, below the lowermost disconformity. This assignment is based on the overall lithologic similarity of these beds to the lower part of Bed II elsewhere. Particularly significant in this correlation is a bed of siliceous earthy claystone (loc. 99), which is common in the lower part of Bed II to the northeast (for example, locs. 42-44). Beds III-IV (und.) and the Masek Beds unconformably overlie this section of Bed II.

Bed II crops out along the western margin of Olbalbal both north and south of the gorge. Exposures to the north (for instance, loc. 1) compare closely in lithology and thickness with those of Bed II between the Second Fault and the mouth of the gorge. Only the Lemuta Member continues south of the gorge for an appreciable distance. Tephra deposits below the Lemuta Member are at most 60 cm thick at locality 157 and have not been found farther to the south. Beds above the Lemuta Member are only 6.2 m thick at locality 157 and likewise have not been found to the south.

The 32-meter section of Bed II to the northeast (loc. 200) comprises only the lower part of Bed II, up to and including the Lemuta Member. An ash-flow tuff 9 m thick directly underlies the Lemuta Member, which is 14.6 m thick. A 3-meter thickness of Bed II is exposed 5 km north of the gorge (locs. 201, 202). It conformably overlies Tuff IF and represents the lowermost part of Bed II, below the Lemuta Member. The section is chiefly claystone, but the lower 2 m contain interbedded mafic tuff and two horizons of chert nodules.

AGE OF BED II

Potassium-argon dating of Bed II has proved unsatisfactory, and the only meaningful date is from Tuff IIA. This tuff is everywhere reworked and is commonly contaminated. G. H. Curtis has dated the biotite in three samples, each from a different locality (Curtis and Hay, 1972). The youngest date, on the least reworked sample, is 1.71 m.y.a. (KA 2320), which may be close to the true age. Other samples gave 1.95 m.y.a. (KA 1761) and 2.08 m.y.a. (KA 1863). The date of 1.0 to 1.1 m.y.a. given in Evernden and Curtis

(1965, KA 664, 664R) was almost certainly obtained from a sample of the Ndutu Beds where they unconformably overlie Bed II near the Second Fault. The other Bed II date of 0.50 m.y.a. (KA 405) in Evernden and Curtis (1965) cannot be accepted because the sampled bed is not known. Although SHK is given as the sample locality, all of the tuffs there are reworked and contaminated by grains from the metamorphic basement. A sample from slightly above the level of Tuff IID at SHK, dated as an experiment, gave an age of 80 m.y.a. (Curtis and Hay, 1972). An unpublished date of 1.93 m.y.a. (KA 2346R) was obtained by Curtis (1971, personal communication) from obsidian fragments of the ash-flow tuff below the Lemuta Member 15 km northeast of the gorge. Despite a relatively high content (24 percent) of radiogenic argon, the date is clearly too old.

The magnetic stratigraphy of Bed II, supplemented by K-Ar dates in Bed I, points to an age of about 1.70 m.y.a. for the base of Bed II, as noted earlier (see Fig. 11). Within Bed II, normal polarity of the Olduvai event is recorded by an ash-flow tuff which underlies the Lemuta Member (loc. 200). The lowermost of twelve samples from the Lemuta Member is normal, and the others are reversed (Table 12). The normal polarity of this one tuff sample probably should not be accepted as wholly reliable without further sampling in view of the erratic results given by some of the sedimentary rocks collected higher in Bed II. Thus the top of the Olduvai event (~1.65 m.y.a.) lies either at the base or within the lowermost 75 cm of the Lemuta Member.

Highly variable but dominantly reversed polarity was measured on seventeen samples of sedimentary rocks above the Lemuta Member. A single bed can give reversed, intermediate, or unstable polarity (Table 12). Part if not most of the magnetic variability is caused by continued growth of hematite crystals after deposition. Thus reversed polarity would be acquired in growth during the reversed portion of the Matuyama epoch, but normal polarity would be superimposed in growth during the Brunhes epoch. Polarity measurements by A. Brock and A. Cox show that reversed rocks extend at least to the base of Tuff IVB, in the middle of Bed IV, and only normal rocks have been found

Table 12. *Polarity Measurements on Samples of Bed II*

Field/lab no.		Stratigraphic position, lithology	Polarity	Measured by
RH 1	(a)	Below Lemuta Member		
		Ash-flow tuff, loc. 200	N	B
	(b)	Lemuta Member		
FB 124		Eolian tuff, zeolitic; 60 cm above base lower unit, 2.7 m thick, loc. 27	N	B
74-I-7C		Eolian tuff, calcareous nodule; 75 cm above base lower unit, 2.7 m thick, loc. 27	R	B
74-I-7D		Eolian tuff, calcareous nodule, 2 m above base lower unit, 2.7 m thick, loc. 27	R	B
FB 123		Eolian tuff, zeolitic; 2.6 m above base lower unit, 2.7 m thick, loc. 27	R	B
FB 122		Eolian tuff, zeolitic, 90 cm above base upper unit, 1.8 m thick, loc. 27	R	B
FB 121		Eolian tuff, calcareous, 1.4 m above base upper unit, 1.8 m thick, loc. 27	R	B
RH 2		Eolian tuff and limestone at top Lemuta Member, loc. 27	R	B
74-1-6B		Eolian tuff, zeolitic; 60 cm above base lower unit, 3.0 m thick, loc. 33*a*	R	B
RH 24		Eolian tuff, calcareous nodule, 1.8 m above base lower unit, 3.0 m thick, loc. 33*a*	R	C
RH 23		Eolian tuff, calcareous nodule, 2.9 m above base lower unit, 3.0 m thick, loc. 33*a*	R	C
74-1-6E		Eolian tuff, zeolitic, from top of lower unit, loc. 33*a*	R	B
RH 21		Eolian tuff, calcareous, near middle of upper unit, 3 m thick, loc. 33*a*	R	C
	(c)	Between Lemuta Member and Tuff IID		
FB 125		Limestone nodule 2 m above Lemuta Member, loc. 27	R	B
FB 126		Reddish-brown mudflow deposit, 7.6 m above Lemuta Member, loc. 27	R	B
FB 127		Reddish-brown hard claystone, 2.6 m below Tuff IID, loc. 29	N	B
ML 3		Nodular limestone 60 cm below Tuff IID, loc. 44	R?	B
RH 25		Limestone about 1.2 m above the Lemuta Member, loc. 33*b*	R	C
RH 26		Limestone about 1.6 m above RH 25	I (closer to R than to N)	C
73-9-18D		Laminated sandy limestone, 45 cm below bird-print tuff, loc. 80	R	B
	(d)	Above Tuff IID		
ML 4		Reddish-brown hard claystone about 1 m above Tuff IID, loc. 44	R	B
FB 111		Same bed as ML 4, collected nearby	I	B
C31, C33		Same bed as ML 4	R	C
ML 5		Reddish-brown hard claystone about 1.5 m above Tuff IID, loc. 7	I	B
FB 115		Top of same bed as ML 5, loc. 7	R	B
FB 118		Bottom of same bed as ML 5, loc. 7	U	B
FB 119		Top of same bed as ML 5, loc. 6	R	B
FB 120		Below FB 119 in same bed, loc. 6	I	B
ML 6		Pale orange limestone, topmost bed in Bed II, loc. 45	R?	B
ML 7		Limestone near top of Bed II, loc. 44	U	B

NOTE: N = normal, R = reversed, I = intermediate, and U = unstable. The "?" refers to specimens that do not agree well, or have a low magnetic intensity. B refers to Andrew Brock and C to Allan Cox. All results are based on polarity measurements both before and after partial demagnetization.

at higher levels. Thus the Brunhes-Matuyama boundary at 0.7 m.y.a. lies at or somewhat above the middle of Bed IV. On the basis of relative stratal thickness between the top of the Lemuta Member (~1.6 m.y.a.) and Tuff IVB (~0.7 m.y.a.), in the eastern part of the gorge, the top of Bed II should be about 1.1 million years old.

The structural history of Beds II and III provides another means of getting a date for the top of Bed II. The disconformity between Beds II and III is a result of earth movements, and Bed III records displacements along several faults which had been inactive during the deposition of Beds I and II. It seems reasonable to correlate the disconformity and fault displacements at Olduvai with the beginning of a major phase of Rift-Valley faulting to the east, which Macintyre et al. (1974) dated at 1.15 to 1.2 m.y.a. by K-Ar analyses of volcanic rocks associated with the earliest faulting. This phase of faulting caused large-scale displacement along the fault which extends from Lake Natron to Lake Eyasie. The date of 1.15 to 1.2 m.y.a. based on structural history is close to that of 1.1 m.y.a. inferred from other evidence, and an age of 1.15 ($\pm$0.10?) m.y.a. is accepted here for the topmost deposits of Bed II.

LITHOFACIES AND ENVIRONMENTS

Bed II comprises seven lithofacies representing six major types of depositional environment: alluvial fan, alluvial plain, lake, lake margin, eolian and a lake-stream complex. The widespread, lowermost disconformity marks a major environmental change, and only the lake facies is found both above and below (Fig. 23). The paleogeography prior to the disconformity was essentially the same as in Bed I, and lake deposits in the center of the basin are bordered by lake-margin deposits. The lowermost of the lake-margin deposits interfinger eastward with alluvial-fan deposits; the uppermost lake-margin deposits interfinger eastward with eolian tuffs.

Above the disconformity, lake deposits are bordered by mixed fluvial and lacustrine deposits which differ vastly from the lake-margin deposits below the disconformity (see Table 13). Because of major lithologic differences, the eastern and western fluvial-lacustrine deposits

are classed as separate lithofacies. The eastern fluvial-lacustrine deposits interfinger eastward with fluvial sediments termed the eastern fluvial deposits. The western fluvial-lacustrine deposits extend westward to the margin of the basin.

Lake Deposits

Lake deposits are present in several different areas and in these varied locales they represent most of the stratigraphic thickness of Bed II. They are most widespread beneath the lowermost disconformity, and in this lower part of Bed I they extend westward from the Fifth Fault, crop out several places in the Side Gorge, and are exposed in two small areas 5 km north of the Main Gorge. Lake beds are also found at higher levels to the west of the Fifth Fault, and they underlie the highest disconformity to the east of the Fifth Fault. Despite certain features in common, the Bed II lake deposits of different areas and stratigraphic levels vary both lithologically and mineralogically.

1. *Lake deposits to the west of the Fifth Fault.* Lake beds beneath the lowermost disconformity extend westward 8.5 km from the Fifth Fault. They are 70 cm to 5.5 m thick and are chiefly greenish claystones which intergrade with reworked tuffs and sandstones near the western margin of the facies. Claystones typically contain little detrital sand and silt, and they have substantial amounts of authigenic minerals: calcite, dolomite, pyrite (now oxidized), and K-feldspar. In addition, fluorite (CaF_2) is in about half of the claystones analyzed by x-ray diffraction. Calcite occurs as euhedral crystals of varied form, the one analyzed sample of which is isotopically heavy (see Fig. 15, Chap. 4). Calcareous claystones grade into calcite-crystal limestones. Tuffs are most common near the western margin of the facies, where they generally form massive, lenticular beds. Their composition and texture suggest that they are composed of redeposited tephra eroded from Tuff IF to the west. Farther east in the facies are thin, widespread tuffs, dominantly tephrite or basalt in original composition. However, the most widespread tuff, 30 cm to 1.2 m above the base of Bed II, is nephelinite with sodic augite, a type not found elsewhere in Bed II at this level. The westernmost tuffs are altered chiefly to zeolites, and the others are altered chiefly to K-feldspar. Molds of radiating sprays of trona crystals are common in tuffs and limestones of the westernmost two-thirds of these deposits. Chert is found west from the Fifth Fault for a distance of 3.2 km. It is most common as nodules but can form lenticular beds as much as 10 cm thick. Chert

66

Table 13. *Composition of Lithofacies in Bed II*

Lithofacies and locale	Claystone	Sandstone	Conglomerate	Tuff	Carbonate rocks	Siltstone	Claystone-pellet aggregate	Siliceous earthy claystone & sandstone	Chert	Remarks
Lake deposits:										
1. Below lowest disconformity to west of Fifth Fault	70%	3%	<1%	22%	5%	0	0	0	≪1%	
2. Above lowest disconformity to west of Fifth Fault	92	3?	0	5	0	0	0	0	≪1	
3. In Side Gorge below disconformity	10	0	0	~5?	~25?	0	~10?	0	≪1	Zeolitite forms approximately 50% of these deposits
4. Between Fifth Fault and FLK	90	5	0	5	0	0	0	0	0	
Lake-margin deposits:										
1. Western part of Main Gorge	42	10	2	46	0	0	<1	0	0	
2. Eastern part of Main Gorge	64	0	0	15	3	0	<1	18	0	
3. Near Kelogi	73	<1	<1	22	<1	0	0	5	0	
Alluvial-fan deposits	6	1	4	89	0	0	<1	0	0	Includes tephra deposits coarser than tuff
Eolian deposits	2	<1	<1	88	10?	0	<1	0	0	Includes a very small amount of dolomite
Eastern fluvial-lacustrine deposits:										
Overall	50	34	1	6	6	1	~1?	1	≪1	Includes a small proportion of dolomite
1. Main Gorge	52	33	1	7	6	1	~1-2?	<1	0	
2. Side Gorge	44	36	1	2	8	2	~2?	4	<1	
Western fluvial-lacustrine deposits	10	60	8	5	12	2	~2?	0	<1	Limestone and dolomite are subequal
Alluvial deposits	72	7	2	6	2	0	0	0	0	Includes 9% mudflow deposits

is most common 2 to 3 m above the base of Bed II in either one or two levels. The one analyzed nodule is isotopically heavy ($\delta O^{18} = 38.7°/oo$).

Lake deposits above the lowermost disconformity are as much as 3.7 m thick, and they thin westward and pinch out 2.4 km west of the Fifth Fault. These deposits are chiefly claystone, and similar claystones lie farther to the west but are subordinate to sandstones and are included within the western fluvial-lacustrine facies. Claystones of the lake facies can be shades of green or brown, and they may be silty or sandy. Euhedral calcite crystals are less common than in the lacustrine clays at lower levels, and authigenic K-feldspar is absent or rare. By contrast, analcime is widespread, and other zeolites (erionite and clinoptilolite) are found in a few beds.

Tuffs are thin, evenly bedded, and mafic in original composition. Vitroclasts are entirely altered, principally to zeolites. Thin, laminated calcareous sandstones are widespread below the bird-print tuff and on the south side of the gorge and are thick enough to include within the western fluvial-lacustrine deposits. They contain footprints of shore birds at RHC (loc. 80). Chert nodules occur widely between the lower sandstone unit and the thin, laminated sandstones above. These nodules lie in cream-colored zeolite-rich laminae. An analyzed nodule has a δO^{18} value of $34.5°/oo$.

2. Lake deposits east of the Fifth Fault. As much as 10 m of lake deposits are known to underlie the uppermost disconformity between the Fifth Fault and FLK (loc. 45). These deposits are principally claystone and

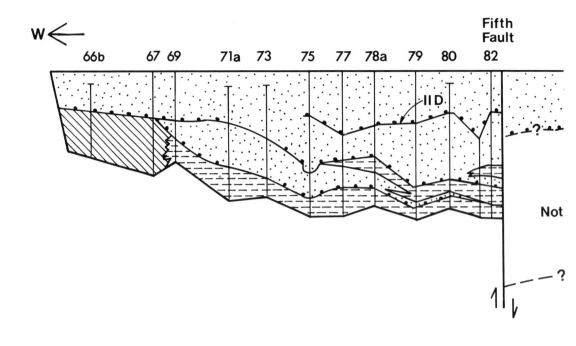

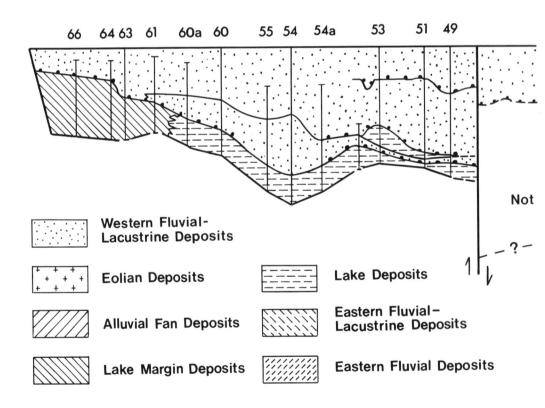

68

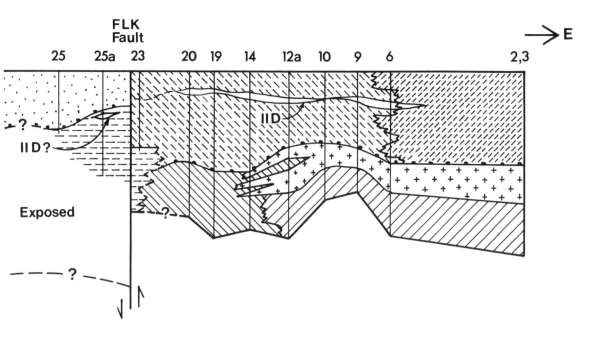

a. NORTH SIDE OF GORGE

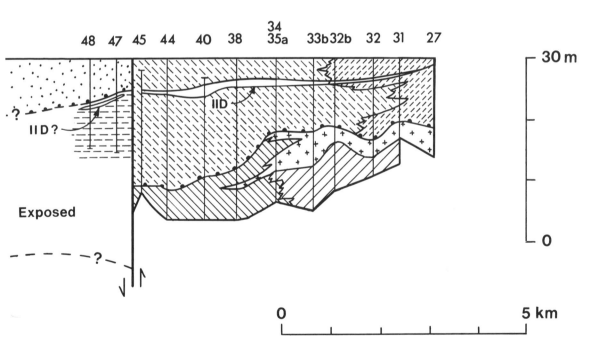

b. SOUTH SIDE OF GORGE

Figure 23. Cross-section showing lithologic facies of Bed II as exposed along the Main Gorge. Section is reconstructed with the inferred original top of Bed II taken as horizontal. Vertical lines represent measured stratigraphic sections.

69

resemble in many respects the lake beds between the lower and middle disconformities to the west of the Fifth Fault. Claystones are dominantly greenish but may be shades of gray and brown. Desiccation cracks were noted at a few horizons near the eastern margin of the facies (locs. 25, 47). Calcite crystals are common in most claystones, and they may be euhedral or rounded. Authigenic analcime is almost invariable present, averaging about 9 percent of the claystones as determined by x-ray diffraction (see Hay, 1970a). K-feldspar and other zeolites were identified in a few samples. Some of the easternmost sandstones contain appreciable sand, and these grade into clayey sandstones. The sandstones are of two types: (1) well-sorted medium-grained, and dominantly of volcanic detritus, and (2) ill-sorted, coarse-grained, and consisting largely of angular quartz clasts, some of which are as much as 1 cm in diameter. The quartz was probably from the nearby inselberg Naibor Soit. There are a few altered tuffs, largely or entirely mafic. Casts and molds of trona and gaylussite are common in the uppermost 3 m of the lacustrine sequence. Small, delicate fish bones and spines are in the sand fraction of nearly all samples, and ostracods were noted in a few.

3. *Lake deposits in the Side Gorge.* Lake deposits crop out over a distance of 1.3 km in the Side Gorge, where they form the lower 4.6 m of Bed II. These are unusual lacustrine deposits consisting largely of zeolitite and zeolitic limestone. Claystone constitutes only about 10 percent of the exposed sections. The disconformity at the top of the facies is not well developed, and in places these deposits grade upward into the eastern fluvial-lacustrine facies. Tuff IIA is taken here rather arbitrarily as the topmost stratum of the lacustrine facies.

The zeolitites and zeolitic limestones are massive, cream-colored, and contain rootmarkings or rootcasts. Zeolite minerals generally constitute 30 to 70 percent of the zeolitites, and the remainder includes detrital sand, altered vitroclasts, claystone pellets, oolites, calcite crystals (commonly rounded), clay, and fine-grained calcite and dolomite. The zeolitites grade into zeolitic limestones, and in a few places (for example, loc. 88) claystone pellets are the largest single component. As noted elsewhere, Tuff IIA is a bed of tuffaceous zeolitite in this area. Molds of gaylussite(?) were noted in limestones beneath Tuff IIA in locality 91. All of the claystones contain substantial amounts (7 to 20 percent) of zeolites, and small amounts of K-feldspar are in some of the analyzed samples. Chert nodules occur widely and are found at varying levels (see Hay, 1968). They are most widespread in Tuff IIA, although they are nowhere abundant in this bed. Two nodules from Tuff IIA gave δO^{18} values of $+34.5°/\text{oo}$ (Stiles et al., 1974).

Rootholes and rootcasts are suggestive of marshland vegetation. Small bones and spines of fish are in many of the samples, and molds of ostracods and small gastropods were noted in a few beds of zeolitite.

4. *Lake deposits north of the gorge.* The 3-meter thickness of Bed II exposed 5 km north of the gorge (locs. 201, 202) is included within the lake deposits. This section consists of olive claystones with interbedding mafic tuffs and a channel filling of augite-rich sandstone or reworked tuff. Claystones contain a wide variety of sand-size materials: calcite crystals (rounded or euhedral), mafic tephra, detrital sand, claystone pellets, oolites and fossils. Small amounts (≤ 5 percent) of finely crystalline authigenic zeolite are in the claystones. The tuffs are thin, evenly bedded, and altered to zeolites. Scattered chert nodules are in a bed of tuffaceous claystone 50 to 70 cm above the base of Bed II, and horizons with abundant nodules lie 1.2 and 1.5 m above the base.

The fossils comprise ostracods, charophytes, and small fragments of bone. The charophytes were identified by Howard Schorn (1972, personal communication), who comments as follows:

> The presence of charophytes is recorded by sinistrally spiralled, 5-celled, calcified oogonia (or gyrogonites) void of the apical coronula cells. The oogonia (or gyrogonites) are formed by deposition of $CaCO_3$ within the 5 cells that sinistrally envelop the developing female reproductive body of members of the Family Characeae. There are some six genera of extant Characeae. The greater majority inhabit clear, quiet freshwater bodies, although there are brackish water species.

These lake deposits of different areas accumulated in a single lake of fluctuating level and generally decreasing size. It was saline, alkaline, and generally similar in its sediments to the Bed I lake. The lake for the lower part of Bed II was about the same size and shape as the lake of Bed I. Salinity was highest in the western half of the lake, as indicated by widespread trona molds in this area. Lake deposits in the Side Gorge represent a zone transitional to the lake-margin environment. The origin of the massive zeolitites, although not fully understood, involved saline, alkaline solutions of sodium carbonate. Zeolites resulted from reaction of these solutions with aluminosilicate materials among which volcanic glass was a minor constituent. The major reactant(s) may have been clay (amorphous?) or aluminosilicate gel of the type formed today around hot springs at Lake Magadi, Kenya (Eugster and Jones, 1968). Saline, alkaline lake water is also recorded by rare gaylussite(?) molds and by chert nodules. Salinity was lower here than in the central and western parts of the lake as indicated by differ-

70

ences in authigenic minerals and in δO^{18} values for chert nodules. Intermittent flooding by fresh water or exposure to the air is indicated by dissolved ostracod tests and gastropod shells, and by etching (that is, solution) of sand-size calcite crystals in some beds. Rootmarkings were made by marshland vegetation, which indicates fresh or brackish water. Lake deposits 5 km north of the gorge (locs. 201, 202) have mineralogic features indicative of saline, alkaline lake water, but the charophytes point to fresh or brackish water. The stream channel records subaerial exposure.

The lake was abruptly reduced in both size and salinity following the lowermost disconformity. Authigenic K-feldspar is rare in the lake deposits above, and evidence of trona is found only in a rather thin zone in the upper part of Bed II. The δO^{18} value of $+34.5°/oo$ for a chert nodule above the disconformity compares with that of $+38.7°/oo$ for a nodule in the same area below it. The lake fluctuated in level, as indicated by sandstone interbeds in the claystones. Some areas were exposed to the air, as shown by desiccation cracks. The coarse, angular quartz of the ill-sorted sandstones was very likely transported from Naibor Soit over exposed mudflats by sheetwash or small streams.

This stage of the Bed II lake presents a geochemical anomaly, inasmuch as salinity would be expected to increase in a lake of decreasing size. Possibly fault movements caused the lake to decrease in size by creating fractured zones that allowed lake water to flow underground to a lower base level in the hydrologic system. Underground leakage would reduce the degree of evaporative concentration and hence the salinity of the lake. Eugster (1970) has inferred a hydrologic system of this sort for the Rift Valley in Kenya, where several lakes drain through alluvium and underground passageways to Lake Magadi, the lowest point in that part of the Rift Valley.

Lake-Margin Deposits

Lake-margin deposits are a recognizable facies only beneath the lowermost disconformity, and they border the lake deposits on the east, west, and south. They have many lithologic features in common with the lake-margin deposits of Bed I.

1. *Eastern lake-margin deposits.* Lake-margin deposits extend eastward a distance of 3 km from FLK-NN (loc. 45b) and VEK (loc. 85). These deposits are 2.5 to 9.7 m thick and intergrade eastward with alluvial-fan and eolian deposits. The westernmost outcrops have some mineralogic features usually diagnostic of Bed II lake deposits, and the transition to the lacustrine facies presumably lies only a short distance west. These eastern lake-margin deposits are 82 percent claystone, 15 percent tuff, and 3 percent limestone, siliceous earth, and zeolitite. Most of the claystones appear waxlike, but others have an earthy luster, caused by a substantial proportion of small particles of biogenic opal.

The more common, waxy claystones are chiefly yellow to brown and commonly contain rootmarkings, usually of marshland vegetation. Silicified fragments of plant stems are in many beds to the west of locality 41. Claystones generally have rather little ($\leqslant 5$ percent) detrital silt and sand. Detritus of sand size is principally volcanic but may include minerals of basement origin. A very few subangular clasts of Engelosin phonolite about 1 cm long were found in one bed of sandy claystone (loc. 42b). Microscopic study shows that one-third to one-half of the claystones contain small particles of biogenic opal. Generally the opal particles amount to only a few percent or less, but they may exceed 10 percent, which is about the point at which the claystones develop an earthy luster and are classed as *siliceous earthy claystones.*

Beds of siliceous earth as much as 30 cm thick are found in the lower 2 to 3 m of Bed II along the east side of VEK (loc. 85) and in a small excavation near the south end of the main FLK site (loc. 45). These beds are white to cream-colored and commonly contain silicified roots and stem fragments. With increasing clay, the siliceous earths grade into siliceous earthy claystones, which constitute most of the facies in a few places (for example, locs. 39, 40). They form massive cream-colored beds 30 cm to 2 m thick that invariably have siliceous rootcasts. The bulk of biogenic opal in both the earths and earthy claystones is in the form of diatoms, spore cases of chrysophyte algae, phytoliths, and silicified plant remains. The earthy claystones contain a small percentage of detrital silt and sand, tephra, and claystone pellets. The thick bed at VEK also contains angular and subangular clasts of basement rock and lava 1 to 2 cm in diameter.

Tuffs are most common in the eastern part of the facies and characteristically form massive, rootmarked lenticular beds. All are reworked, and all but one are dominantly mafic. The exception is a tuff at the base of Bed II in HWK (near locs. 42, 43) that consists of trachytic materials eroded from Tuff IF. Many tuffs contain particles of opaline silica, and opal-bearing tuffs grade into siliceous earthy claystones. Tuff IIA, the most widespread of the tuffs, is dominantly a medium-gray eolian tuff as far west as HWK, where it grades into cream-colored siliceous earthy claystone. In its western-most exposure it is tuffaceous zeolitite similar to that of Tuff IIA in the Side Gorge. The eolian tuff is massive, varies from 30 cm to 2 m in thickness, and exhibits lenticular bedding of an unusual sort in which the lower surface is relatively even and the upper surface undulates. The thicker parts of the undulating bed probably represent low dunes. The eolian tuff has the texture of a rather well-sorted medium-grained sand, generally with a clay matrix. Some of the rock fragments have been rounded and polished by wind transport. Vitroclasts are altered and the tuff cemented by zeolites in a few places (for example, locs. 19, 38). The upper tongue of the Lemuta Member grades westward from eolian tuff into both siliceous earthy claystone (locs. 38-40) and soft, chalklike limestone (locs. 15a, 19).

Limestone occurs most widely as thin, nodular beds and lenses of white or pale gray color. They are finely crystalline and relatively pure calcite. With decreasing size, the nodular beds grade into nodular concretions. A few beds are of relatively soft, cream-colored limestone having a chalklike appearance. These contain admixed clay, biogenic opal, or tuffaceous materials.

Authigenic zeolites vary inversely with biogenic opal and unaltered vitroclasts in the eastern lake-margin deposits. Zeolites are almost invariably present in the opal-free claystones, which are most common near the eastern and western margins of the facies. Similarly, zeolites are characteristic of tuffs with altered vitroclasts and are absent or rare in tuffs with fresh vitroclasts. The authigenic minerals clinoptilolite, kenyaite ($NaSi_{11}O_{20.5}(OH)_4.3H_2O$), and chalcedonic quartz are found in the eastern lake-margin deposits at VEK (loc. 85; Hay, 1970a). This mineralogic assemblage has a significant degree of geochemical similarity to that of the lake deposits 650 m distant in the Side Gorge.

Coarse, vertical rootmarkings suggesting marshland vegetation are common throughout the facies. Rootmarkings probably of grass or brush are common, however, and they are the dominant type near the eastern margin. Widespread lacustrine plant fossils are diatoms and spores of chrysophyte algae. Phytoliths and silicified plant remains are also widely distributed. The phytoliths are quite varied and probably represent both grasses and nongrass vegetation (see Rovner, 1971). The facies contains abundant faunal remains, among which water- and swamp-dwelling forms predominate (M. D. Leakey, 1971a). Avian and crocodile remains are abundant in the lower 1.5 m of one excavation at HWK (sites 48a, 48b), and deinothere skeletons were excavated both at the base of Bed (loc. 44, site 46b) and 1.8 to 2.1 m above the base (loc. 45a, site 40g). Hippo and crocodile teeth were noted in siliceous earthy claystones at widely scattered localities. The basal clays of Bed II contain four species of fresh-water molluscs (M. D. Leakey, 1971a p. 253), and small fish bones are common in the sand fractions from claystones of various levels.

2. *Western lake-margin deposits.* Lake-margin deposits to the west of the Fifth Fault lie between lake deposits and the western margin of the basin, a distance of nearly 3 km. These beds are 2.6 to 9.8 m thick, and measured sections are 46 percent tuff, 42 percent claystone, 10 percent sandstone, and 2 percent conglomerate. Claystones are generally gray or brown, rootmarked, and rather sandy. Siliceous earthy claystones and siliceous rootcasts are absent, but microscopic examination was not made for small particles of biogenic opal. Zeolites are in some beds. Tuffs are reworked and form lenticular beds, most of which are rootmarked. Tephra is dominantly trachytic and resembles that of Tuff IF, from which it was very likely derived. Mafic vitroclasts are, however, in some tuffs and sandstones, particularly near the top of the facies. Vitroclasts are generally fresh, but a tuff was found wholly altered to zeolite at one place near the eastern margin of the facies (loc. 61). Sandstones are chiefly of basement detritus but may contain tephra, either primary or reworked from older deposits. Conglomerates are generally a mixture of basement detritus with clasts of the Naabi Ignimbrite. Although faunal remains are not rare, few have been collected and rootmarkings are the principal source of ecological information. These suggest that marshland was greatly subordinate to grass and brush.

3. *Southern lake-margin deposits.* Lake-margin deposits crop out over a north-south distance of 1.4 km at Kelogi (locs. 99-101), to the south of stratigraphically equivalent lake deposits. These are as much as 6.1 m thick and are roughly 76 percent claystone and 24 percent tuff. They also include very small amounts of conglomerate and sandstone. Claystones are dominantly brown, rootmarked and somewhat sandy. Zeolites are in at least a few of these beds. A bed of siliceous earthy claystone was found at one place (loc. 99). Most of the biogenic opal is in phytoliths, but there are also spores of chrysophyte algae, which are the only definitive evidence for lacustrine conditions in this area.

Tuffs are reworked and either form massive, rootmarked beds or else they are thin-bedded, obviously stream-worked deposits, some of which fill channels. Tephra is indistinguishable from that of Bed I, from which it was probably derived. Vitroclasts in the northernmost tuffs can either be fresh or altered to zeolites, whereas those to the south (loc. 101) are wholly altered.

Lake-margin deposits of these three different areas accumulated on low-lying terrain intermittently flooded by the lake. The lake-margin terrain included large areas of marshland and was similar in most respects to the lake-margin terrain of Bed I. The eastern area was flooded more frequently than other areas, as indicated by sediments and fossils. The proportion of claystone is highest here, and the content of sand in the claystones is lowest. Biogenic opal is vastly more abundant here, as are rootmarkings of marshland vegetation, vertebrate remains, and archeological materials. Low dunes of wind-worked tephra, represented by Tuff IIA, spread over the lake-margin terrain from the east.

As one might expect, the lake water flooding the eastern marginal terrain decreased in salinity from west to east, away from the center of the lake. The evidence for a salinity gradient is in both the alteration pattern of vitroclasts and biogenic opal and in the assemblage of authigenic minerals at VEK (loc. 85). Alteration of glass and diatoms in the easternmost deposits reflects the transition to eolian and fluvial environments. It should be noted that the maximum extent of lacustrine flooding cannot presently be defined accurately in deposits to the west and south of the gorge, and the most distal deposits (for instance, locs. 66a, 101) may lie slightly outside the zone flooded by the lake.

Alluvial-Fan Deposits

Alluvial-fan deposits form the lowermost part of Bed II in the eastern 6 to 7 km of the Main Gorge. They also crop out along the western margin of Olbalbal both north and south of the gorge. These deposits are 3 to 14.3 m thick in the gorge, and they thin abruptly to 1.7 m to the south (loc. 157). A maximum thickness of 18 m is exposed 15 km to the northeast at locality 200, which is only about 8 km from the southwestern margin of the basin of Lake Natron. Measured sections of the facies are 89 percent tephra deposits, 6 percent claystone, 4 percent conglomerate, and 1 percent of sandstone. The tephra deposits are principally reworked tuffs and lapilli tuffs, and ash-flow tuffs form the remainder. Tephra deposits together with conglomerates form the entire thickness of the facies in the easternmost exposures; claystones are restricted to the area west of the Second Fault, and they are most common near the western margin of the facies.

Reworked tuffs and lapilli tuffs are crudely stratified and may be cross-bedded. They consist largely of trachytic tephra, principally pumice, but they also commonly include a small percentage of mafic vitroclasts. Many samples contain a small to moderate percentage of clay-coated, sand-size clasts, some of which are rounded. The vitroclasts in nearly all areas are altered to zeolites. Rootmarkings are rare in the easternmost deposits and they increase in number westward. Ash-flow tuffs have been found in the gorge (loc. 3a), along the western margin of Olbalbal (near loc. 1a), and 15 km to the northeast (loc.

200). They are massive trachytic deposits 1.2 to 8.5 m thick that consist chiefly of pumice but include lava clasts of Olmoti type. Pumice is altered to zeolites in all of these deposits. The deposit to the northeast (loc. 200) is distinguished by welding of pumice and by obsidian clasts as much as 8 cm in diameter.

Claystones are brown to reddish-brown and are generally tuffaceous. Claystone pellets are common, and a few samples identified as claystone in the field proved on microscopic examination to be claystone-pellet aggregates. Rootmarkings are both widespread and abundant. Analcime of sugary appearance can be seen with the hand lens in most samples.

Conglomerates form lenticular beds, some of which fill steep-walled channels as much as 2 m deep. Clast size decreases westward, with maximum diameters of about 60 cm in the eastern part of the facies. East of the Second Fault, clasts appear to be entirely from Olmoti; westward, they include welded tuff and lava from Ngorongoro and Sadiman. Channel alignments were measured at several localities, and rather consistent directions were obtained from nine channels to the east of the Third Fault. Eight of the nine channels are aligned between N32°W and N95°W, averaging N56°W. Measurements farther north are N70°W (loc. 1) and N35°W (loc. 200). Nine channels measured to the west of the Third Fault are generally shallow and vary widely in orientation. They point to meandering streams and are too few to obtain a meaningful average direction of stream flow.

Except for rootmarkings suggestive of grass and brush, the alluvial-fan deposits have yielded no evidence of fauna or flora.

These alluvial-fan deposits of Bed II represent a continuation of Bed I sedimentation in the same area. The deposits originated largely in explosive eruptions of Olmoti, and most of the tephra was redeposited by streams on an alluvial fan of low gradient. There are, however, a few features of the alluvial fan not recorded by the Bed I fan deposits. Mafic tephra in the Bed II fan deposits probably points to explosive eruptions from a volcano other than Olmoti. Sediment was now transported by wind on a significant scale, as shown by pelletoid tuffs and claystone-pellet aggregates, and saline, alkaline soils are suggested by reddish-brown, rootmarked claystones with analcime.

Eolian Deposits

Eolian deposits form the bulk of the Lemuta Member and are present throughout the eastern 7.5 km of the gorge. They are exposed discontinuously along the western side of Olbalbal for at least 4.8 km southeast and 7 km north of the gorge, and they crop out 15 km to the northeast (loc. 200). These deposits interfinger westward with claystones of the lake-margin facies over an east-west distance of about 1 km. Eolian tuffs grade downward into alluvial-fan deposits, and the contact between the two facies is taken at the base of the lowermost bed of eolian tuff. By this definition, some beds of water-worked tuff similar to that of the alluvial-fan deposits are included within the eolian deposits. This facies is generally 3 to 7 m thick in the gorge, and its thickest sections are to the north, at localities 1a (14.5 m) and 200 (13.7 m).

Measured sections of the facies average 88 percent tuff, 10 percent limestone, 2 percent claystone, and less than a percent each of sandstone, conglomerate, and claystone-pellet aggregate. At least 95 percent of the tuffs are wind-worked, or eolian, and the remainder are water-worked. The eolian facies corresponds to the Lemuta Member as far west as Long K (loc. 38), and the upper and lower units of the Lemuta Member are useful subdivisions in respect to lithology and origin. The lower unit contains most of the water-worked tuff and conglomerate, and the upper unit contains nearly all of the limestone and dolomite.

Eolian tuffs are typically massive or crudely bedded, rootmarked, and pale yellowish-brown. Eolian tuffs of the lower unit are on the whole poorly sorted and may contain isolated pebbles of lava or pumice. This tephra is dominantly trachytic and mineralogically similar to that in the underlying alluvial-fan deposits. There is also a small to moderate percentage of mafic tephra, principally augite and vitroclasts. Clay is common as pellets, pelletoid coatings, and matrix material. Eolian tuffs on the upper unit are better sorted overall, and the clasts are mostly of medium sand size. Mafic nephelinite tephra predominates, and it may form all of the deposit. Some trachytic tephra is commonly present, and a small mount of basement detritus was noted in a few samples. Claystone is common as pelletoid coatings. Cementing materials form 20 to 40 percent of the tuffs, which are consequently rather well indurated. Zeolites are the principal cement and were formed by alteration of vitroclasts and nepheline. Calcite is a major cementing material in the upper unit and may be accompanied by dolomite. The tuffs were altered at the land surface, penecontemporaneous with deposition. This is suggested by modern examples and is documented by pebbles of zeolitic tuff from the lower unit in conglomerates of the upper unit. Likewise, eolian-tuff pebbles of the upper unit are in the overlying conglomerates. The alteration and cementation are easy to recognize in the field and are one of the major criteria for distinguishing eolian tuffs of the lake-margin facies from those of the eolian facies.

Limestones generally form massive, nodular beds of pale gray or yellowish-brown color. Some of them are tuffaceous, and these may grade into eolian tuffs. Laminated calcretes are in the upper 1.5 m of the facies in the vicinity of the Second Fault (loc. 28). A few of the limestones comprise angular fragments of tuffaceous limestone in a calcareous matrix.

Conglomerates are restricted to the eastern part of the facies and have not been found west of locality 33b. Those of the lower unit are nearly everywhere similar to those of the alluvial-fan deposits in the same places. However, south of the gorge (loc. 2b) they include detritus from Ngorongoro and Sadiman in addition to Olmoti clasts. The rare conglomerates of the upper unit contain rounded clasts of Bed II eolian tuff in addition to Olmoti clasts.

Claystones are most common in the western half of the facies and to the south of the gorge (loc. 157). They are reddish-brown at locality 157 and are gray and brown to the west. The sand fraction of the southernmost claystone (loc. 157) is principally of detritus from Ngorongoro and Sadiman.

Most of the rootmarkings are suggestive of grass, but large, vertical root holes of marshland vegetation are locally common in the western part of the facies. The skull and horn cores of an alcelaphine antelope (Cat. no. 208) were collected from a lens of coarse-grained, reworked tuff in the lower unit at locality 1a. A few isolated fragments of bone and teeth were found elsewhere in the facies.

The eolian facies consists largely of wind-worked materials deposited on and adjacent to an alluvial fan. Eruptions of Olmoti either had ceased or were greatly reduced, and detritus from Ngorongoro and Sadiman was spread northward by streams beyond the earlier limit recorded in the underlying alluvial-fan deposits. Trachytic materials of the lower unit are from Olmoti and may be entirely older ejecta, eroded by wind from both stream channels and interfluve areas of the higher, eastern part of the fan. Nephelinite tephra was supplied by explosive eruptions from an unknown source. Eolian tuffs have many features in common with the present-day deposit of wind-worked tephra on the plain north of the gorge, and the two probably have a similar origin. The more important similarities are in grain size and sorting, pelletoid coatings, and claystone pellets. These Bed II eolian tuffs may, like the modern deposits, represent the low trailing ridges, later modified by wind, that were left by dunes in moving westward. The dunes of Bed II were stabilized by marshland vegetation near the eastern margin of the lake. The content of augite in Tuff IIA provides strong supporting evidence for this picture. Augite constitutes about 15 percent, on average, of the constituent particles of Tuff IIA in the eolian facies, whereas it averages about 30 percent and commonly reaches 50 percent of Tuff IIA in the lake-margin facies. Augite increases downwind in a similar fashion in the Holocene dune deposits on the plain to the north of the gorge. The active dunes near the western margin of the dune deposits average 60 to 75 percent augite! The nature and origin of this enrichment are discussed in the chapter on Holocene sediments.

The climate was relatively hot and dry, and soluble salts were concentrated at the land surface by evapotranspiration. This is documented by pedogenic limestones and by the extent and penecontemporaneous nature of zeolitic alteration. Grassland vegetation is suggested by root-markings, and the rarity of vertebrate fossils may reflect either a sparse fauna or unfavorable conditions for preserving bones.

The Lemuta Member very likely accumulated over at least a few tens of thousands of years. Particularly significant is evidence that the lower unit was altered before the upper unit was deposited. This is shown by clasts of altered, zeolitic tuff from the lower unit in the unaltered basal bed of the upper unit at locality 35. A deposit of eolian tuff probably requires a minimum of about 10,000 years for extensive alteration, as shown by study of the Namorod Ash and the Naisiusiu Beds. Thus the lower unit represents a time span on the order of 10,000 years or more. The upper unit is thicker and altered to at least the same extent as the lower unit, suggesting a time span of 20,000 years or more for both units.

Western Fluvial-Lacustrine Deposits

The western fluvial-lacustrine deposits are a thick, widespread facies exposed discontinuously between the Fifth Fault and the western margin of the basin, a distance of 9 km (Fig. 24). Here it is generally 10 to 20 m thick but thins to about 5 m near the western margin. That part of the facies above the uppermost disconformity extends as much as 3.8 km farther east, where it intergrades with the eastern lake-margin deposits. The underlying part of the facies, exposed only to the west of the Fifth Fault, presumably interfingers eastward with lake deposits. This facies is underlain by lake deposits everywhere except near the western margin, where it overlies lake-margin deposits. The basal contact of the western fluvial-lacustrine deposits is a disconformity in most places and gradational elsewhere (for example, locs. 54, 55, 72a, 73).

This facies is unlike all others of Bed II in its high ratio of sandstone to claystone. It comprises about 60 percent sandstone, 12 percent carbonate rocks, 10 percent claystone, 8 percent conglomerate, 5 percent tuff, and perhaps 2 to 3 percent claystone-pellet aggregate. There is also a very small percentage of siltstone and chert. The upper and lower parts of the facies, separated in many places by the middle disconformity of Bed II, differ considerably in lithology and origin and will be considered as separate subfacies. The lower subfacies contain chiefly lacustrine and lake-margin sediments; the upper subfacies is dominantly fluvial or mixed fluvial-lake-margin sediments.

The lower subfacies is a crudely lens-shaped body as much as 9.4 m thick. To the east it thins and interfingers with purely lacustrine sediments; to the west it thins and either pinches

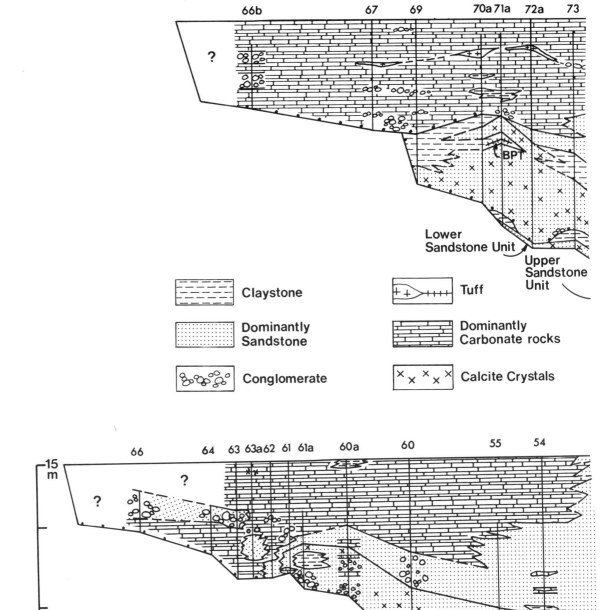

W ←

66b   67   69   70a 71a   72a   73

BPT

Lower Sandstone Unit

Upper Sandstone Unit

	Claystone		Tuff
	Dominantly Sandstone		Dominantly Carbonate rocks
	Conglomerate		Calcite Crystals

66   64 63 63a 62 61 61a   60a   60   55   54

-15 m

0   3km

-0

b. South Side of Main Gorge

Figure 24. Generalized cross-sections of the western fluvial-lacustrine deposits of Bed II. Lacustrine claystones are marked with *lc*, and *IID* represents Tuff *IID*.

76

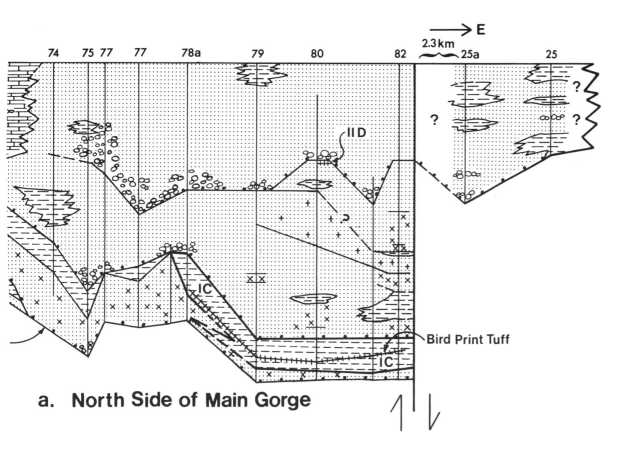

a. North Side of Main Gorge

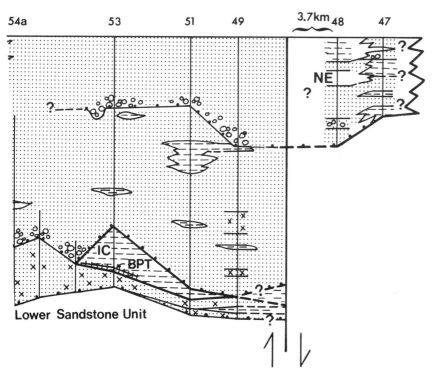

out (locs. 67-69) or interfingers with the lowermost deposits of the upper subfacies (locs. 61-62). Sandstones form at least three-fourths of the subfacies, and most of the remainder is claystone, found principally at the top of the subfacies on the north side of the gorge. There are some conglomerates, principally on the south side of the gorge in the western part of the subfacies. Beds of carbonate rocks occur widely and increase in proportion from east to west. The small percentage of tuffs does not show any particular distribution pattern. Casts and molds of trona have been found in several places, mostly in the lower part of the deposit.

Sandstones of this lower subfacies were earlier described as a stratigraphic unit termed the *augitic sandstone,* which generally comprises two units separated by an erosional surface. The lower unit is 15 cm to 1 m thick and differs lithologically from the upper, which is as much as 8.5 m thick. Sandstone of the lower unit is characteristically poorly sorted, and much of it is clayey. Detritus is quartzose and is typically a bimodal mixture of sand and granules. Fragments of Naabi Ignimbrite are found in the western sandstones, and clasts of Engelosin phonolite are in the easternmost sandstones. A small proportion of augitic tephra is widespread. Most of the sandstones contain abundant coarse crystals of calcite, both euhedral and rounded. The euhedral crystals of one sample (loc. 49) are isotopically heavy, with $\delta O^{18} = +30.8°/_{oo}$ and $\delta C^{13} = +5.03°/_{oo}$ (J. R. O'Neil, 1973, personal communication). Trona casts were found in locality 55. Abraded chert nodules from the underlying lake beds are in some of the easternmost sandstones.

Sandstones of the upper unit of the lower subfacies are generally medium-grained, well sorted, and form beds 10 cm to 1 m in thickness. The beds may have a wide lateral extent and are commonly laminated, cross-bedded, or ripple-marked. Rootmarkings are not rare, and bird prints are abundant on bedding planes in a few places. Most sandstones contain substantial proportions of basement detritus, rounded calcite crystals, and volcanic materials — principally mafic tephra. Augite probably constitutes 5 to 10 percent of the sandstones, on the average, giving them a greenish-gray color which generally serves to distinguish them from sandstones of the overlying subfacies. Oolites or claystone pellets are found in a few samples examined microscopically. Sandstones range from unconsolidated to well cemented by calcite. Fragments and abraded nodules of chert are present locally at the base of the sandstone unit. Trona casts are in the base of the upper sandstone in locality 73.

Claystones are dominantly greenish, commonly pale olive, and resemble the lacustrine clays to the east. The zeolite clinoptilolite is a major constituent of claystones between the upper and lower sandstone units near the eastern margin of the subfacies. Casts of trona are in equivalent claystones to the west (loc. 54).

Beds and nodules of chert are at various horizons in the eastern part of the subfacies. A bed of green chert is near the base of the lower sandstone unit at localities 80 and 82a. A sample of this chert gave a δO^{18} value of $+34.8°/_{oo}$ (J. R. O'Neil, 1973, personal communication). Nodules are widespread in the clinoptilolite-rich claystone between the upper and lower sandstone units, and nodules are in sandstone of the upper unit at locality 78b.

Conglomerates form lenticular beds, some of which fill shallow channels. They are dominantly of pebbles, most of which are of basement rock and Naabi Ignimbrite derived from the west. Clasts of basement rock are of types found in the Mozambique belt (see Fig. 1), and coarse-grained quartzite is most abundant. The one measured channel is oriented northwest-southeast.

The carbonate rocks are mostly aphanitic, hard, and white to yellowish-gray. They typically form massive, nodular beds 10 to 30 cm thick, which can either be dolomite or limestone. These rocks were formed in muds and sands at shallow burial depths, by processes discussed more fully in the description of the upper subfacies. The remaining carbonate rocks are lacustrine and comprise a thin, laminated bed of dolomite above the basal sandstone unit and calcite-crystal limestones, found in both the upper and lower sandstone units. Tuffs are fine- to medium-grained, and those interbedded with claystone are generally thin (2-10 cm) and evenly laminated. Tuffs interbedded with sandstone are reworked, sandy, and grade into tuffaceous sandstones. Tephra is almost exclusively mafic and is dominantly tephrite in composition. One exception is a sodic nephelinite tuff near the base of the upper sandstone unit in locality 54. Vitroclasts are altered, chiefly to palagonite and zeolites.

Faunal remains have been found in several places but are most common in conglomerates near the western margin of the facies. The only mammalian fossil identified is a hippo skull (loc. 61a). Fragments of ostrich eggshell occur in sandstones of the upper unit at localities 53, 54, and 60.

This lower subfacies of the western fluvial-lacustrine deposits records principally lacustrine and semilacustrine environments. The claystones, chert beds and nodules, and euhedral calcite crystals point to saline, alkaline lake water relatively undisturbed by waves and currents. Salinity was high, at least locally, as shown by trona and by isotopic measurements. The ill-sorted mixture of detritus in the lower sandstone unit is suggestive of transport and deposition on mudflats, either by sheetwash or small streams. The abraded calcite crystals in these sandstones are evidence of intermittent flooding and wave action. The upper sandstone unit is a mixture of lacustrine and fluvial sediment (see Table 15). Oolites, abraded calcite crystals, and ripple marks indicate shallow, wave-agitated water, and the bird prints and rootmarkings are evidence of subaerial exposure. Fluvial features are not rare and include cut-and-fill (scour) structures and medium-scale festoon cross-bedding. The broad zone of lacustrine sandstones can be explained by fluctuating lake level on nearly flat lake-margin terrain. Sand-size materials, supplied either by ash-falls or by streams, would be reworked by waves and currents over a broad zone during a rise in lake level.

The upper subfacies is between 3 and 18 m thick between the Fifth Fault and the western margin of the basin. Much of this variation reflects the irregular surface eroded into Bed II at the base of Beds III-IV (und.). This deposit is about 55 percent sandstone, 20 percent carbonate rocks, with the remainder comprising chiefly conglomerate, tuff, and claystone. Carbonate rocks are the dominant rock type in the western 5 to 6 km of the subfacies (Fig. 24). Conglomerates are generally coarsest and most abundant near the western margin, but they are wide-spread to the east above the uppermost disconformity. Claystone forms the highest proportion of the subfacies near its eastern margin.

The sandstones are of several types. Probably 10 to 20 percent are clearly fluvial, and between 2 and 5 percent have features diagnostic of lacustrine or lake-margin environments. The majority of sandstones either have a mixture of fluvial and lacustrine features, or else they lack textures and structures diagnostic of either environment. The fluvial sandstones are found throughout the subfacies near the western margin of the basin. To the east, they are most common above the uppermost disconformity. Lake-margin sandstones have been found only below the uppermost disconformity and within 1.5 km of the Fifth Fault. Sandstones with a mixture of fluvial and lacustrine features are most common either below the uppermost disconformity to the west of the Fifth Fault or above the disconformity to the east. Several features are common to most of the sandstones of these three groupings. They are dominantly medium-grained, yellowish-gray, and weakly cemented by calcite or zeolites, or by both. Detritus is dominantly of basement origin and derived from the west. Augite-bearing mafic tephra is in many sandstones, and the vitroclasts are altered to palagonite and zeolites.

The fluvial sandstones may be coarse-grained, and some are conglomeratic. They generally form lenticular beds 30 cm to 1.5 m that may have rootmarkings and medium- to large-scale festoon cross-bedding. Only rarely does the grain size decrease systematically upward as in deposits of meandering streams. Sorting is generally poor, and a matrix of silt and clay is commonly present. Some beds contain claystone pellets or abraded calcite crystals.

The lacustrine sandstones are well sorted, fine- to medium-grained, and laminated or rippled. They may contain interbeds of pale olive claystone. Either claystone pellets or abraded calcite crystals can be a major constituent. The lake-margin sandstones are poorly sorted, commonly clayey, and contain abundant euhedral calcite crystals or crystal aggregates. Trona replaced by calcite was noted in one bed (loc. 49). Another bed, 4 m above the base of the subfacies (loc. 82a), is a bimodal mixture of fine- to medium-grained sand from the west and coarse sand and granules from eastern sources, including Engelosin. These poorly sorted sandstones are similar in most respects to the poorly sorted sandstones of the lower subfacies that were deposited on mudflats.

The sandstones of mixed and indeterminate origin form beds 30 cm to 1.5 m thick that commonly occur in sequences 3 to 5 m thick. Individual beds can be traced for considerable distances where exposures are good (Plates 3, 4). Thin, even stratification is the most common structure, but cross-bedding is not rare, and some beds are massive. Rootmarkings are not uncommon. Sorting varies widely but is generally moderate to good. Scattered pebbles or angular clasts of tuff are in some of the more poorly sorted sandstones. Sand-size grains of Engelosin phonolite were noted in a few beds. Abraded calcite crystals and claystone pellets are almost invariably present, and they may form as much as 30 percent of the sand fraction. Calcite oolites are in a few samples examined microscopically.

Conglomerates are typically fluvial and form lenticular beds, some of which fill channels as much as 2 m deep. Clasts are chiefly of basement rocks and the Naabi Ignimbrite, but tuff and carbonate rocks derived from Bed II are not rare, and they predominate in some conglomerates. Average clast size decreases from west to east, with the exception of tuff clasts, which are coarsest in locality 80. The nine measured channels range over $125°$ in orientation, from $S75°W$ to $N20°E$, averaging $N45°W$. The stream channels were probably sinuous, judging from the wide range in orientation, although the channel fill does not have the textures and structures typical of meandering streams.

Claystones can be lacustrine, fluvial, or of indeterminate origin. Lacustrine claystones are found principally below the disconformity in the vicinity of the Fifth Fault. These are typically waxlike and pale olive, and they may contain euhedral calcite crystals. The sand-silt fraction is dominantly of basement origin except near the eastern margin, where volcanic detritus is about equally abundant. Fluvial claystones are at varying levels toward the west, and they are found above the uppermost disconformity near the eastern margin of the subfacies. These are typically brown, rootmarked, sandy, and are associated with sandstones and conglomerates of

fluvial origin. Claystones of indeterminate origin are most common in the eastern part of the subfacies, above the uppermost disconformity. These are most commonly brown, waxlike, and may or may not have rootmarkings. Illite is generally the only clay mineral, as based on diffractometer analyses of the fraction finer than $2 \mu m$ in seven samples. Analcime constitutes between 5 and 10 percent of the analyzed samples, fluvial, lacustrine, and indeterminate.

Tuffs are characteristically reworked and contaminated with detrital sand, claystone pellets, and rounded calcite crystals. They may grade into sandstone, commonly in the same bed. As with the sandstones, some beds are clearly fluvial, whereas others have both fluvial and lacustrine features. Rare sand-size grains resembling Engelosin phonolite are in a few samples, principally from the vicinity of the Fifth Fault. All of these reworked tuffs are dominantly or entirely mafic, and glass is altered to palagonite and zeolites.

A few tuffs have features suggesting deposition in ponded water. Tuff IID is laminated and uncontaminated in the one locality (80) where it is found in place. An uncontaminated mafic tuff above the level of Tuff IID is evenly laminated or ripple-marked in several places over an east-west distance of 2 km (locs. 67, 70a, 72a). Vitroclasts are altered, and the tuffs are cemented by zeolites. Clasts of these two tuffs are in conglomerates only slightly younger than the tuffs, indicating that the tuffs were altered shortly after they were deposited, presumably in saline, alkaline lake water.

The few siltstones are pale brown and generally micaceous. One bed, near the base of the subfacies in locality 80, has gaylussite crystals replaced by calcite. These siltstones were deposited in ponded water, which was locally an alkaline brine.

Claystone pellets are widespread through the subfacies and are particularly common above the uppermost disconformity. Most of the pellets are in sandstones, but several sandy claystone-pellet aggregates have been identified. Most of the aggregates are massive, rootmarked, and form beds 30 cm to 1 m thick. One bed contains a substantial proportion of pellets that have abraded calcite crystals as nuclei. A few aggregates are thin and interlaminated with sandstone. The massive, rootmarked aggregates probably accumulated on the land surface, whereas the laminated aggregates were deposited in shallow water.

The carbonate rocks occur almost entirely as nodular beds of aphanitic limestone and dolomite, and the remainder are thin sandy calcite-crystal limestones. The aphanitic carbonate rocks form nodular beds 10 to 60 cm thick that are white to yellowish-gray and very pale orange. They are most commonly interbedded with sandstones, and most of them contain between 5 and 25 percent of detrital sand. Some are root-marked, and many contain curved fractures partly filled by relatively coarse calcite. Field staining and laboratory work indicate that about equal proportions of these beds are of calcite (that is, limestones) and dolomite. Some beds are mixtures of calcite and dolomite. Most of these carbonate rocks probably were formed at shallow depth in the zone of contact between calcium-bearing surface runoff and alkaline ground water rich in carbonate ions. Rootmarkings suggest that some beds represent caliche soils. The dolomite points to high salinities, which could be produced by upward capillary movement and evaporation near the margin of a saline lake of fluctuating level, where ground water should be at least intermittently saline. Dolomite caliche of probably similar origin has been found at the margin of a saline lake in west Texas (Friedman, 1966).

The calcite-crystal limestones occur principally in the eastern part of the subfacies below the disconformity where they are associated with lacustrine claystones and lake-margin sandstones. Another was noted near the top of the subfacies to the west (loc. 67), which is significant in showing the broad extent of saline, alkaline lake water.

Ostracods and remains of small fish have been found in several places. Ostracods and small delicate scales and bones of fish are common in a laminated claystone-pellet aggregate beneath the uppermost disconformity in locality 80. Broken ostracod tests have been found in a claystone and a few sandstones near the eastern margin of the subfacies (locs. 25, 25a, 47). Small, delicate fish bones are in some of the siltstones and lacustrine claystones. Small pelecypods with both valves intact were noted in conglomeratic sandstone in the lower part of the subfacies near the western margin of the basin (loc. 67). Small,

unidentified bone fragments are in many sandstones.

Remains of mammals are widespread but have not been systematically studied. They are particularly common in conglomerates and sandstones above the uppermost disconformity. Buffalo skulls have been collected from this upper part of the subfacies at localities 77 (Kar K, site 3) and 25 (FK, site 12b). A primate skull (Colobus sp.) and an elephant tooth were found above the disconformity in MLK (loc. 53, site 33). The mandible of a white rhino (Ceratotherium simum), collected near locality 68a, very likely comes from this facies, but the horizon is unknown.

In summary, these western sediments are unique in their high content of sandstone, and they record a complex shifting pattern of fluvial, lake-margin, and lacustrine environments. The lower subfacies has the highest proportion of lake-margin and lacustrine sediment. Most if not all of the sandstones of this subfacies interfinger eastward with lacustrine clays. The lower part of the upper subfacies, between the middle and upper disconformities, consists chiefly of sandstones that either have a mixture of fluvial and lacustrine features or else lack features diagnostic of either environment. These interfinger westward with fluvial detrital sediments and carbonate rocks of water-table and caliche origin. Probably they interfinger eastward with lacustrine sediments in an area not now exposed. These sandstones were probably deposited by streams at the lake margin and reworked by the lake during periods of rising lake level. Deposits above the uppermost disconformity contain a relatively high proportion of fluvial deposits, which extend to the eastern margin of the facies. There is evidence for impermanent saline, alkaline ponds or small shallow lakes, but none for a relatively stable body of water in any one place above the uppermost disconformity. Water-table and caliche carbonate rocks are widespread and are the dominant type of rock in the western part of the basin. Wind was a significant agent in sediment transport, particularly for the upper part of the facies, as indicated by claystone pellets and grains of calcite coated with clay. Presumably much of the associated sand-size detritus was redeposited by wind.

The different patterns of sedimentation in the three different intervals of the facies are very

Table 14. *Criteria for Depositional Environments of Claystones in Eastern Fluvial-Lacustrine Deposits*

	Lacustrine	Lacustrine mudflat	Marshland	Fluvia
Bedding	Massive or laminated	Generally massive	Massive	Massive to thinly bedded; commonly lenticular
Texture, appearance	Generally waxlike and pale olive, with little sand and silt	Generally waxlike and olive gray to brown; commonly fractured in paleosol; variable content of sand and silt	Yellowish-gray to cream-colored, with earthy luster; variable content of sand and silt	Generally waxlike and brown, with a moderate to high content of sand and silt
Biogenic features	Delicate remains of small fish are common; ostracods present locally	Delicate remains of small fish are locally common; ostracods rare; root-markings common.	Remains of fish; diatoms, algal encysting cases, and rare charophytes, silicified plant remains and rootcasts	Scattered faunal remains, rootmarkings common.
Miscellaneous compositional features	Illite generally predominates in clay-mineral fraction; many samples have calcite crystals, euhedral or abraded	Either montmorillonite alone or a mixture of montmorillonite, illite, and mixed-layer clay; claystone pellets can be common; abraded calcite crystals rare	Montmorillonite and illite; claystone pellets common; abraded calcite crystals can be common	Montmorillonite is clay mineral in unaltered claystones; claystone pellets can be common
Associated sediments	Lacustrine and lake-margin sandstones; calcite-crystal limestones; laminated tuffs	Lake-margin sandstones; nodular limestone; claystone-pellet aggregates	Lake-margin sandstones	Fluvial sandstones and conglomerates; nodular limestone

likely a result of fault displacements which rather abruptly changed the paleogeography. The lenticular shape of the lower subfacies is also a result of deformation, as discussed more fully in the environmental synthesis of Bed II.

Eastern Fluvial-Lacustrine Deposits

The eastern fluvial-lacustrine deposits form the upper part of Bed II over a distance of 7 to 8 km in the Main and Side gorges. The facies is 11 to 24 m thick, with the thickest sections in the Side Gorge and nearby areas of the Main Gorge (Fig. 25). The thinnest sections are over a "rise" or slightly domed area near the eastern margin of the facies (locs. 9-12a, 32b-35a). This facies intergrades and interfingers eastward with the eastern fluvial deposits. Its basal contact is taken at the widespread disconformity over the Lemuta Member, and in the Side Gorge where the disconformity is not readily apparent, the contact is taken at the base of the main chert unit.

The eastern fluvial-lacustrine deposits are about 51 percent claystone, 34 percent sandstone, 6 percent each of tuff and limestone, and the remaining 3 percent comprises conglomerate, siltstone, and claystone-pellet aggregate.

Chert nodules are in the lowermost part of the facies in the Side Gorge. These deposits record a shifting pattern of sedimentation in which lacustrine and lake-margin sediments alternate with fluvial and eolian sediments over broad areas. This facies is of particular interest because it contains most of the archeological materials and faunal remains of Bed II. Consequently it will be described in more detail than the other facies of Bed II. To simplify this description, the major rock types will be categorized by their inferred depositional environment.

Claystones are classified into four environmental types. Clays are classed as lacustrine if they were deposited in lake water and were buried before being exposed at the land surface. Semilacustrine claystones were deposited in lake water but were exposed as a mudflat before burial. Still others were deposited in marshland intermittently flooded by lake water. Finally, many claystones were deposited by streams, either in a channel or on a floodplain. Criteria for identifying claystones of these different environments are given in Table 14. Claystones of the lacustrine mudflat can be difficult to distinguish from those of fluvial origin, particularly where they have weathered for an appreciable period of time. Moreover, the two categories overlap to some extent, as for example a mud-

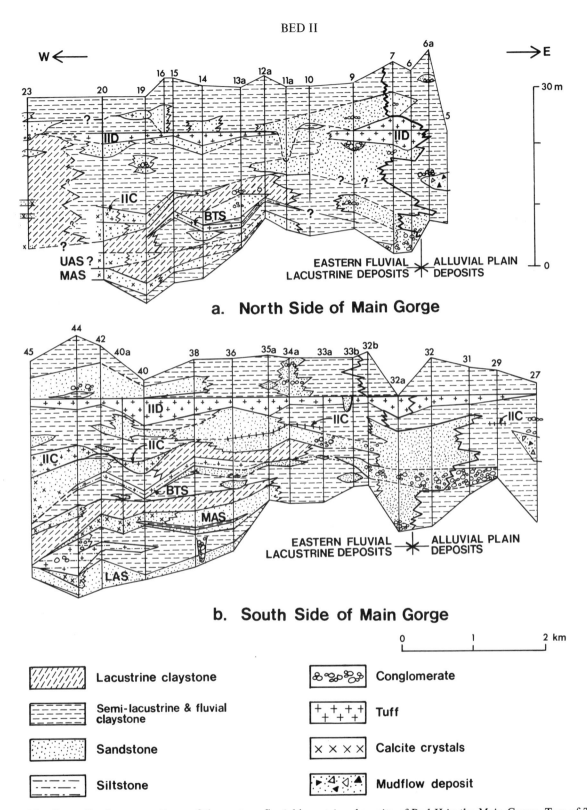

a. North Side of Main Gorge

b. South Side of Main Gorge

Lacustrine claystone

Semi-lacustrine & fluvial claystone

Sandstone

Siltstone

Conglomerate

Tuff

Calcite crystals

Mudflow deposit

Figure 25. Generalized cross-sections of the eastern fluvial-lacustrine deposits of Bed II in the Main Gorge. Top of Tuff IID is taken as a horizontal datum. Abbreviations are as follows: LAS, lower augitic sandstone; MAS, middle augitic sandstone; UAS, upper augitic sandstone; IIB, Tuff IIB; IIC, Tuff IIC; IID, Tuff IID; and BTS, brown tuffaceous siltstone. The alluvial-plain deposits are termed eastern fluvial deposits elsewhere in this chapter.

Table 15. *Criteria for Depositional Environments of Sandstones in Eastern Fluvial-Lacustrine Deposits*

	Lacustrine	Lake margin	Fluviatile
Shape of bed	Broad, sheetlike, generally thin ($\leqslant 1m$)	Sheetlike; thin to moderately thick (1-2 m)	Elongate; either sheetlike or lenticular in cross-section
Bedding structures	Ripple marks, lamination, small-scale cross-bedding	Commonly massive; may be crudely laminated	Cross-bedding, crude lamination, massive bedding
Texture	Generally well-sorted; medium grain size, rarely with granules and small pebbles	Generally poorly to moderately sorted; commonly bimodal, with granules and small pebbles in matrix of medium sand	Medium- to coarse-grained; poorly to moderately sorted
Composition	Abraded calcite crystals common; oolites common locally; claystone pellets generally rare	Abraded calcite crystals; common locally; claystone pellets common and larger claystone clasts not rare	Claystone pellets commonly present and locally abundant
Matrix, cement	Coarse calcite cement common	Clay matrix common, with or without zeolite/calcite cement	Silt-clay matrix; zeolite cement in some beds
Mineralogic alteration	Alteration generally minimal; glass can be unaltered	Highly variable alteration	Moderate to high degree of alteration; glass entirely altered in most samples
Biogenic features	Small abraded bone fragments	Faunal remains and artifacts locally common; rare ostracods, diatoms; rootmarkings common; siliceous rootcasts and silicified plant remains present locally	Faunal remains, artifacts locally common; rootmarkings generally common

flat subject to flooding by both lake and stream. Probably about half of the claystones can be readily classified in the field as to environmental type. Most of these are of either the lake or lacustrine mudflat. Siliceous earthy claystones of marshland origin are the least common, and are found almost exclusively above Tuff IID.

Lacustrine claystone forms the highest proportion of the facies at PLK (loc. 23), which has the westernmost exposures of the facies in the Main Gorge. To the south and east, most of the lacustrine claystone is found in two sheetlike deposits, one overlying the bird print-tuff and the other between the brown tuffaceous siltstone and Tuff IIC. The upper part and the eastern margin of these sheets show evidence of exposure at the surface before burial. A third sheetlike deposit with widespread evidence of exposure is found between Tuffs IIC and IID. It was identified as lacustrine principally on the widespread content of illite in the clay-mineral fraction. In the Side Gorge, lacustrine claystone overlies the upper augitic sandstone and occurs widely beneath Tuff IID.

Sandstones are subdivided into lacustrine, lake-margin, and fluvial types, which are generally easier to distinguish than types of claystone (Table 15). Lacustrine sandstones have features diagnostic either of shallow, wave-agitated water or of a beach. They can grade into lake-margin sandstones, which include both fluvial sandstones slightly reworked by the lake in rising to

a higher level and lacustrine sandstones reworked by streams during a drop in lake level. Lake-margin sandstones generally have features such as rootmarks resulting from subaerial exposure. Some of the lake-margin sandstones have silicified rootcasts of marshland vegetation. Sandstones classed as fluvial were deposited in stream channels.

Lacustrine sandstones are found only below Tuff IIC, with the possible exception of PLK (loc. 23). They are most common as sheetlike deposits in the westernmost exposures of the facies in the Main Gorge. Within a single sheet, lacustrine sandstone grades eastward into lake-margin sandstones. In the Side Gorge, the upper augitic sandstone is a similar sheetlike deposit of lacustrine and lake-margin origin. Fluvial sandstones, although found at many levels, are most abundant above the brown tuffaceous siltstone. Most of the fluvial sandstone in the Main Gorge is in the form of elongate sheetlike bodies which seem to be oriented approximately east-west. These sandstones were deposited in a main drainageway from the volcanic highlands to the lake (see Figs. 33-37, 39).

Conglomerates form lenticular beds, some of which fill steep-walled stream channels. A few deposits are more or less sheetlike, and probably represent a series of lenticular beds at about the same level (for example, locs. 32-33*b*). The coarser conglomerates consist largely of cobbles and small boulders; the finer-grained conglomerates are of pea-size pebbles. Conglomerates are fluvial, with the possible exception of a few pebble layers within lacustrine sandstones, which may have been deposited on a beach or lacustrine bar.

Clasts are of volcanic rock, basement rock, and of detritus eroded from Bed II. For convenience in characterizing the clast distribution pattern, detritus from Olmoti and Ngorongoro is designated the eastern suite, that from Sadiman, Lemagrut, and Kelogi the southern suite, and the northern suite comprises Engelosin phonolite and quartz, pink feldspar, and granite gneiss derived from the northwest. The northern suite can also include clasts of the Naabi Ignimbrite. Volcanic detritus of the eastern and southern suites predominates in all of the coarser- and in most of the finer-grained conglomerates. Clasts of limestone from Bed II are widespread and are particularly common in the Side Gorge, where they are the dominant clast type in many of the finer-grained conglomerates. Clasts of eolian tuff from the Lemuta Member are in most of the conglomerates overlying the Lemuta Member and have been found as much as 7.6 m above the Lemuta Member. Clasts of other types of tuff from Bed II occur widely and predominate in a few conglomerates.

Clasts of volcanic rock show the expected type of distribution pattern, in which Lemagrut detritus is most abundant in the westernmost conglomerates and Olmoti detritus is most abundant in the easternmost. Ngorongoro detritus predominates over a broad central zone, and Sadiman detritus is most abundant between the zones richest in Lemagrut and Ngorongoro debris. There is, however, considerable intermixture, and Lemagrut detritus has been identified in some of the easternmost conglomerates (loc. 6), and clasts from Olmoti have been found as far west as localities 13a and 34.

Granules and small pebbles of the northern suite are a minor component of many conglomerates consisting largely of volcanic detritus from the east and south. These clasts of the northern suite are characteristically angular or subangular and from 4 mm to 1.5 cm in average diameter. The few conglomerates composed largely of northern-suite clasts are thin and lenticular, suggesting transport by sheet flood, small streams, or by both. One example is a layer 2 to 3 cm thick consisting largely of quartz

clasts below Tuff IIC in TK (loc. 19). Beneath Tuff IID at DK (loc. 13) is a small, narrow channel filled with pebbles of quartz and pink feldspar.

The conglomerates with clasts of both northern and southern (and/or eastern) suites probably represent channel deposits of the main drainageway, fed by tributary streams from both the volcanic highlands to the southeast and the basement terrain to the north and northwest. The angularity of most clasts from the north is a striking feature, which is explained by transport during brief periods, either in small streams or by sheetwash over mudflats, as a result of flash floods.

Dispersed clasts of pebble and cobble size are a widespread feature in this facies. The clasts are dispersed in and upon paleosols of the lacustrine mudflat, in siliceous earthy claystone of marshland origin, and in massive sandstones, principally of lake-margin origin. Clasts are commonly of both northern and southern (and/or eastern) suites, but in a few places they are entirely of northern derivation. These dispersed clasts have several explanations. Isolated pebbles or cobbles may represent manuports, or clasts transported by hominids. Transport by sheet flood or small streams can account for granules and small pebbles on mudflat paleosols and possibly in lake-margin sandstones. Finally, pebbles originally deposited in a layer overlying unconsolidated sediment may have been dispersed by the trampling or burrowing of animals and by the churning effect of roots.

Twenty-two channel alignments were measured in this facies (Table 16). The channels range in depth from 15 cm to 4.6 m, and they are filled with conglomerates and sandstone. They are most nearly adequate for paleogeographic purposes in the Side Gorge and the nearby part of the Main Gorge. Here they suggest a shift from northward flow during the deposition of Tuff IIB to northwesterly flow shortly prior to Tuff IIC. At higher levels, streams appear to have shifted to a northeasterly flow direction. These measurements are utilized in paleographic maps for different stratigraphic subdivisions of the facies (Figs. 27, 32-37).

Tuffs reflect the same range of environments as do the sandstones and claystones. Thin, laminated lacustrine tuffs are widely interbedded with lacustrine claystones, and the lower part of

Table 16. *Channel Alignments Measured in the Eastern Fluvial-Lacustrine Deposits of Bed II*

Stratigraphic unit or interval	Locality	Measurement	Reliability of measurement		
			Fair	Good	Excellent
Lower augitic sandstone	43*a*	N70°W		X	
Tuff IIB	30	N20°W ± 10°		X	
	44	N15°W ± 15°		X	
	44	N5°W		X	
	85	N10°E ± 10°		X	
	85	N5°E	X		
	85	N15°W			X
	88	N0°E		X	
	88	N15°W		X	
Brown tuffaceous siltstone to Tuff IIC	12	N65°E	X		
	45	N30°W	X		
	88	N35°W		X	
	91	N15°W			X
Tuff IIC to Tuff IID	89	N13°E		X	
	93	N15°E		X	
	93	N15°E		X	
	93	N0°E ± 20°	X		
Above Tuff IID	23	N57°W ± 10°		X	
	33*b*	N22°E ± 5°		X	
	94	N45-50°E			X
	94	N30°E		X	
	94	N30°E		X	

Tuff IID is lacustrine over a large area. Tuff IIC locally contains siliceous rootcasts and root channels of marshland vegetation. The augitic sandstone units contain a high proportion of augitic tephra that was redeposited by streams and by lacustrine waves and currents. It is not generally possible without microscopic study to distinguish samples of these augitic sandstone units that are dominantly tephra (that is, reworked tuffs) from those that are dominantly older detritus (tuffaceous sandstones). Very likely most of the sandstones with 75 percent or more of augite can be properly classed as reworked tuffs.

Limestones are chiefly of water-table or caliche origin. These form nodular beds that grade with decreasing size into nodules. Nodular beds are generally 15 to 30 cm thick and no more than a few tens of meters in lateral extent. This limestone is aphanitic, hard, white to yellowish-gray, and is commonly sandy or rootmarked, or both. All of the samples studied mineralogically are calcite with the exception of dolomite from BK (loc. 94). Nodules and nodular beds are widespread in the uppermost 1 to 3 m of Bed II. At lower levels they are most common in claystones of the mudflat or lake-margin environment.

Perhaps a quarter of the limestones are lacustrine. Most of these are characterized by a high content of augite and rounded calcite crystals, with or without oolites, and they are clearly a deposit of shallow, wave-agitated lake water. Nearly all of these limestones are restricted to lacustrine parts of the augitic sandstone marker units. Fine-grained, zeolite-bearing limestones are in the lower part of the facies in the Side Gorge. These were precipitated chemically, probably in saline, alkaline lake water.

Claystone-pellet aggregates occur widely in association with eolian siltstones, with lacustrine and fluvial sandstones, and with claystones of lacustrine and mudflat environments. Thin layers are interbedded with lacustrine sandstones near the eastern margin of the middle augitic sandstone, and with lacustrine claystones in SHK (locs. 90 and 91). Poorly defined layers of claystone-pellet aggregate are widely interbedded with claystone beneath Tuff IID in the Side Gorge (locs. 88, 91). The thickest beds, as much as 1.2 m thick, are massive and lie above Tuff IID (Plate 10; locs. 6, 7, 20, 89). Claystone pellets are common although subordinate to detrital sand in many lake-margin sandstones, and they form a substantial proportion of the siliceous earthy claystones above Tuff IID.

Most of the siltstones are massive, root-marked, and contain a high proportion of silt-sized particles of claystone. These are of eolian origin and are generally rich in zeolites as a result of alteration in a saline soil or mudflat environment. Orange to yellowish-brown zeolitic siltstones form a substantial proportion of Tuff IIB, and they are the principal constituent of the brown tuffaceous siltstone marker bed. In a few places there are lacustrine siltstones with flakes of mica and small, delicate fish remains.

Chert nodules of the main chert unit are found undisturbed at PEK (loc. 88a) and FC (loc. 89), and they are slightly reworked at MNK (loc. 88) and SHK (loc. 91). Undisturbed nodules are at three levels in 2.4 m of lacustrine sediments at PEK, and reworked nodules and chert fragments are scattered through as much as 1 m of limestone and sandstone above the main chert unit to the west (locs. 88-91). The chert nodules are isotopically moderately heavy, with δO^{18} ranging from 32.0 to 34.3°/oo in seven samples from the main chert unit (Stiles et al., 1974). Thin layers of zeolitite are interbedded with lacustrine claystones in the lower 1.2 m of the facies at PEK, and zeolitite locally forms the matrix for chert nodules in the main chert unit (locs. 88, 88a, 89).

The eastern fluvial-lacustrine facies contains a record of at least three rather lengthy periods of high lake level separated by episodes of low level. The lowermost flooding is represented by the middle augitic sandstone and overlying claystone (see Fig. 25). The brown tuffaceous siltstone and underlying fluvial sandstones and conglomerates were deposited at a time of low level. Claystones above the brown tuffaceous siltstone were deposited when the lake level had risen again. The overlying conglomerates and sandstone were deposited at a time of lower level. The interval between Tuffs IIC and IID represents either one or two episodes of advance and withdrawal of the lake. The lacustrine lower part of Tuff IID very likely records a brief period of time when the main drainageway was flooded. These large-scale events are treated more fully in the environmental history and geologic history of Bed II.

The lower part of the facies, up to the level of the bird-print tuff, was deposited in an area being actively folded, resulting in highly variable paleogeography and in great variation in sediment thickness over short distances (Fig. 26).

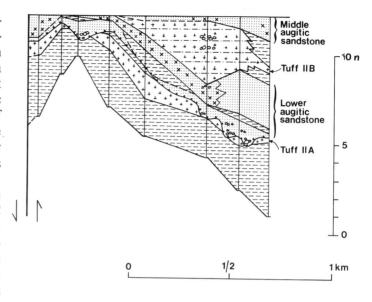

Figure 26. Cross-section of the lower part of Bed II up to the level of the bird-print tuff between localities 42, (HWK-E), at right, and 45b (FLK-NN), at left. Bird-print tuff is taken as horizontal datum at top.

The paleogeographic development of this interval is described here in five stages.

The earliest stage (Fig. 27a) is represented in the Main Gorge by the lower part of the lower augitic sandstone and in the Side Gorge by the main chert unit and the overlying limestone with reworked chert nodules. Sediments of this age apparently do not extend east of the HWK area (loc. 42). The lake covered the present area of the main chert unit (locs. 88a-91) and extended as far northeast as FLK-NN (loc. 45b). The lake water was moderately saline, as indicated by δO^{18} values of chert nodules and by lacustrine zeolitites. During a drop in lake level, nodules were slightly reworked in the southernmost exposures (locs. 88, 90, 91), and the main chert unit was eroded away in one place (loc. 88), which was utilized as a chert factory (or workshop) site. The main chert unit and the workshop site were buried by lacustrine limestone during a rise in lake level. During this time, the lake never fully withdrew from a small embayment to the east (loc. 88a). Equivalent deposits

87

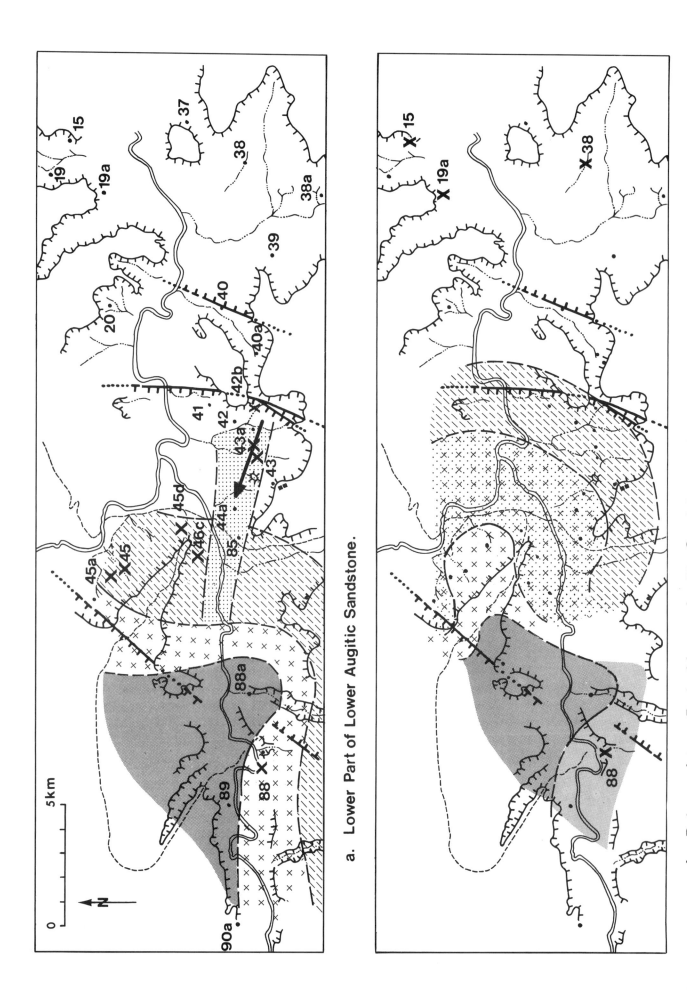

a. Lower Part of Lower Augitic Sandstone.

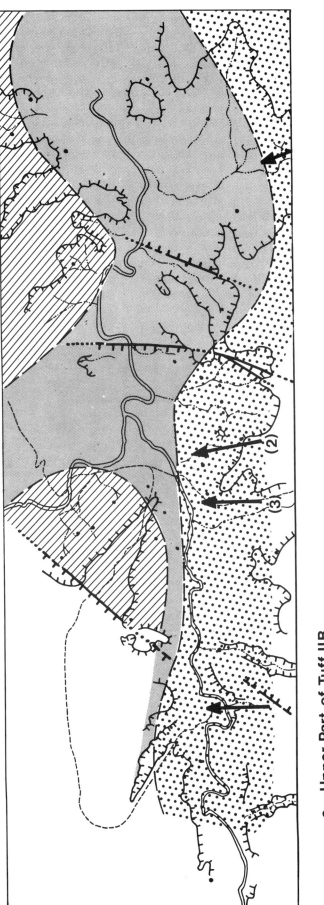

c. Upper Part of Tuff IIB.

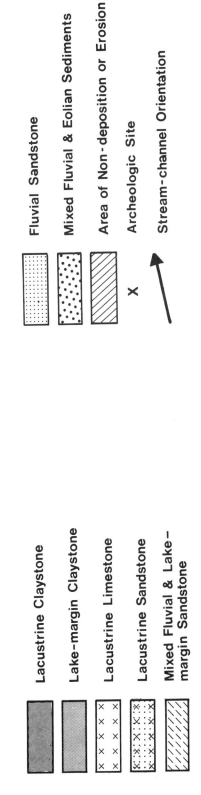

Lacustrine Claystone

Lake-margin Claystone

Lacustrine Limestone

Lacustrine Sandstone

Mixed Fluvial & Lake-margin Sandstone

Fluvial Sandstone

Mixed Fluvial & Eolian Sediments

Area of Non-deposition or Erosion

X Archeologic Site

Stream-channel Orientation

Figure 27. Paleogeography and sedimentation for three stratigraphic units of the eastern fluvial-lacustrine deposits near the junction of the Main and Side gorges. Map *a* represents the lower part of the lower augitic sandstone. Map *b* represents the upper part of the lower augitic sandstone, map *c* the upper half of Tuff IIB.

farther east are lake-margin sediments (locs. 45, 45a, 45c), weathered clays of paleosols (locs. 45d, 46c), and conglomeratic sandstones of fluviatile origin (locs. 42, 42a, 42b, 43, 44). The conglomeratic sandstone is in the form of an elongate body oriented east-west, which appears to represent the channel deposits of streams which flowed westward into the lake. The one channel measurement (loc. 43a) is N70°W, which fits with this interpretation.

The second stage (Fig. 27b) is represented by the upper part of the lower augitic sandstone. It was deposited as the lake spread eastward, and augite-rich lacustrine sandstone and limestone are found between FLK-NN (loc. 45b) and Castle Rock (loc. 44). In the Main Gorge these lacustrine sediments are bordered on the east by dominantly fluvial sandstones, which thin eastward and include sandy claystones that are either floodplain or mudflat deposits. In the area of the Side Gorge this stratigraphic interval is represented by claystones that are either purely lacustrine (loc. 88a) or were exposed as mudflats (locs. 88, 89). Equivalent deposits are very thin or missing at SHK (locs. 90, 91).

The third stage (Fig. 27c) is represented by Tuff IIB, which in most places marks an abrupt change from lacustrine and lake-margin to fluviatile and eolian sedimentation. Sediments of this age are absent in the FLK area but otherwise have about the same distribution as those of the underlying stratigraphic interval. The fluvial and eolian sediments of Tuff IIB crop out in an east-west belt between SHK (loc. 91) and HWK-E (loc. 42). Conglomerate is widespread in the upper part of the unit, where much of it fills well-defined, commonly steep-walled stream channels. The deepest channel, at VEK (loc. 85), is 4.6 m deep. Seven channel measurements are between north-south and N20°W, and thus indicate northward-flowing streams. Detritus from Sadiman and Lemagrut constitutes at least 90 percent of the clasts in conglomerates. Most of these fluvial and eolian sediments are root-marked, highly altered, and cemented by zeolites. In the vicinity of HWK (loc. 43a) and VEK (loc. 85) they are commonly orange to reddish-brown. These highly altered deposits probably were exposed for a rather lengthy period to a saline, alkaline soil environment. Contemporaneous deformation may account for pinching out of these deposits in the vicinity of FLK, for

the cutting of the deeper channels, and possibly for the high degree of subaerial alteration.

The fluvial-eolian sediments interfinger eastward with claystones which seem to occupy an east-west embayment whose axis lay slightly to the north of the alluvial deposits. The lithology of the claystones suggests that they were deposited in lake water and intermittently exposed as mudflats. The lacustrine embayment persisted in the Side Gorge at PEK (loc. 88a) through deposition of the lower half of Tuff IIB. The embayment had disappeared when the upper half of Tuff IIB was deposited.

The fourth stage is represented by the middle augitic sandstone, a sheetlike body of lacustrine sandstone which occupies a large eastern embayment in the same general area as that of the preceding stage. The sandstone pinches out to the west, in the vicinity of localities 85 and 45d, and does not appear to be represented by sediments in the area of FLK and most of the Side Gorge. It probably pinches out on the north side of the Main Gorge (locs. 13a-14) and interfingers with lake-margin claystones on the south side (locs. 34-35a). It may be represented by fluvial conglomerates and sandstones farther to the east (for example, loc. 32a). This sandstone sheet provides an ideal example of sedimentation in shallow, wave-agitated water. The western part comprises cross-bedded and laminated calcareous sandstones and calcite-crystal limestones in which oolites are widespread and are locally abundant (locs. 19, 40, 41). Interbedded claystone contains abundant ostracods, flakes of mica, and remains of small fish (loc. 39). Farther east the sandstones are laminated and well sorted but lack oolites and abraded calcite crystals. Thin interbeds of claystone pellets are near the eastern margin of the sheet, indicating nearby mudflats. The vertical sequence of sandstone types within the sheet (for instance, at loc. 41) records a progressive deepening and presumably an eastward spreading of the lake.

The fifth stage of paleogeographic development is represented by the bird-print tuff which was deposited in a small, shallow pond or lacustrine embayment roughly 700 m across in a northwest-southeast direction (Fig. 28). It is reworked and contaminated by claystone pellets around the margin of the embayment (locs. 42, 45, 45a, 45b). In the central area it is laminated and graded, and although it shows no evidence

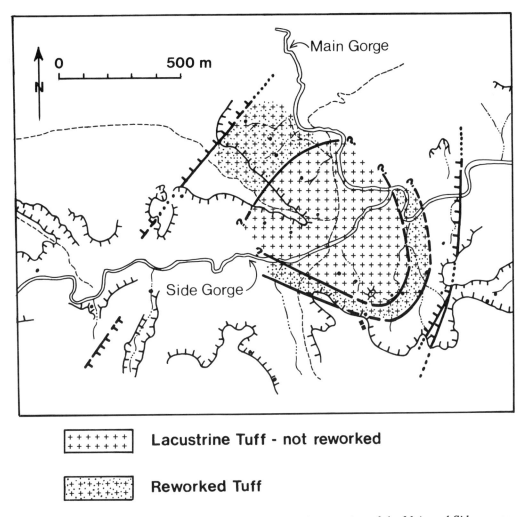

Lacustrine Tuff - not reworked

Reworked Tuff

Figure 28. Map showing nature of bird-print tuff near the junction of the Main and Side gorges.
Locality numbers are given in Figure 6.

of reworking, bird prints of plovers(?) are locally abundant (for example, loc. 44), indicating either very shallow water or temporary exposure. The tuff was very likely deposited at nearly the same level everywhere, and if it is taken as a horizontal datum (Fig. 26), a cross-section of the underlying part of Bed II shows the extent to which this area was deformed following erosion of the disconformity at the base of the facies. The disconformity is 6.8 m higher at FLK-N (loc. 45a) in this reconstruction than it is at HWK (loc. 43), and lacustrine limestones of the upper part of the lower augitic sandstone (locs. 45, 45c) are topographically 3 to 6 m higher than equivalent sandstones of fluvial origin to the east (locs. 42, 43). This deformation

probably also accounts for the pinching out of Tuff IIB and the middle augitic sandstone in the vicinity of FLK. These stratigraphic relationships suggest anticlinal folding, with a northeast-southwest axis passing through FLK (loc. 45), about 200 m from a major fault. Very likely the folding accompanied displacements along the fault. Deposits to the south appear to have been deposited in an east-west trough or synclinal flexure (Fig. 29).

Rootmarkings are the principal evidence as to the nature of the flora of the eastern fluvial-lacustrine deposits. Most of the rootmarkings are finely textured and represent grass and possibly brush. Coarse vertical root channels and rootcasts of marshland vegetation are found in many

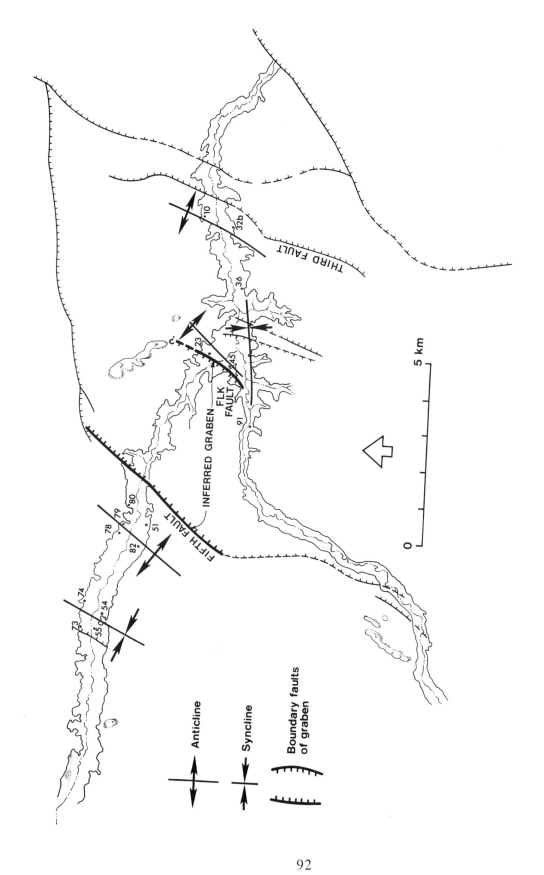

Figure 29. Structural features active during the deposition of Bed II above the lowermost disconformity.

92

Table 17. *Proportions of Specimens in Selected Taxa from Excavated Sites in the Eastern Fluvial-Lacustrine Deposits*

	Locality	Site	Chelonia	Crocodile	Proboscid	Hippo	Equid	Bovid
Main Gorge	HWK	48c levels 3-5	1.4	0	0	3.2	8.4	76.1
	TK	19a	0	0	0	5.0	15.0	50.0
	TK	19b	0	0	1.3	1.3	26.5	44.3
	TK	19d	0	0	0	12.8	33.7	45.3
Side Gorge	MNK	71a	76.7	3.7	0.4	1.3	0.9	13.4
	MNK	71c	2.8	3.1	0.2	12.8	20.9	45.9
	FC	62a	5.2	13.8	1.7	12.1	31.0	31.0
	FC	62b	2.2	0	0	18.7	18.7	51.6
	BK	66	2.4	2.6	0.7	2.0	18.0	53.2

NOTE: Data are taken from M. D. Leakey (1971a, p. 257).

places but are widespread at only a few horizons. They are particularly abundant in the siliceous earthy claystone above Tuff IID. Opal phytoliths of the type found in grasses (Rovner, 1971) were noted in thin sections of siliceous earthy claystone. Raymonde Bonfille has begun a study of the pollen in Bed II, and this should provide important information as to the types of grass, brush, and marshland vegetation. Lacustrine floral elements observed microscopically include diatoms, chrysophyte algal spores, and charophytes. The diatoms and algal spores were found in the siliceous earthy claystones and sandstones. Charophytes were noted in a sample of siliceous earthy claystone representing Tuff IID at BK, and in two samples of lacustrine claystone from PLK.

Ostracods are the only invertebrate forms noted. They are most common at PLK (loc. 23), where they are in about half the samples. They are abundant at one place (loc. 39) in claystone within the middle augitic sandstone. Ostracods were also found in several samples collected from the lower part of the facies in the Side Gorge (for example, loc. 88a).

The vertebrate assemblage is varied, comprising fish, reptiles, birds, and mammals (see L. S. B. Leakey, 1965). Forms favoring open savannah and riverine conditions (equids, antelopes, and hippos) are much more common, relative to swamp-dwelling and lacustrine forms, than they are below the disconformity at the base of the facies. Not only are the same faunal elements represented in different proportions above and below the disconformity, but several new forms appear at Olduvai for the first time in

deposits immediately above the disconformity. Examples include *Pelorovis oldowayensis* and *Damaliscus sp. nov.* (Gentry and Gentry, in press). Alcelaphines are the dominant bovid group, and the bovid fauna overall is suggestive of relatively dry, open savannah (Alan Gentry, 1973, personal communication). Remains of the white rhino (*Ceratotherium*), presently a grazer, are much more abundant than are remains of the black rhino (*Diceros*), a browser.

The analysis by M. D. Leakey of materials on excavated sites (1971a) clearly shows the dominance of bovids and equids and the wide occurrence of hippos above the disconformity (Table 17). Remains of crocodile and chelonia are much more abundant on sites in the Side Gorge than in the Main Gorge, and remains of proboscids and hippos may be somewhat more common in the Side Gorge. Bovids and equids are represented about equally in sites of the Main and Side gorges. Noteworthy concentrations of single species are at SHK (loc. 91), where remains of many gazelles were found in the same deposit, and at BK (loc. 94), where 24 *Pelorovis* skeletons were embedded in claystone within a stream channel.

Many vertebrate fossils were noted in the course of geologic studies. Probably most common are scattered equid teeth in conglomerates and sandstones. Faunal remains recovered in the excavation at PLK (loc. 23) include hippo and suid teeth, tortoise scutes, crocodile vertebrae, ostrich eggshell, and a *Theropithecus* maxilla (M. D. Leakey, personal communication, 1970). Catfish (clariid) remains and *Pelorovis* horn cores were uncovered in an earlier excavation at the top of Bed II and PLK (M. D. Leakey, 1971a).

Fish remains from archeologic excavations at SHK and BK are chiefly of clariid fishes but include a few of percoid fishes (Greenwood and Todd, 1970). Bones, spines, and scales of small fish are in the sand fraction of many lacustrine claystones and lake-margin sandstones throughout the facies. The samples submitted to P. H. Greenwood were too fragmentary for generic identification.

Bird remains from several excavated sites, principally in the Side Gorge, have been studied by P. Brodkorb. Swimming birds (cormorants, pelicans and grebes) are indicated by remains from HWK, DC, BK, SHK, FC, and PEK. Heron bones are from SHK and BK, and SHK also yielded flamingo bones (Brodkorb, 1972, personal communication).

In summary, the faunal and floral remains represent varied habitats, which accord with the environments inferred on purely geologic evidence. The most striking feature is the abrupt increase in proportion of forms favoring open savannah at the base of the facies. This change was probably caused indirectly by faulting, which modified the shape of the lake and greatly reduced the fringing marshland and correspondingly increased the area of grassland.

Eastern Fluvial Deposits

Eastern fluvial deposits constitute all of Bed II above the Lemuta Member in the downstream 4 km of the Main Gorge. These deposits interfinger and intergrade westward with the eastern fluvial-lacustrine deposits over a distance of about 1.5 km. Fluvial deposits can be traced northward 7 km along the western margin of Olbalbal, and a 6.2-meter thickness of fluvial deposits is exposed to the south of the gorge (loc. 157). The facies is disconformably overlain by Bed III as far east as localities 4 and 27; farther east, it is unconformably overlain by the Masek Beds.

The eastern fluvial deposits are about 72 percent claystone, 11 percent mudflow deposits, 7 percent sandstone, 6 percent tuff, and 2 percent each of conglomerate and limestone. The boundary between the eastern fluvial and eastern fluvial-lacustrine deposits is drawn primarily on

the basis of color and lithologic differences in sandstones and claystones. An additional, mineralogic criterion is the presence or absence of basement detritus. The eastern fluvial deposits have only volcanic detritus, whereas the eastern fluvial-lacustrine deposits are a mixture of basement and volcanic detritus.

Claystones are brown to reddish-brown, root-marked, and massive to crudely stratified. Most of them have a relatively high content of detrital silt and sand, and they grade into clayey sandstones. The silt and sand fractions are of volcanic detritus, principally anorthoclase, rock fragments, and augite. Claystones have been modified to varying extents by chemical reactions. The least modified of the claystones are soft and brown or reddish-brown; the most altered are hard and bright reddish-brown. The soft brown claystones generally contain 10 to 15 percent zeolites, whereas the hard reddish-brown claystones are generally 30 to 40 percent zeolites, generally analcime and either chabazite or phillipsite. The zeolites occur as disseminated crystals or as a cement, and they fill fractures and line cavities such as root channels. Montmorillonite is the dominant or the only clay mineral in the soft brown claystones, whereas illite is the dominant type in the hard, zeolite-rich claystones. The zeolites were formed chiefly by reaction of montmorillonite with alkaline solutions of sodium-potassium carbonate (Hay, 1970a):

$$\text{montmorillonite} + K^+ + Na^+ + CO_3^= \rightarrow \text{illite} + \text{analcime} + SiO_2 + H_2O + CO_2$$

The reddish-brown color is largely attributable to dehydration of yellowish-brown hydrated ferric oxide to anhydrous ferric oxide (hematite). These alterations were very likely penecontemporaneous with deposition, in view of the fact that hard, reddish-brown claystones are interbedded with soft brown claystones.

The mudflow deposits are massive, commonly resistant beds 1 to 9.5 m thick which are composed of blocks, chiefly of lava, in an unsorted clayey matrix. One deposit, 1 to 2 m thick, is in the westernmost exposures (locs. 5, 6a, 27), and two totaling 7 m are found nearer the mouth of the gorge (loc. 3). The westernmost deposit probably lies stratigraphically slightly below Tuff IIC, judging from its relative position between the Lemuta Member and Tuff IID (Fig. 22b, locs. 27, 29). Mudflow deposits crop out discontinuously along the western margin of Olbalbal between the mouth of the gorge and locality 1, 7 km to the north. The sand fraction comprises rock fragments, crystals, and zeolitic pumice lapilli. The crystals comprise chiefly anorthoclase and augite but include small amounts of altered olivine, sphene, perovskite, and biotite. Much of the augite is bright green and unusually coarse. The clay matrix ranges from soft, yellowish-brown, and poor in zeolites to reddish-brown, hard, and rich in zeolites. Most of the zeolitic alteration was probably penecontemporaneous with deposition, but the uppermost mudflow near the mouth of the gorge was intensely altered during or shortly before the deposition of the Masek Beds (Hay, 1970a).

Lava blocks decrease in size from east to west, and the largest are 1.2 m long at locality 3 and only about 25 cm long near the Second Fault. They coarsen, on average, in a northward direction, and the largest block ($2 \times 3.8 \times 3.8$ m) was measured near locality 1. Coarse, porphyritic trachyte of Olmoti type is the dominant lava and forms the largest blocks. The trachyte is, however, of two varieties, which are restricted to different areas. The trachyte in the Main Gorge weathers pale gray and contains large oligoclase phenocrysts in a coarse groundmass of alkali feldspar, hornblende, sodic augite, and zeolites. Trachyte to the north (locs. 1-1a) weathers greenish-gray and contains large phenocrysts of anorthoclase in a coarse groundmass of deeply pleochroic hornblende, aegirine, alkali feldspar, and zeolites. Coarse, porphyritic olivine basalt of Olmoti type forms at least several percent of the blocks in all of the deposits. This basalt contains abundant large phenocrysts of bright green augite and altered olivine. Perhaps 5 percent of the blocks are of trachytic welded tuff, and a much smaller percentage is of zeolitic lapilli tuff. In the deposits near the mouth of the gorge are rounded clasts of waxlike, dark reddish-brown claystone as much as 10 cm in diameter.

Sandstones generally form lenticular beds 50 to 100 cm thick which are massive or crudely bedded and commonly rootmarked. They are medium- or coarse-grained and commonly have a clay matrix, which is generally reddish-brown. Most of the volcanic debris in the sandstones is from Olmoti, judging principally from the rock fragments. Many if not most of the sandstones are cemented by zeolites.

The only widespread tuff is Tuff IID, which is reworked, rootmarked, and contaminated with detrital sand. A reworked trachytic pumice tuff lies above Tuff IID in a few places (locs. 31, 32), and sandy, reworked mafic tuff lies above Tuff IID in CK (loc. 7). Tuff IIC was noted only in locality 29, where it forms a thin, lenticular bed. Vitroclasts are wholly altered, and all of these tuffs are cemented by zeolites. The tuffs are commonly a bright reddish-brown.

Limestone occurs as nodules and nodular beds of limited extent. They are aphanitic and sandy, and in the Main Gorge they are generally pale gray or brown. Those found along the western margin of Olbalbal are generally pale reddish-brown. All are of water-table or caliche origin.

Conglomerates occur at various levels but are most widespread at and near the base of the facies. They are lenticular in cross-section and fill narrow channels in a few places. Clasts range from pebbles to boulders as much as 60 cm in average diameter. A few conglomerates consist largely of large cobbles and boulders (for instance, loc. 26b). Clasts are chiefly of the same rock types found as blocks in the mudflow deposits. Ngorongoro detritus is comon in the westernmost conglomerates (for example, loc. 31) and forms a small proportion of the clasts as far east as locality 26b. Clasts of eolian tuff from the Lemuta Member are not rare as far east as locality 26b. There is in addition a very small percentage of green nepheline phonolite pebbles which superficially resemble those derived from Sadiman. The two pebbles studied microscopically have, however, a coarser and more feldspathic groundmass than the Sadiman phonolites.

Channel alignments suggest that streams flowed northwest, then changed to a more westerly direction near the margin of the facies. The two deepest and best-defined channels, oriented about N65°W, are cut in the mudflow deposit (loc. 27). Two rather poorly defined channels at locality 31 appear to be oriented N85°W and N60°E. Although the latter measurements are not reliable, they fit with the inferred main east-west drainageway nearby in the eastern fluvial-lacustrine deposits.

Finely textured rootmarkings suggestive of

grass are widespread and are the only type of plant fossil. Vertebrate remains are rather rare, and the only form identified is *Beatragus antiquus*, an alcelaphine bovid, from locality 3.

The claystones, sandstones, tuffs, and conglomerates have all the characteristics of fluvial sediments, and the high proportion of claystone suggests an alluvial plain of very low gradient. Olmoti was the source of the mudflows and of most of the detritus in the stream deposits. The coarse nature of many conglomerates deposited on this nearly flat surface points to transport and deposition by flash floods. The mudflows originated by slumping of older rocks and soil, not as a result of explosive eruptions. They probably originated high on the flanks of Olmoti in view of the fact that they had sufficient momentum to carry large blocks at least a few kilometers over the alluvial plain.

ENVIRONMENTAL SYNTHESIS AND GEOLOGIC HISTORY

The paleogeography of the lower part of Bed II was similar in most respects to that of the upper part of Bed I. The widespread disconformity above this lower part of Bed II marks a paleogeographic change in which the size of the lake and adjacent marshland was greatly reduced. The disconformity resulted from earth movements, the most important element of which was a graben, or subsiding trough, between the Fifth Fault and the fault a short distance west of FLK. Subsidence was accompanied by gentle anticlinal folding on the upthrown sides of the faults bounding the graben (Fig. 29). These anticlines were eroded prior to deposition of the lowermost augitic sandstone, both east and west of the graben. The lower part of Bed II was also warped upward along an axis west of and roughly parallel to the Third Fault. This uplift resulted in erosion of the upper part of the Lemuta Member on the south side of the gorge (locs. 32a-33b). Downwarping accompanied the uplifts, as shown by anomalously great thicknesses of sediment in two troughs, or synclinal flexures. One trough has an axis slightly east of, and parallel to the westernmost fault; the other, near the junction of Main and Side gorges, is elongated east-west, and has an axis oblique or perpendicular to all faults in this area.

The anticline and syncline between the FLK fault and KK (loc. 41) coincide closely with areas of thickening and thinning of Tuff IF (Fig. 16). The anticline of Bed II is located where Tuff IF is thinnest and was exposed intermittently above lake level, whereas the syncline includes the area where Tuff IF is thickest and shows no evidence of exposure. Moreover, Tuff IF and the bird-print tuff of Bed II are purely lacustrine over very nearly the same area (Fig. 28). From this comparison, I conclude that Tuff IF records slight earth movements of the same type that substantially modified the basin during the deposition of Bed II between the disconformity and the bird-print tuff.

The graben subsided further after the bird-print tuff was deposited, and the adjacent fault blocks were tilted downward toward the graben. This tilting is inferred from the inclined attitude of the bird-print tuff and the underlying lacustrine sediments in a reconstruction of Bed II with the inferred original top taken as horizontal (Fig. 30). The same conclusion follows from a reconstruction taking Tuff IID as horizontal. The two disconformities to the west of the Fifth Fault are probably a result of two relatively rapid stages of subsidence. It is not clear as to why these two disconformities cannot be found in deposits of Bed II to the east of the graben.

Vertical displacements to the east or south of the gorge seem to be indicated by the stratigraphic distribution of eolian tuff clasts from the Lemuta Member. Tuff clasts are most abundant in conglomerates directly overlying the Lemuta Member, as would be expected from erosion of the disconformity in the area of the gorge. They also have been found as much as 7.6 m above the disconformity (loc. 32) and as little as 2.3 m beneath Tuff IID (locs. 32, 33a). Nowhere in the gorge does the Lemuta Member stand sufficiently high for erosion to supply the tuff clasts only a short distance below Tuff IID. Tuff clasts seem to be most abundant in conglomerates with the highest proportion of debris from Ngorongoro. Consequently, it seems likely that the Lemuta Member was uplifted, probably along faults, to the south of the gorge and possibly near the western margin of Olbalbal.

With this structural framework as a background, the history of Bed II will now be developed from time-equivalent subdivisions of Bed II.

Bed II Up to the Lowest Disconformity

This lowermost unit of Bed II comprises four facies: lake, lake-margin, eolian, and alluvial-fan deposits. These facies represent a lake basin bordered on the east by an alluvial fan of pyroclastic materials discharged from Olmoti. Eolian deposits accumulated on the fan surface and on the adjacent lake-margin terrain. The Olmoti tephra represents a continuation of the Bed I eruptive phase. Mafic tephra, probably from distant volcanoes, is widespread as a minor constituent of Bed II below the level of the Lemuta Member, and the Lemuta Member contains a large volume of augitic nephelinite tephra from an unknown volcano, probably to the east or northeast. Mozonik volcano, in the basin of Lake Natron to the northeast, has an appropriate composition, but it is younger than the Lemuta Member (Isaac and Curtis, 1974).

The lake basin can be subdivided into three zones: a perennial saline lake, a surrounding zone flooded for extended periods by saline alkaline lake water, and an outer zone flooded only infrequently, at times of highest water level. The perennial lake was about 9 km long and 5 km wide, and the outer zone was about 24 km long and 16 km wide, covering an area roughly eight times as large as that of the perennial lake (Fig. 31; see also Fig. 38).

The perennial lake was highly saline, as shown by authigenic minerals and by isotopic measurements. Salinities were highest in the western part of the lake. Lake water of the adjacent zone was generally less saline, although chert nodules, trona molds and casts, and rare K-feldspar testify to intermittent (or localized areas of) brine. Lake deposits in the Side Gorge belong to this zone and contain a high proportion of unusual zeolitite deposits. The outermost, or lake-margin zone, was flooded by fresh to brackish water and included large areas of marshland. This terrain supported a rich fauna of water- and swamp-dwelling forms.

Tephra of the alluvial fan to the east was mostly redeposited by streams, but some was emplaced by hot ash flows to form ignimbrites. This alluvial fan, like that of Bed I, had relatively sparse vegetation, at least seasonally. In the latest stages of fan growth, sand-sized particles of Olmoti tephra was extensively reworked by wind to form an extensive eolian deposit which was spread over the lake-margin terrain at least a kilometer beyond the foot of the alluvial fan. The overlying eolian tuffs are composed largely of tephra produced in contemporaneous eruptions. Some of the tephra was reworked into dunes, which migrated westward over the fan and adjacent lake-margin terrain. Wind-deposited ash on the alluvial fan was altered in a saline, alkaline soil environment to form zeolitic eolian tuffs. The eolian tuffs are estimated to have accumulated over a minimum of 20,000 years.

Disconformity to Bird-Print Tuff

The interval between the disconformity and the bird-print tuff comprises lake deposits, eastern and western fluvial-lacustrine deposits, and probably a small proportion of eastern fluvial deposits. These facies were deposited in a basin, vastly changed by earth movements, in which the area occupied by lake and marshland was greatly reduced (Fig. 32). Augitic ash of tephrite composition was showered widely over the basin and was reworked to form large volumes of sandstone extraordinarily rich in augite.

Clasts of Engelosin phonolite in sediments above the disconformity near the Fifth Fault (locs. 49, 82a) are significant in determining the paleogeography to the north of the gorge. The clasts are too large and angular to have been readily transported by wind. If, as seems likely, they were carried by running water, they are evidence that the surface sloped gently southward or southwestward between Engelosin and the beds with Engelosin detritus. On this assumption, the area flooded by lake probably did not extend northeast of the inselberg Naibor Soit.

Lacustrine sediments of the perennial lake in the graben are nowhere exposed, but deposits outside the graben reveal a lake that was chemically rather similar to the earlier and larger lake of Bed II. Although K-feldspar is absent, trona indicates that salinities were high, at least in small brine pools. Chert was formed widely, and it has relatively uniform δO^{18} values (+32.0 to 34.8°/oo), suggestive of moderate salinity. A sample of euhedral calcite crystals is isotopically heavy and gave δO^{18} and δC^{13} values very close to the average of crystals from Bed I. Domi-

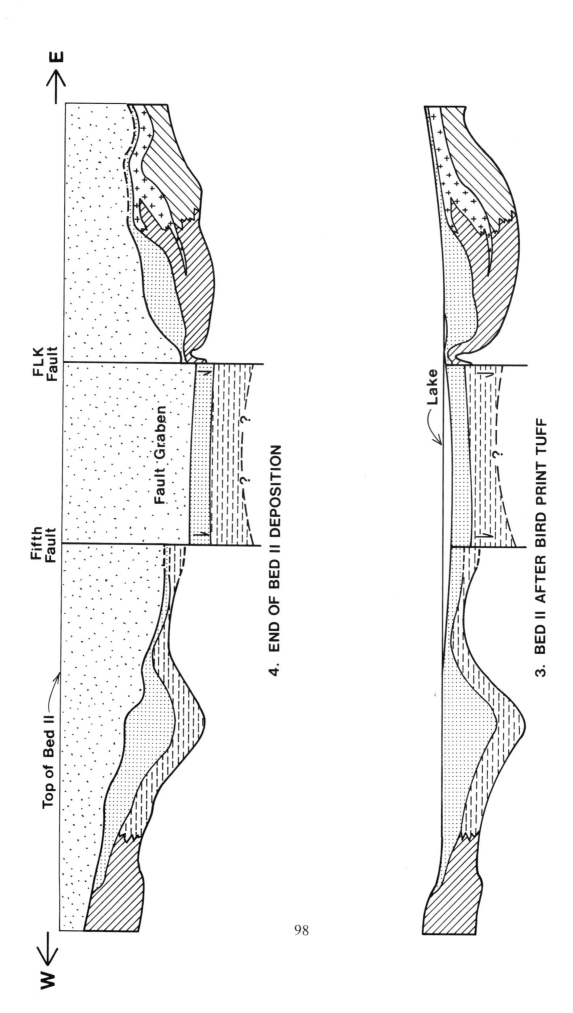

E

W

Top of Bed II

Fifth Fault

FLK Fault

Fault Graben

?

4. END OF BED II DEPOSITION

Lake

?

3. BED II AFTER BIRD PRINT TUFF

98

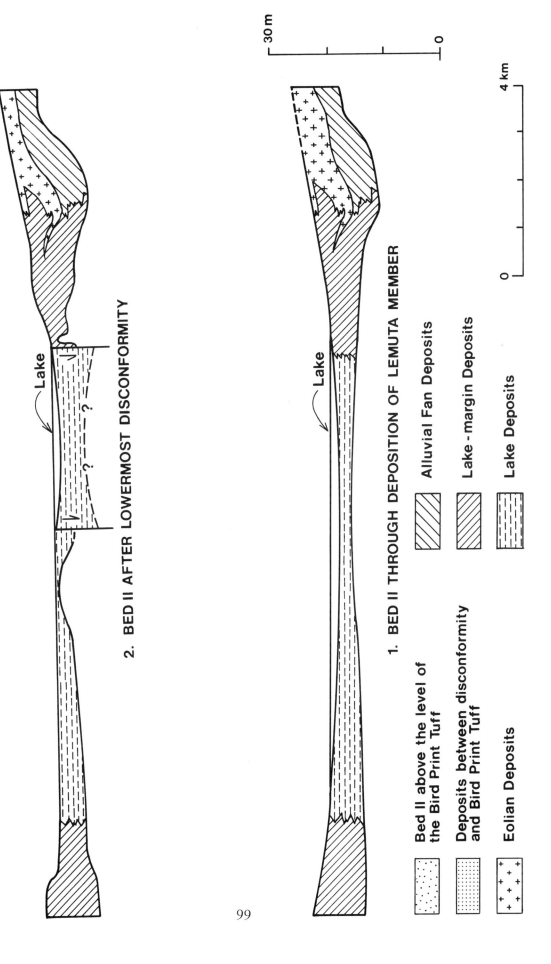

30 m

0

4 km

0

2. BED II AFTER LOWERMOST DISCONFORMITY

Lake

1. BED II THROUGH DEPOSITION OF LEMUTA MEMBER

Lake

Alluvial Fan Deposits

Lake - margin Deposits

Lake Deposits

Bed II above the level of the Bird Print Tuff

Deposits between disconformity and Bird Print Tuff

Eolian Deposits

Figure 30. Geologic development of Bed II as based on measured sections on the south side of the Main Gorge between localities 31 and 66. Amount of graben subsidence is purely conjectural.

99

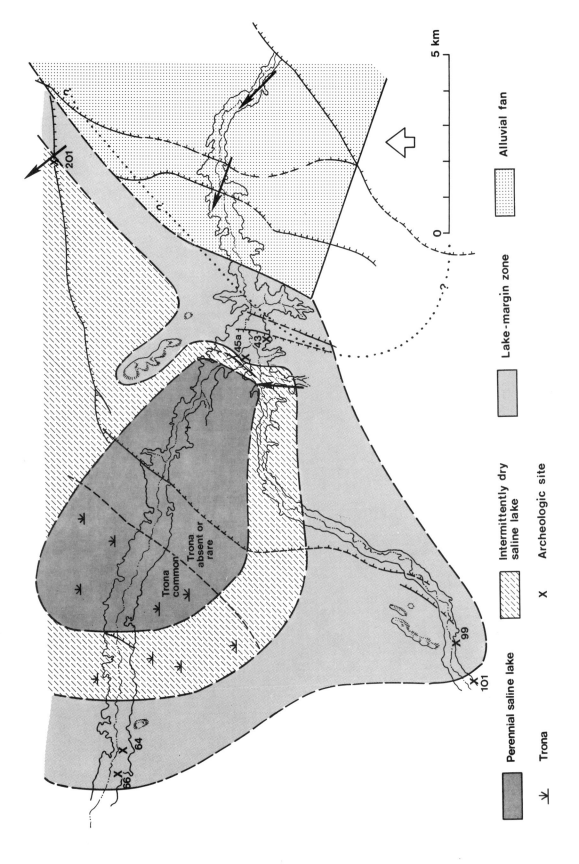

Figure 31. Paleogeography of Bed II in the area of the gorge for the interval between the base of Bed II and the lowermost disconformity. Arrows indicate flow direction of streams as based on channel measurements.

100

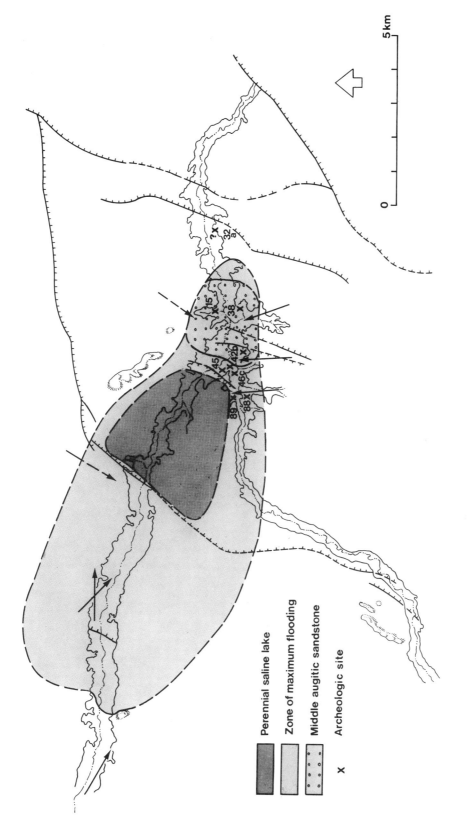

Perennial saline lake

Zone of maximum flooding

Middle augitic sandstone

x Archeologic site

0   5 km

Figure 32. Paleogeography of Bed II in the area of the gorge for the interval between the lowermost disconformity and the bird-print tuff. The paleogeography of this interval to the east of the perennial lake is unusually variable, and more detailed information is given in Figure 27. Solid arrows indicate flow direction of streams as based on channel measurements; dashed arrows represent flow direction inferred from clast composition.

101

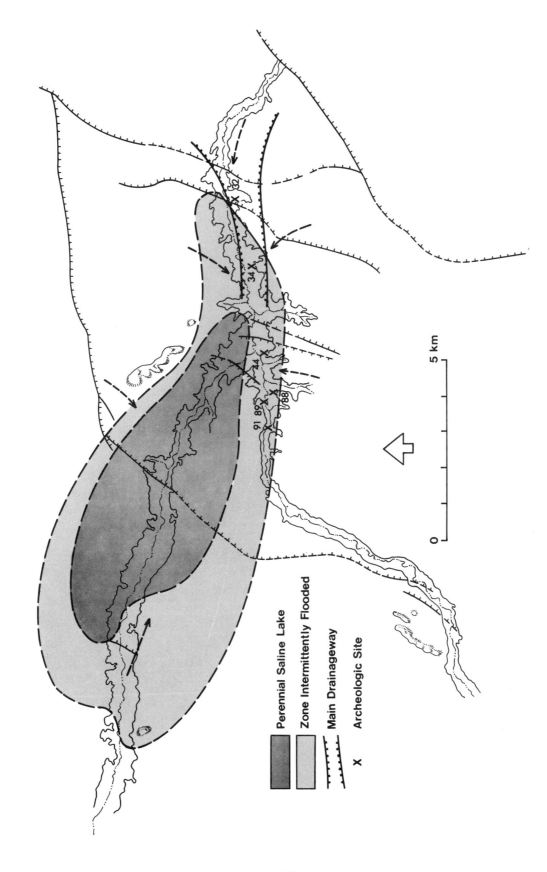

Perennial Saline Lake

Zone Intermittently Flooded

Main Drainageway

X Archeologic Site

5 km

0

Figure 33. Paleogeography in the area of the gorge for the interval represented by the upper augitic sandstone. Dashed arrows indicate direction of stream flow inferred from clast composition.

102

nantly lacustrine sandstones form a broad zone around the margin of the lake, both northwest and southeast of the graben. The paleogeography along the southeastern margin of the graben varied greatly at different levels in this interval, at least partly as a result of contemporaneous folding (Fig. 27). Eastern fluvial deposits, if correctly assigned to this interval, are found on the south side of the gorge to the east of the uplifted area near the Third Fault. Clasts in conglomerates are from both the northern and eastern suites, suggesting that the conglomerates were deposited in a main channelway which drained westward into the lake, across the uplifted area.

Bird-print Tuff to Brown Tuffaceous Siltstone

The interval between the bird-print tuff and the brown tuffaceous siltstone contains the same facies as the underlying interval (lake, eastern and western fluvial-lacustrine, and eastern fluvial), and the overall paleogeography remained essentially the same (Fig. 33). This interval records first an expansion and then a retreat of the lake. The main drainageway to the east end of the lake is clearly documented by fluvial conglomerates and sandstones deposited during retreat of the lake. Augitic tephra was contributed to the basin, at least during the lower part of this interval.

Lacustrine clays were deposited far to the east by the expanding lake in the early stages of this period. The claystones lack mineralogic evidence of high salinities, and the water was probably fresh to brackish. Lacustrine clays also were deposited far to the west, but their original extent is uncertain because they were eroded away in many places beneath a disconformity to the west of the Fifth Fault. These clays of the western area have mineralogic indicators of moderate or high salinity. The lower salinity along the eastern margin of the lake suggests that this part of the lake received a greater volume of fresh river water than did the western part. Augitic tephra was reworked and concentrated along the southern margin of the lake to form the upper augitic sandstone.

The lake retreated from both the eastern and western shorelines, and the present reconstruc-

tion of the lake at times of low level is based on correlation of the brown tuffaceous siltstone with the disconformity eroded in lacustrine claystones to the west of the Fifth Fault. This correlation is tentative and based on the fact that both features mark the first major withdrawal of the lake following the bird-print tuff. This major retreat of the lake very likely resulted from subsidence of the graben.

Eastern fluvial deposits of the interval are chiefly reddish-brown zeolitic claystones but include a deposit of coarse conglomerate which was probably deposited in the main drainageway.

Brown Tuffaceous Siltstone to Tuff IIC

This interval comprises the same facies found below the brown tuffaceous siltstone, and the paleogeography was basically the same (Fig. 34). This records another expansion of the lake, and lacustrine claystones were deposited nearly as far east as the Third Fault. Fluvial sands and gravels accumulated in the main drainageway during retreat of the lake. Farther east, coarse mudflows swept down from the slopes of Olmoti and spread westward over a nearly flat alluvial plain (see Fig. 39). Streams deposited sandstones to the west of the lake in the graben, and carbonate rocks of caliche and water-table origin were formed on the westernmost part of the basin. Small amounts of mafic tephra were supplied to the basin, and evidence of earth movements is lacking.

Deposits of the perennial lake are dominantly claystones with mineralogic evidence for high salinity. Fresh or brackish water is suggested by claystones deposited near the eastern end of the lake at times of high level. Lacustrine sandstones and limestones composed of abraded calcite crystals (loc. 20) appear to mark the eastern shoreline of the lake at low level.

Clasts of Engelosin phonolite and other detritus of the northern suite are unusually widespread and abundant in deposits of this interval. They are found not only in deposits of the main channelway, but in sandstone to the west of the graben (loc. 82a). Much of this detritus is too coarse to have been transported by wind, and its wide distribution seems to indicate that it was

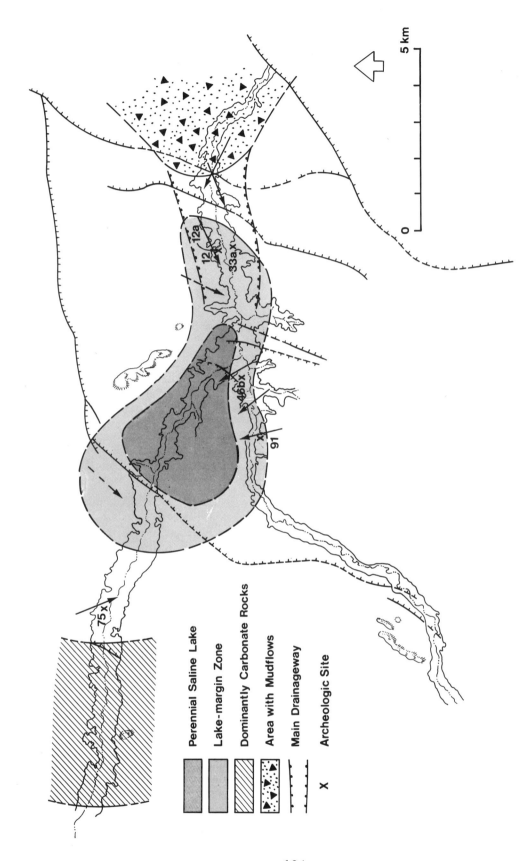

Perennial Saline Lake
Lake-margin Zone
Dominantly Carbonate Rocks
Area with Mudflows
Main Drainageway
X Archeologic Site

5 km

Figure 34. Paleogeography of Bed II in the area of the gorge for the interval between the brown tuffaceous siltstone and Tuff IIC. This interpretation assumes that the brown tuffaceous siltstone is correlative with the middle disconformity to the west of the Fifth Fault. See Figure 32 for meaning of arrows.

104

carried by streams from the north which meandered widely over a surface sloping gently to the south or southwest.

Channel alignments for this interval help to document the drainage pattern as inferred from the distribution of clast types and elongation of sand bodies of the main drainageway, in which were measured three channels (N65°E, N60°E, N85°W), suggestive of streams flowing west and west-southwest. Two measurements in the Side Gorge and one at FLK (loc. 45) fit with the inferred location of the perennial lake in the graben.

Tuff IIC to Tuff IID

This interval comprises Tuff IIC and the overlying deposits up to the base of Tuff IID. The paleogeography for this interval remained about the same as that for the deposits below Tuff IIC (Fig. 35). The lake was restricted to the graben at times of low level, and at times of high level it spread rather far to the east within the main drainageway. Sands and gravels were deposited in the drainageway at times of low level. Sediments of the perennial lake are chiefly claystone and contain mineralogic evidence of high salinity. Fresh-water marshland fringed the southeast margin of the lake at the time Tuff IIC was deposited. Mafic tephra was showered widely over the basin and is most abundant in deposits to the west of the graben. A relatively small amount of trachytic tephra was deposited in the area southeast of the graben.

A paleogeographic problem is presented by clasts of the northern suite, including Engelosin phonolite, in a stream-channel conglomerate at DC (loc. 93), on the south side of the lake. Perhaps the northern-suite clasts were transported westward by wave action along a beach on the south side of the lake to the mouth of a stream flowing north, where they were mixed with detritus from Lemagrut and Kelogi.

Tuff IID

Tuff IID is a product of several dominantly phonolitic trachyte eruptions over a relatively brief time. Thin interbeds of claystone (for example, loc. 38) reflect short periods when no ash fell. Where the eruptive sequence appears to be most complete (loc. 40), the lowermost layers are relatively rich in crystals and rock fragments, and mafic vitroclasts are not rare.

The tuff is thickest and its lower part is lacustrine over a large elongate area within the main drainageway to the east of the lake (Fig. 36). Thus, a sizable area of the drainageway must have been flooded when the first showers of ash fell. If, as seems likely, Tuff IID is represented to the west of FLK by a clayey, zeolitic tuff interbedded with lacustrine clays (Fig. 23), a perennial lake still occupied the graben where Tuff IID was deposited. By analogy with earlier stages (for example, Fig. 34), the lacustrine area to the east would represent an embayment from the perennial lake in the graben. Casts and molds of trona are locally abundant in Tuff IID(?) within the graben, providing clear evidence of deposition either in brine pools or on an exposed salt flat. Lacustrine tuff of the eastern embayment is nearly everywhere altered to zeolites, showing that it was deposited largely in saline, alkaline lake water. The upper part of the tuff in the embayment is reworked and root-marked, suggesting that the embayment had disappeared.

Tuff IID in the Side Gorge is thin, discontinuous, and contaminated by claystone pellets and detrital sand. It overlies lacustrine claystone and claystone-pellet aggregates. In this area it probably was deposited on mudflats and reworked by the wind.

Tuff IID is laminated and lacustrine to the west of the Fifth Fault at RHC (loc. 80), where it occurs both in place and as clasts. Tuff clasts elsewhere to the west of the Fifth Fault are of massive rather than laminated tuff, suggesting that the lacustrine tuff at RHC represents a rather small area of ponded water.

Tuff IID to top of Bed II

The perennial lake in the graben disappeared shortly after Tuff IID was deposited (Fig. 37). This event is recorded by the disconformity separating lacustrine claystones from sandstones of the western fluvial-lacustrine facies (locs. 25, 25a 47, 48). This disconformity is a surface of considerable erosion to the west of the Fifth Fault. Relatively small ponds were present in the

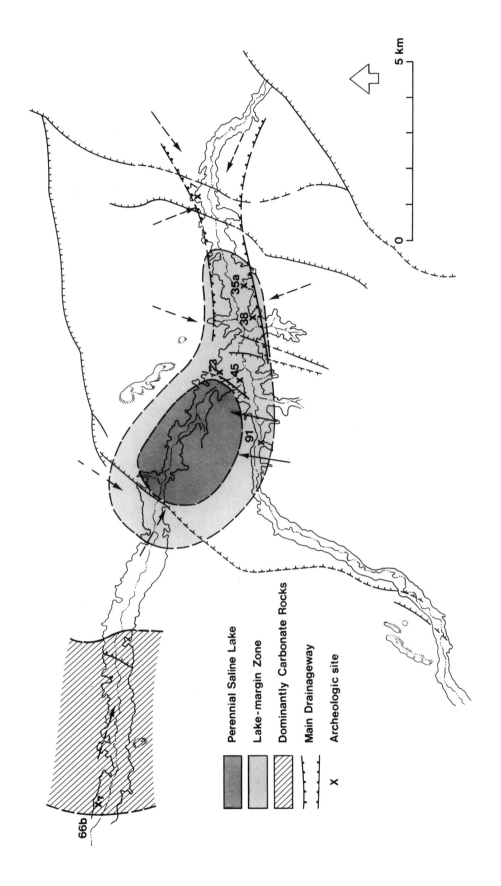

Perennial Saline Lake

Lake-margin Zone

Dominantly Carbonate Rocks

Main Drainageway

X   Archeologic site

5 km

Figure 35. Paleogeography of Bed II in the area of the gorge for the interval between the base of Tuff IIC and the base of Tuff IID. The maximum shoreline to the west of the Fifth Fault is poorly defined and may have been exceeded at infrequent intervals. See Figure 32 for meaning of arrows.

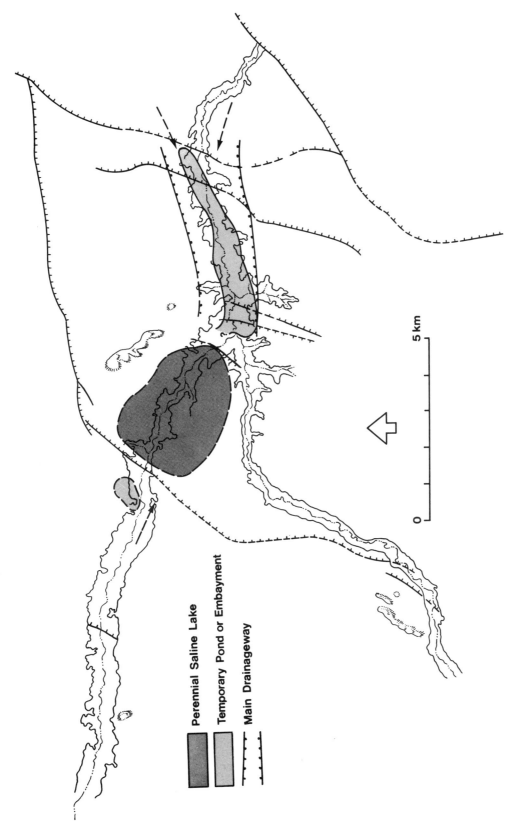

Perennial Saline Lake

Temporary Pond or Embayment

Main Drainageway

5 km

0

Figure 36. Paleogeography for Tuff IID. Arrows indicate inferred direction of stream flow.

107

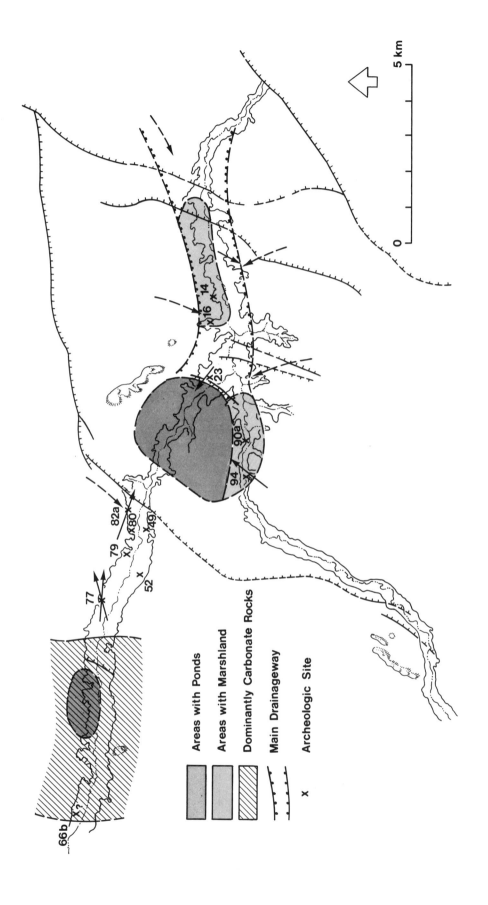

Areas with Ponds
Areas with Marshland
Dominantly Carbonate Rocks
Main Drainageway
x Archeologic Site

Figure 37. Paleogeography for Bed II in the area of the gorge above Tuff IID. See Figure 32 for meaning of arrows.

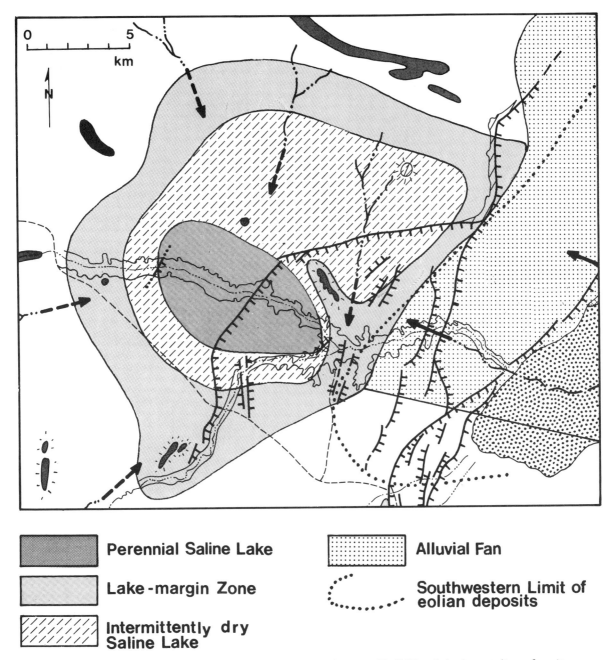

Perennial Saline Lake

Lake-margin Zone

Intermittently dry Saline Lake

Alluvial Fan

Southwestern Limit of eolian deposits

Figure 38. Regional paleogeography of the part of Bed II between Tuff IF and the lowest disconformity. See Figure 32 for meaning of arrows.

graben as shown by lenticular beds of waxlike claystone, some of which contain ostracods (for instance, loc. 25). The main east-west drainage-way is represented by channel fillings of sandstone and by areas of marshland. In the area of the Side Gorge was another, probably larger area of marshland. Clasts of Engelosin phonolite are locally common in sediments of this interval

(loc. 15, 16), showing that the basin axis had not shifted to some point north of the gorge. Very likely the lowest point was in the graben near its southeastern margin. As evidence, detrital sediments of the graben are of western origin except along the eastern margin, where detritus of eastern origin is intermixed. Stream-channel measurements support this inferred paleogeogra-

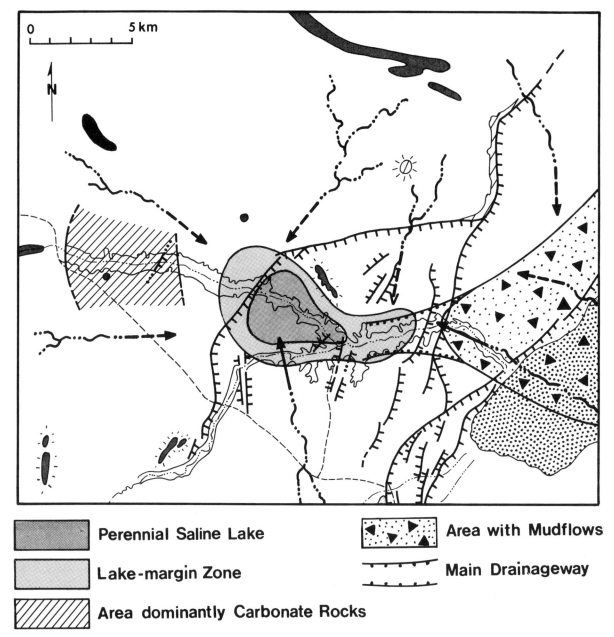

Perennial Saline Lake

Lake-margin Zone

Area dominantly Carbonate Rocks

Area with Mudflows

Main Drainageway

Figure 39. Regional paleogeography of the part of Bed II between the brown tuffaceous siltstone and Tuff IIC. This interpretation assumes that the brown tuffaceous siltstone is correlative with the middle disconformity to the west of the Fifth Fault. See Figure 32 for meaning of arrows.

phy. Well-defined channels at BK (loc. 94) are oriented N30°-50°E, which represents a significant eastward shift from the channel direction (N10°E) measured beneath Tuff IID nearby at DC. A channel with debris of easterly origin at PLK is oriented about N60°W, which accords with streams flowing into the graben.

Fault movements very likely caused the perennial lake to disappear and resulted in erosion of the widespread disconformity. Faulting may have destroyed the lake by creating or opening fractured zones that allowed lake water to flow underground to a lower level in the hydrologic system (as noted earlier in this chapter).

Mafic tephra was showered widely over the

basin several times between the deposition of Tuff IID and the end of Bed II. Trachytic tephra falls are recorded by two tuffs in the southeastern part of the area.

Climate of Bed II

A semiarid climate prevailed during the deposition of Bed II. The average climate probably resembled that of the upper part of Bed I more closely than it did the lower fossiliferous levels. This comparison is suggested by the fauna and is supported by a variety of geologic features. The fauna, at least above the lowermost disconformity, is dominated by forms suggestive of relatively dry, open savannah. However, the abrupt increase in proportion of these forms at the disconformity reflects the paleographic change caused by faulting rather than a rapid change in the climate. Nowhere in Bed II have been found urocyclid slugs, which require damp conditions in evergreen forests. The following geologic features are evidence of an arid or semiarid climate:

1. The perennial lake was highly saline and did not have an outlet. The lake of Bed II below the level of the disconformity was more saline than the lake of Bed I, as judged from the common occurrence of trona. Isotopic measurements are unfortunately inadequate for a satisfactory comparison, although it is perhaps significant that a chert nodule of Bed II is isotopically heavier than a nodule of Bed I in the same area. Lake deposits above the disconformity cannot be satisfactorily compared with deposits of the Bed I lake because the size, depth, and to some extent the salinity of the lake were determined by earth movements as well as by climate.

2. Wind-transported sediment is both widespread and locally abundant in the eastern and western fluvial-lacustrine facies, and it forms nearly all of the eolian facies. The eolian deposits are particularly significant in showing that vegetation on the alluvial fan and adjacent lake-margin terrain was at least seasonally insufficient to prevent large-scale movement of sand-size sediment by the wind.

3. Three lithofacies exhibit abundant evidence of zeolitic alteration at the land surface, demonstrating that saline, alkaline soils were widespread. Wind-worked tephra was altered and cemented to form eolian tuffs mineralogically similar to those formed in late Pleistocene and Holocene time in the Olduvai region.

4. Coarse conglomerates were deposited on a relatively flat surface in the eastern fluvial and eastern fluvial-lacustrine facies. These are suggestive of flash floods caused by torrential downpours in an arid or semiarid climate.

5. Mudflows, found in the eastern fluvial facies, are generated most readily in an arid or semiarid climate with torrential downpours.

6. Caliche limestone and dolomite are abundant in the western part of the basin, and they point to a climate in which evapotranspiration exceeds rainfall.

Features suggestive of a drier than average climate are found at four levels in Bed II. Eolian tuffs of the Lemuta Member reflect a relatively dry episode that lasted at least 20,000 years. The tuffs were deposited and altered before the basin was changed by earth movements, and they very likely represent a regional climatic event. The lower tongue of eolian tuff, consisting of older materials eroded by the wind, is the best evidence for a period of desiccation. The upper tongues are in accord with a dry climate, although they are not equally good evidence inasmuch as their sediment was supplied largely by falls of tephra. Tuff IIB contains a high proportion of zeolitic eolian siltstone, and it has relatively deep channels filled with conglomerates that are unusually coarse for Bed II in this area. This association of features clearly indicates that the local environment was much drier than that which prevailed when the overlying and underlying sediments were laid down. Tuff IIB was deposited in a part of the basin which was being deformed, hence the features suggestive of aridity may reflect a temporary paleogeographic change caused by earth movements. The brown tuffaceous siltstone and the associated zeolitic alteration at a higher level may likewise represent either a regional climatic or local paleogeographic change. Finally, caliche limestones and calcareous nodules are widespread in the topmost part of Bed II and are particularly abundant in the eastern fluvial-lacustrine facies. Evidence of standing water or marshland is extremely rare in the uppermost deposits. These features of the uppermost interval suggest a drop in the water table which could have resulted either from earth movements or climatic change.

HOMINID REMAINS AND ARCHEOLOGIC MATERIALS

Sixteen of the hominid fossils collected at Olduvai Gorge are very probably from Bed II (M. D. Leakey, 1971a, p. 234; in press, 2). A few additional fossils found on the surface may possibly be from Bed II. Remains clearly assignable to *Homo habilis* (H-13, 16, 37, 41) have been found only below the level of Tuff IIB (Table 18), whereas remains of *Homo erectus* (H-9, 36) have been found only near the top of Bed II, at or above the level of Tuff IID. *Australopithecus cf. boisei*, by contrast, occurs both high and low in Bed II (H-3, 20, 26, 30, 38). Four of the hominid fossils are probably from the eastern lake-margin deposits (H-16, H-17, H-20, and probably H-30), and the remainder are from the eastern fluvial-lacustrine deposits.

Bed II has thus far yielded sixty-three archeologic sites (Table 18; Appendix B) which represent three cultural types: Oldowan, Developed Oldowan A and B, and Acheulian. These cultural types are characterized by M. D. Leakey as follows:

It [the Oldowan] is characterised by choppers of various forms, polyhedrons, discoids, scrapers, occasional subspheroids and burins, together with hammerstones, utilised cobbles and light-duty utilised flakes. In Bed II there is evidence for the existence of two industrial complexes whose contemporaneity, in the broad sense, has been confirmed at a number of different sites. One is clearly derived from the Oldowan and has been termed Developed Oldowan, while the second must be considered as a primitive Acheulian. In the Developed Oldowan A from Lower Bed II Oldowan tool forms persist, but there is a marked increase in spheroids and subspheroids and in the number and variety of light-duty tools. In the upper part of Middle Bed II and in Upper Bed II a few bifaces are also found in Developed Oldowan assemblages, but they form such a negligible proportion of the tools that it has been considered unjustifiable to assign this industry to the Acheulean. It has, therefore, been termed Developed Oldowan B, to distinguish it from the preceding phase (A) which does not include bifaces. Sites where bifaces amount to 40 percent or more of the tools have been classed as Acheulean [1971a, pp. 1-2].

The Oldowan in Bed II is found from the base up nearly to the level of Tuff IIB (loc. 88, MNK,

Table 18. *Stratigraphic Position of Archeologic Sites and Hominid Fossils in Bed II*

Stratigraphic interval	Number and localities of archeologic sites	Number and localities of hominid fossils[a]
Base of Bed II to top of Lemuta Member	*9 sites:* loc. 43 (sites 48a, 48b), 45a (40f, 40g), 64, 66, 99, 101, 201	*3 fossils:* 43 (H-20), 45a (H-16, H-17?)
Lower augitic sandstone and equivalent	*10 sites:* 15, 38(a), 42b, 43 (48c), 43a, 45(a), 45a (40h), 45d, 46c (44a), 88	*3 fossils:* 42b (H-41), 45d (H-37?), 46a (H-40?)
Between lower augitic sandstone and Tuff IIB	*2 sites:* 88 (71a, 71b)	*3 fossils:* 88 (H-13, H-14, H-15)
Between Tuff IIB and upper augitic sandstone	*2 sites:* 32a?, 89 (62a)	
Upper augitic sandstone and equivalent	*7 sites:* 32?, 34?, 44a, 88 (71c), 89 (FC-E), 89 (62b), 91a	*1 fossil:* 89 (62b)
Between brown tuffaceous siltstone and Tuff IIC	*7 sites:* 12 (23a), 12a, 33a, 46b (44b), 75, 91 (68a, 68b)	
Tuff IIC to Tuff IID	*8 sites:* 7 (27a), 23 (a-c), 35a, 38(b), 45 (42), 91 (68c)	
Above Tuff IID	*17 sites:* 14, 16 (19a-e), 23 (d, e, 15), 49, 52, 77 (3), 79 (4), 80 (5b), 82a (6), 90a, 94 (66)	*4 fossils:* 46 (H-9?), 90a, (H-36, H-38), 94 (H-3)
Above Lemuta Member, position uncertain	*1 site:* 66b	*2 fossils:* 88 (H-32), 89 (H-26)

[a] Archeologic sites and hominid identification number in parentheses. Archeologic data are from Appendix B. "?" refers to a fossil or site whose stratigraphic position is uncertain. It pertains principally to surface finds of hominid fossils where the source bed is not clearly established.

site 71a). The Developed Oldowan A is found only in the lower augitic sandstone. The lowest site with Developed Oldowan B lies immediately below the upper augitic sandstone (loc. 89, FC, site 62a), and the uppermost sites are above Tuff IID (loc. 16, TK, sites 19b, d; loc. 94, BK, site 66). The only described Acheulian site is at EF-HR (loc. 12, site 23a), and this site very

Table 19. *Paleogeographic Location of Oldowan, Developed Oldowan, and Acheulian Sites in Bed II*

Industry	Lake margin	Inland	Indeterminate
Oldowan	Loc. 43 (HWK-E, site 48*a*) Loc. 88 (MNK, site 71*a*) Loc. 45*a* (FLK-N, site 40*f*)		
Developed Oldowan A	Loc. 43 (HWK-E, site 48*c*) Loc. 46*c* (FLK-S, site 44*a*) Loc. 88 (chert workshop)		
Developed Oldowan B	Loc. 45 (FLK, site 42) Loc. 88 (MNK, site 71*c*)? Loc. 89 (FC-west, sites 62*a*, 62*b*) Loc. 91 (SHK-west, sites 68*a-c*)		Loc. 16 (TK, sites 19*b, d*) Loc. 94 (BK, site 66) Loc. 25 (FK)
Acheulian		Loc. 7 (CK, site 27*a*) Loc. 12 (EF-HR, site 23*a*) Loc. 12*a* (DK-EE) Loc. 32 (Elephant K) Loc. 49 (site 35) Loc. 52 (MLK-east, site 34) Loc. 77 (Kar K, site 3) Loc. 79 (site 4) Loc. 82*a* (site 6)	

NOTE: Classification of archeologic sites is taken from M. D. Leakey (1971*a*) except for the sites at locs. 12*a* and 88 which were found (chert work shop) after her volume was in press.

likely lies above the brown tuffaceous siltstone. M. D. Leakey (1971*a*) has classified eight other sites as Acheulian (locs. 7, CK; 25, FK; 32, Elephant K; 49; 52, MLK-E; 77, Kar K; 79; and 82*a*). One of these (loc. 32) is in deposits that may well be stratigraphically equivalent to the upper part of the upper augitic sandstone. Another site (CK, loc. 7) lies between Tuffs IIC and IID, and the remaining six sites are stratigraphically higher than Tuff IID. On the basis of a high percentage of bifaces, another site (loc. 12*a*) found in the course of geologic work is clearly Acheulian. This site overlies the brown tuffaceous siltstone. M. D. Leakey has recently pointed out (in press, 1) that bifaces of Acheulian type are found in site 71*c* at MNK, which lies within the upper augitic sandstone. Thus, the oldest known bifaces of Acheulian type at Olduvai Gorge are only slightly younger than the oldest Developed Oldowan B ($\sim$1.4 to 1.5 m.y.a.).

The eastern fluvial-lacustrine deposits contain forty-six of the sixty-three known sites, and the remainder are distributed as follows: western fluvial-lacustrine deposits – 8, lake-margin deposits – 8, and lake deposits – 1. No sites have

been found in the alluvial-fan, alluvial-plain, or eolian deposits. At least one isolated artifact was found in all facies, but isolated artifacts, like archeologic sites, are far more abundant in the eastern fluvial-lacustrine deposits than in any other facies of Bed II. The concentration of archeologic materials in the eastern fluvial-lacustrine deposits may partly reflect a relative stable supply of fresh water, for the eastern margin of the lake was relatively fresh at times of high level, and the main drainageway to the east of the lake contained a river which may have flowed throughout the year. This eastern terrain may also have supported a relatively high concentration of animals and edible vegetation.

The paleogeography of hominid activities can be viewed in more detail by relating the archeologic site to inferred shorelines for various stratigraphic intervals of Bed II (see Figs. 27, 31-37). Sites situated a kilometer or less from the lake are classed as *lake-margin,* and those more distant are termed *inland.* The minimum shoreline is taken as a reference point for classifying those sites where the shoreline is not accurately located for the time represented by the site. By this method, some sites can be erroneously

113

classed as inland where the shoreline was at an intermediate position. The result of this analysis is that thirty-eight sites are lake-margin, eighteen are inland, and seven are indeterminate. The indeterminate sites are above Tuff IID, where a perennial lake, if present, was greatly reduced in size. Despite possible errors in paleogeographic assignment, it seems highly significant that nine of the ten Acheulian sites are inland, and the other is indeterminate, whereas seven of the Oldowan B sites are lake-margin, and the others are indeterminate (Table 19). The correlation between industry and paleogeography is close, even if site 71c, at MNK, is classed as atypical Acheulian rather than Developed Oldowan B.

6

BEDS III AND IV

STRATIGRAPHY AND DISTRIBUTION

Beds III and IV are distinguishable stratigraphic units, as described by Reck (1951), only in the eastern parts of the Main and Side gorges (Figs. 40, 41). The cliffs at JK (loc. 14) can serve as a type section for both Beds III and IV. The contact between Bed II and Bed III is in most places disconformable and easy to locate. Beds II and III can, however, be very difficult to separate to the east, where both are chiefly reddish-brown in color (for example, locs. 5-7, 27-32b). The basal tuff of Bed III is in some places useful in locating the contact where the disconformity is not recognizable (locs. 27-32). In some other places, the contact was located only by correlating through a series of measured sections. Beds III and IV are generally easy to distinguish as far west as FLK (loc. 45) and JK (loc. 14), where Bed III interfingers over a broad zone with sediments similar to those of Bed IV. The interfingering was first demonstrated at JK through a series of excavations (Kleindienst, 1964). Farther west, where Bed III is lithologically indistinguishable from Bed IV, the two units are combined into Beds III-IV (und.). Between the fault at FLK and the Fifth Fault, the contact with Bed II is commonly difficult to locate and can be gradational. To the west of the Fifth Fault, the contact with Bed II is generally sharp, and it is in many places an erosional surface.

Beds III and IV, either separate or undivided, are exposed over a smaller area than are Beds I and II. They pinch out a short distance east of the Second Fault and are not found along the western edge of Olbalbal either north or south of the gorge. They do, however, extend to the western margin of the basin and as far south as Kelogi.

Bed III

Bed III is dominantly a reddish-brown deposit, chiefly of volcanic detritus. It is about 85 per-cent claystone, and most of the remainder is sandstone and conglomerate. Its thickness ranges from 4.5 to 11 m and varies systematically where deposited on different fault blocks, pointing to contemporaneous fault movements. Within the same fault blocks, it is thicker on the south side of the Main Gorge than it is on the north (Fig. 41). Four different tuffs were found at more than one locality in the eastern part of the Main Gorge (Table 20). They are numbered from 1, the oldest, up to 4, the youngest. Only the lowermost has aided appreciably in correlating, and the rest have thus far been recognized in only a few places.

Tuff 1, at the base of Bed III, is a fine- to medium-grained vitric trachyte tuff which is gray to reddish-brown. It is 15 to 60 cm thick and can be found in most places between localities 27 and 32b. It is distinguished by pumice with abundant small tabular crystals of sodic plagioclase which give the pumice a micaceous appearance under the hand lens. Tuff 2 is a gray to reddish-brown fine-grained vitric trachyte tuff, 15 to 30 cm thick, which lies 1.5 to 2.4 m above the base of Bed III (locs. 14, 27). It is characterized by small crystals of biotite. Tuff 3 is a coarse-grained orange-brown vitric tuff 30 to 45 cm thick that lies about 2.6 m below the top of Bed III in locality 27. It consists largely of mafic (basaltic?) scoria particles 1 to 4 mm in diameter and altered to palagonite. Augite and altered olivine are in the scoria. Tuff 4 is a yellow to pink fine-grained vitric tuff of trachyte or trachyandesite composition. It is 30 to 60 cm thick and lies 1.2 to 1.5 m below the top of Bed III (locs. 33b, 36). A few other thin trachyte tuffs were noted elsewhere in Bed III (for example, loc. 95a), but as yet they can be neither correlated nor placed in stratigraphic position relative to the numbered tuffs. Of mineralogic significance is a nephelinite or phonolite tuff about 3 m above the base of Bed III at locality 22. It probably lies at roughly the same level as tuff 2.

Table 20. *Minerals in Tuffs of Beds III-IV*

Stratigraphic position and location of tuff	Olivine	Augite	Sodic augite	Brown hornblende	Biotite	Ilmenite	Melanite	Perovskite	Sphene	Plagioclase	Anorthoclase	Nepheline
Bed III, tuff 1 (locs. 9, 27-32b)	−	+	+[a]	xx	−	−	−	−	−	xx	−	−
Bed III, tuff 2 (locs. 7, 14, 27)	−	+[a]	+	−	xx	−	−	−	+	xx	−	−
Bed III, tuff 3 (locs. 4, 27)	xx	xxx	−	−	−	−	−	−	−	−	−	−
Bed III, tuff 4 (locs. 33b, 36)	−	+	−	x	−	−	−	−	−	xx	xx	−
Bed III, tuff 3 m above base, loc. 22	−	+[a]	xx	+	−	xx	+	+	x	xx[a]	−	?
Bed IV, Tuff IVA (locs. 17, 44b, 45, 95)	−	−	−	−	xx	−	−	−	−	−	xx	+?
Bed IV, 1 cm tuff 1.7 m above base of Bed IV east of JK (loc. 14)	−	+	xx	−	−	xx	xx	+	−	−	−	xx
Bed IV, tuff filling fissures 1.5 m below Tuff IVB at JK (loc. 14)	−	+	xx	+	−	+	+	x	x	−	−	+?
Bed IV, Tuff IVB (locs. 14, 36)	−	xx	−	−	−	x	−	−	−	xx	−	−
Beds III-IV (und.), 7.3 m above base at loc. 25	−	+	−	x	+	−	−	−	−	xx	xx	−
Beds III-IV (und.), tuff near base exposed section, loc. 204	−	+[a]	xx	+	−	x	−	−	x	+	−	xxx
Beds III-IV (und.), tuff near top exposed section, loc. 204	−	−	xx	−	−	xx	−	−	xx	−	x	xx

NOTE: Frequency symbols are xxx = abundant, xx = common, x = between rare and common, + = rare, and − = absent.
[a]Minerals are possibly detrital contaminants.

Bed IV

Bed IV is approximately 68 percent claystone, most of which is soft-weathering and gray or brown. The remainder comprises sandstone (19 percent), siltstone (7 percent), conglomerate (4 percent), and tuff (2 percent). The contact with Bed III is generally an erosional surface, and the basal bed of Bed IV is in most places a sandstone or conglomerate. The contact of Bed IV with the Masek Beds is sharp, and in a few places the Masek Beds fill channels eroded into Bed IV or through Bed IV into Bed III. Bed IV generally ranges from 2.4 to 7.3 m in thickness in the Main Gorge, and it is as much as 10 m thick in the Side Gorge. The variation in thickness is a result of fault movements during the deposition of Bed IV as well as postdepositional erosion.

Bed IV contains both basement and volcanic detritus. Only a small amount of volcanic detritus is found in the northwestern exposures (locs. 13, 14), and volcanic detritus increases in abundance both to the east and south. Volcanic and basement detritus can be intermixed in the same beds, or beds consisting of volcanic detritus can be interstratified with beds consisting largely of basement detritus. At VEK (locs. 85-87), for example, a reddish-brown deposit of volcanic sediment is underlain and overlain by gray deposits containing chiefly basement detritus. In the southern part of Long K (for instance, loc. 38b), Bed IV is entirely reddish-brown volcanic sediments indistinguishable from those of Bed III.

Within Bed IV are two marker tuffs, termed Tuffs IVA and IVB. Tuff IVA is the lower of the two and is present at widely separated localities in the Main and Side gorges (locs. 17, 36, 44b, 45, 95). Tuff IVB is found only in the vicinity of WK (loc. 36) and JK (loc. 14). A very small amount of tuff was also found beneath Tuff IVB at JK. This tuff beneath Tuff IVB is of interest primarily in documenting nephelinite volcanism while Bed IV was being deposited.

Tuff IVA is a fine-grained vitric trachyte tuff 15 to 30 cm thick. It is typically laminated and yellowish-gray, but it can be massive and reddish-brown (TK, loc. 17). The principal primary minerals are biotite and feldspar, and analcime is the alteration product of the glass. The tuff lies in the lower part of Bed IV at WK (loc. 36) and HEB (loc. 44b), whereas it is in the middle of Bed IV at FLK (loc. 45) and GTC (loc. 95). This difference suggests that Bed IV accumulated more rapidly below the tuff toward the east than it did toward the west.

Tuff IVB is a fine- to medium-grained crystal-lithic tuff of probable trachyandesite composition which is 15 cm to 2 m thick. It is hard and reddish-brown, and where thickest (loc. 36) it is laminated or thin-bedded and contains thin layers of claystone. Its crystals are chiefly angular fragments of plagioclase and augite originating in the explosive fragmentation of crystalline lava. The tuff contains a high proportion of zeolites (analcime and chabazite), which may be reaction products of fine-grained vitric ash whose texture has been lost in the alteration process.

Siltstone and silty sandstone are widespread in the Main Gorge and have proved useful in correlating. Most of these silt-rich deposits lie in the lower half of Bed IV, and a thick bed underlies Tuff IVB at both JK (loc. 14) and WK (loc. 36) and very likely extends eastward at the same level. A thin siltstone locally overlies Tuff IVB at JK.

Beds III-IV (Undivided)

Beds III and IV are combined in a single unit not only to the west of FLK (loc. 45) and JK (loc. 14) but to the south, near Kelogi, and to the north, along the east-west extension of the Fifth Fault. The deposits near Kelogi resemble Bed III in lithology and are separated by erosional surfaces from Bed II and the Masek Beds. They have a maximum exposed thickness of 7.5 m at locality 98a. Beds III-IV (und.) to the west of FLK and JK are about 62 percent claystone, and the remainder is nearly all sandstone and conglomerate. These are dominantly of basement detritus and are chiefly gray and brown. Hard reddish-brown sediments similar to those of Bed III have not been found west of PLK (loc. 23). Beds III-IV (und.) range in thickness from about 4 to 29 m, with the thickest sections in the graben between the FLK fault and the Fifth

Fault. The variation in thickness is controlled largely by structural position, but it is also determined to some extent by the depth of erosion into Bed II, and the depth to which Beds III-IV (und.) have been eroded beneath the Masek Beds.

Beds III-IV (und.) cannot presently be subdivided into units which can be traced laterally either near Kelogi or to the west of FLK. The disconformity that separates Bed III and IV to the east of FLK may lie within this sequence, but it has not been located, possibly because nearly all of the conglomerates and many of the sandstones have an erosional base. Only a very few tuffaceous beds have been found in Beds III-IV (und.). A thin bed of tuffaceous siltstone at PLK (loc. 23), 4.6 m above the base, is correlated with tuff 2 of Bed III on the basis of mineral composition and relative stratigraphic position. A vitric tuff, trachyte or trachyandesite, was noted at FK (loc. 25), 7.3 m above the base of Beds III-IV (und.). It is cream-colored to pink and is similar to clasts of tuff in a conglomerate 4 m above the base of Bed III at JK (loc. 14).

A few discontinuous exposures tentatively correlated with Beds III-IV (und.) crop out along the east-west extension of the Fifth Fault 5 km north of the gorge (locs. 203-205). The thickest section, 8.2 m, is at locality 204 and is chiefly ostracodal claystone with interbedded sandstone. A tuff and tuffaceous sandstone 30 cm thick are near the base of the section, and a tuffaceous bed 15 cm thick is near the top. Exposures at locality 205, about 6 m thick, are chiefly sandstone and conglomerate, and at locality 203 is a 4-meter thickness of claystone and sandstone. These sediments are correlated with Beds III-IV (und.) on their clast composition, nature of fossilization, and mineral content of tuffs. Clasts of lava from Sadiman and Ngorongoro are incompatible with the paleogeography for Bed II but fit with that inferred for Bed III, a point developed in more detail in the facies analysis. Bones at locality 204 are stained and impregnated with iron (and manganese?) oxide, similar to those of Beds III and IV elsewhere, and are unlike fossilized bones of Bed II. Finally, the tuffaceous material is of nephelinite or phonolite composition and is mineralogically closest to the nephelinite or phonolite tuff in Bed III at locality 22 (Table 20).

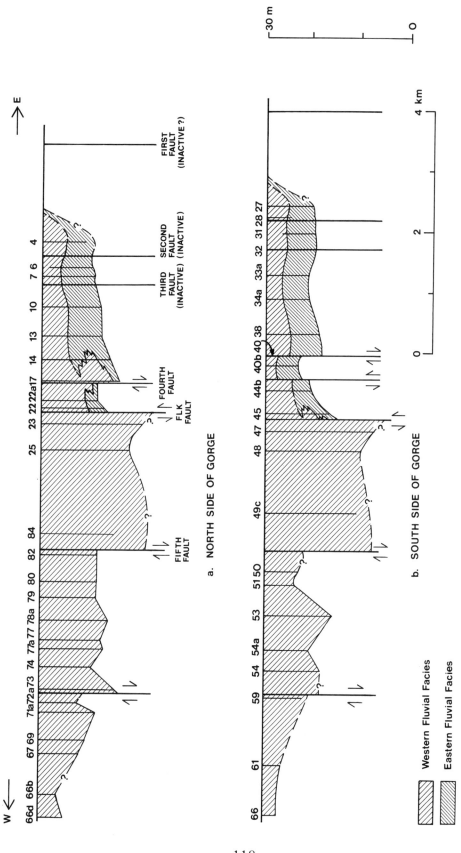

a. NORTH SIDE OF GORGE

b. SOUTH SIDE OF GORGE

Western Fluvial Facies

Eastern Fluvial Facies

Figure 42. Cross-sections showing lithologic facies and thickness of Beds III-IV as exposed along the Main Gorge. Sections are constructed with the top of Beds III-IV taken as horizontal. Vertical lines represent measured stratigraphic sections.

118

AGE OF BEDS III AND IV

Magnetic stratigraphy is presently the only geophysical method to provide information about the age of Beds III and IV. Polarity studies were first undertaken by A. Brock, who made laboratory measurements on eight samples from Bed III, eight from Bed IV, and four from Beds III-IV (und.). F. H. Brown and M. D. Leakey were largely responsible for the field sampling. Most of the samples are of hard, reddish-brown zeolitic claystone, and the remainder are of limestone. Three of the eight Bed III samples have reversed polarity, four are normal, and one is magnetically unstable. The reversely polarized samples were collected both high and low in Bed III, demonstrating that Bed III was deposited during the Matuyama epoch, more than 700,000 years ago. The reversely polarized rocks probably acquired their polarity penecontemporaneous with deposition, as they did their reddish-brown color. The normal polarity of the other samples is probably a result of the continued growth of hematite crystals in a later period of normal polarity. Four of the Bed IV samples have normal polarity, and the other four were magnetically unstable or of low magnetic intensity. These preliminary results led to the tentative conclusion that the Bed III-IV contact coincided approximately with the Matuyama-Brunhes boundary at 700,000 years before the present (Brock et al., 1972).

In 1972, A. Cox initiated a more detailed sampling program in Bed IV and the upper part of Bed III with the aim of locating the Brunhes-Matuyama boundary more precisely. Although this work is not yet completed, the most significant result is the discovery of reversed polarity in three samples from the siltstone beneath Tuff IVB at the hominid excavation of WK (loc. 36). Tuff and calcrete samples from various levels in the Masek Beds, measured both by Brock and Cox, have normal polarity, and presumably they were deposited during the Brunhes normal epoch. Thus, the Brunhes-Matuyama Boundary probably lies within Bed IV, and no lower than Tuff IVB.

As discussed earlier, the contact between Beds II and III is estimated as about 1.15 m.y.a. on the basis of stratal thicknesses in the gorge and dated fault movements to the east. If Tuff IVB is taken as 0.7 million years old, relative stratal thicknesses in Beds III and IV can be used to obtain an age of about 0.80 m.y.a. for the top of Bed III. The top of Bed IV is about 0.60 million years old if one assumes that Bed IV sediments above Tuff IVB accumulated at the same rate as the underlying Bed IV sediments. If the Brunhes-Matuyama boundary lies above Tuff IVB, the age limits of Bed IV should be slightly older than 0.6 to 0.8 m.y.a.

LITHOFACIES AND ENVIRONMENTS

Beds III and IV and Beds III-IV (und.) comprise three lithofacies: western fluvial deposits, eastern fluvial deposits, and fluvial-lacustrine deposits. The eastern and western fluvial facies (Fig. 42) are distinguished primarily on the basis of composition and postdepositional alteration. The western fluvial deposits are chiefly of basement detritus from the west and northwest, and the eastern fluvial deposits are almost entirely of volcanic detritus from the east and south. The western fluvial deposits are dominantly gray and brown, and they generally have been modified to a rather limited extent by postdepositional chemical processes. By contrast, most sediments of the eastern fluvial deposits are bright reddish-brown and contain a relatively high proportion of zeolites. Sediments composed of volcanic detritus intergrade and interfinger over a broad zone with sediments composed of basement detritus. These sediments of the transitional zone are dominantly gray and brown and are included within the western fluvial deposits on the basis of their degree of alteration. Within the western fluvial deposits are also included gray and brown deposits of Bed IV in which the sand fraction is chiefly volcanic detritus. Sediments of the fluvial-lacustrine facies resemble in many respects the sediments of the zone transitional between eastern and western fluvial deposits. They differ from these in their content of lacustrine features, as for example ostracods and abraded calcite crystals.

Eastern Fluvial Deposits

Measured sections of the eastern fluvial deposits are 83 percent claystone, 9 percent sandstone, 6 percent conglomerate, and the remaining 2 percent comprises tuff, limestone, and mud-

Table 21. *Composition of Lithofacies in Beds III-IV and the Masek Beds*

Facies and stratigraphic unit	Claystone	Sandstone	Conglomerate	Tuff	Carbonate rocks	Siltstone	Claystone-pellet aggregate	Mudflow deposits
Beds III-IV								
Eastern fluvial deposits (dominantly Bed III)	83%	9%	6%	<1%	~1-2%	0	<1%	<1%
Western fluvial deposits (all)	62	22	8	<1	tr.	2	1	0
Western fluvial deposits (Bed IV)	67	19	4	2	<1	7	1	0
Western fluvial deposits (Beds III-IV (und.))	61	30	8	<1	<1	0	1	0
Fluvial-lacustrine deposits	50-55	34	5	2	0	1?	5-10	0
Masek Beds								
Western fluvial deposits	76	14	10	0	tr.	0	≤1	0
Eastern fluvial deposits (SW margin of Olbalbal)	~75	~15	~5	0	0	0	~5?	0
Eastern fluvial deposits (Kelogi)	~25	~50	~25	0	0	0	0?	0

flow deposits (Table 21). The figures for claystone and sandstone are only approximate, because a substantial proportion of these sediments are difficult to classify accurately in the field, largely due to alteration. Limestone is found principally in the lower 2 m of Bed III, but other rock types are found at various levels.

Bed III constitutes the bulk of this facies (Fig. 42), and the remainder is found in Bed IV, principally within the Main Gorge, and in Beds III-IV (und.) near Kelogi. In the southern part of Long K (for example, loc. 38b), all of Bed IV consists of typical eastern fluvial deposits. Lenticular beds of reddish-brown claystone and conglomeratic sandstone, typical of this facies, are found near the base of Bed IV near the junction of the Main and Side gorges (locs. 44, 44b, 86, 87) and underneath Tuff IVA in the Side Gorge at GTC (loc. 95).

Claystones are characteristically silty, sandy, massive, rootmarked, and reddish-brown. The massive character probably reflects the churning action of roots and burrowing animals. The amount of detrital silt and sand is generally between 5 and 40 percent, averaging perhaps 25 percent. Vertical fissures filled with sandstone or sandy claystone are in some of the claystones with the least amount of sand and silt (for instance, locs. 38, 40). The fissures are as much as a meter deep, and are as much as 2 cm wide at the top and narrow downward. These represent deep cracks exposed at the surface and filled with sediment during the deposition of Bed III.

Volcanic detritus predominates overall in the silt-sand fraction of the claystones, but a small percentage of basement detritus is not rare, and basement detritus can be common in beds near the western margin of the facies.

Most conspicuous of the postdepositional changes is the reddish-brown color, and another major change is in the high content of zeolites, which form on the average 12 to 15 percent of the claystones. Zeolites may either be disseminated or occur as veinlets, and intersecting veinlets may form boxworks which make the claystones resistant to erosion. Sandy claystones with more than 15 percent zeolites are hard, bright, reddish-brown, and have a bricklike appearance (Hay, 1970a). Calcite is common, principally as veins and nodules. In addition, calcic plagioclase and some or all of the augite grains are generally etched. These reactions took place at shallow depth, soon after deposition. Probably the most convincing evidence of this is provided by clasts of distinctive, zeolite-rich claystone in conglomerate of a Bed III channel eroded through the source bed. In some places, zeolite veinlets cut across the Bed III-IV contact, indicating that they postdate Bed III.

The most common type of sandstone is massive, clayey, and reddish-brown. At a distance, these sandstones resemble the harder of the zeolite-rich claystones. Widespread, particularly near the base of Bed III, are laminated, medium-grained zeolitic sandstones with a high content of reddish-brown claystone pellets (for example, locs. 14, 45, 93). Claystone pellets predominate over detrital sand in some layers. Widespread in the middle and upper part of Bed III are relatively well-sorted medium to coarse uncemented sandstones (for instance, locs. 13, 32b). A massive bed of pelletoid sandstone, deposited by wind, was found near the middle of Bed III at locality 31. Most of the detrital grains are coated by claystone, and sand-sized rock fragments are unusually well rounded. Within all types of sandstone, Olmoti detritus is most common in the easternmost exposures and is rare to the west of DK (loc. 13). Sandstones in the Side Gorge contain minerals from Kelogi (for example, aegirine, quartz) and from the Laetolil Beds (melanite, perovskite) as well as from Lemagrut.

Conglomerates range in clast size from granules to small boulders. The coarser beds are between 30 cm and 2 m thick and commonly fill relatively narrow channels. Grain size decreases from north to south, and whereas lava cobbles are commonly 30 cm long on the north side of the gorge (locs. 38, 40, 44), they rarely exceed 2.5 cm at PLK (loc. 23) or 8 cm between JK (loc. 14) and CK (loc. 6).

Clasts are chiefly lava and welded tuff, but limestone is common, and it predominates in many conglomerates in the lower 2 m of Bed III. Most of the limestone clasts are subangular and have been eroded from the uppermost part of Bed II. At higher levels in Bed III, the limestone clasts are probably from Bed III, as are the clasts of tuff, sandstone, and claystone in some beds. The distribution pattern of volcanic detritus in the Main Gorge is about the same as that in Bed II above the Lemuta Member in the same area; Olmoti detritus predominates in the easternmost exposures and decreases rapidly in a westward direction. Sadiman detritus, the easiest to identify, is most abundant near the junction of the Main and Side gorges, where it forms 20 percent of a fifty-pebble sample that was counted. It was, however, noted in all conglomerates of the Main Gorge excepting those of locality 5. Conglomerate clasts in Beds III-IV (und.) near Kelogi are chiefly of Lemagrut lava but include tuff from the Laetolil Beds and gneiss from Kelogi. Also common here are green nephelinite and phonolite indistinguishable from those of Sadiman. Stream-channel measurements and paleogeographic considerations, discussed in the

following paragraphs, make Sadiman seem an unlikely source, and this green lava was very likely derived either from the Laetolil Beds or from lava which crops out on the western slopes of Lemagrut.

Tuffs of widely varying composition have been found in Bed III (Table 20). Most of these are fine-grained, and a few are delicately laminated, showing that they were deposited in ponded water. The laminated tuffs are localized, suggesting that the ponds were small—possibly those left in the channel of a partly dried stream. All of the tuffs are altered and either replaced or cemented by zeolites.

Clayey mudflow deposits were noted in localities 30, 31, and 38. The deposit at localities 30 and 31 is 1 to 1.2 m thick and contains both rounded and angular clasts of Olmoti detritus, the largest of which is 30 cm long. The other deposit is 60 cm thick and contains angular fragments of Lemagrut lava no more than a centimeter long. Both deposits are hard, reddish-brown, and rich in zeolites.

Limestones are nodular and of caliche or water-table origin. Calcareous concretions are widespread and abundant, and they have a total volume much greater than that of the nodular beds. The concretions are generally either very irregular in shape or are crudely cylindrical, often branching, and on the order of 1 cm in diameter. These may form openworks as much as 60 cm thick. The cylindrical structures probably were formed by precipitation of calcium carbonate in permeable channels such as rootcasts and burrows.

Conglomerate-filled channels are common, and a few contain either sandstone or tuff. Most of the channels are between 30 cm and 1 m deep, but a few are 2 to 3 m deep. Alignments of forty-three channels were measured, and of these, forty are in Bed III between VEK (loc. 86) and the Second Fault (loc. 27). The orientations vary widely but average approximately north-south. Orientations are rather consistent in several localities, yet differ strikingly from those in nearby areas. At VEK (near loc. 86), for example, four channels range from N27°W to N57°W, averaging N48°W, whereas nearby to the east (loc. 44), seven channels range from N20°E to N85°E, averaging N54°E. Channels in both areas are at several different levels in Bed III, and the consistent orientations may indicate that the drainage pattern in this area through the deposition of Bed III was partly controlled by movement of fault blocks. Elsewhere, channel directions can vary widely at the same locality. At locality 31, for example, four steep-walled, deep channels are oriented N35°W, N40°W, N40°E, and N85°E.

Rootmarkings suggestive of grass and brush are widespread in the eastern fluvial deposits. Remains of vertebrates and invertebrates are relatively rare as compared with their occurrence in the western fluvial deposits. Pelecypod shells (*Unio* sp.?) have been found in two places, one in Bed III (loc. 88), the other in Bed IV (locs. 86-87; M. D. Leakey, 1974, personal communication). Hopwood's (1951) faunal list for Bed III very likely refers only to fossils collected from the eastern fluvial deposits. Although some of the identifications may be outdated, the listing serves to indicate that Bed III includes many of the forms represented in the western fluvial deposits. These include crocodiles, hippos, rhinos, bovids, and equids.

The eastern fluvial facies was deposited on a semiarid alluvial plain along the northwestern margin of the volcanic highlands. Both the types of deposits and their alteration suggest an environment much like that of the eastern fluvial deposits of Bed II. The high proportion of claystone is evidence that the alluvial surface had a low gradient. Bedding and textural features of conglomerates point to deposition by braided rather than meandering streams. Many of the streams had relatively straight courses, as shown by consistent channel direction at some localities. Sinuous channels are suggested by measurements in a few other places. The coarser detritus probably was transported by flash floods, in view of the coarse nature of many conglomerates interbedded with claystone.

The many massive, rootmarked claystones can be viewed as a sequence of poorly developed alluvial paleosols. Claystones with deep fissures compare closely with the vertisols found in floodplain clays and channel-filling clay plugs of the Omo River delta (Butzer, 1971).

The calcareous nodules, zeolitic alteration, and reddish-brown color are indicative of a relatively low or fluctuating water table and a saline, alkaline soil environment. Claystone pellets are suggestive of saline mudflats eroded by the wind. The paucity of faunal remains may reflect either poor conditions for fossilization, a relatively small animal population, or a highly seasonal concentration of animals.

Western Fluvial Deposits

The western fluvial deposits are about 62 percent claystone, 27 percent sandstone, 8 percent conglomerate, and the remainder includes tuff, limestone, siltstone, and claystone-pellet aggregate (Table 21). The proportions of claystone, tuffs, and siltstone are highest in the eastern part of the gorge, whereas conglomerate and sandstone increase to the west.

Claystones are most commonly gray and brown but can be mottled or reddish-brown. Most of these are massive and rootmarked, and many are highly fractured as a result of weathering. They generally have between 5 and 35 percent of silt and sand. The silt-sand fraction is almost entirely metamorphic detritus as far east as localities 22 and 45; farther east, volcanic detritus is commonly intermixed and increases in proportion to the south and east. The clay mineral is illite, with or without a small amount of interstratified montmorillonite (Hay, 1970a). Calcite concretions are widespread, and they are most commonly between 1 and 10 mm in diameter.

Claystones in the eastern part of the facies (that is, within Bed IV) differ in several respects from those found to the west. Vertical fissures filled by claystone and sandstone are widespread. These are commonly 1 to 2 m in depth and as much as 5 cm wide at the surface. Wax-like claystones with very little silt and sand are common in this eastern area. Finally, claystones of the eastern area have been altered mineralogically to a greater extent, on average, than have those to the west. Zeolites, principally analcime, generally constitute 8 to 12 percent of the eastern claystones but only 3 to 8 percent of those to the west. Moreover, reddish-brown colors are more common, and quartz and augite are more commonly etched to the east and south.

Sandstones form lenticular beds as much as 2.5 m thick that vary greatly in grain size and sedimentary structures. The sand is almost exclusively quartzose metamorphic detritus, with or without limestone, to the west of PLK (loc. 23) and FLK (loc. 45). Volcanic detritus is commonly intermixed to the east and south (see Kleindienst, in press), and constitutes nearly all of the grains in a few sandstones along the southeastern margin of the facies. Olmoti detritus predominates in the easternmost sandstones and rapidly decreases westward. Volcanic detritus in the Side Gorge is exclusively from Lemagrut and the Laetolil Beds, and it is associated with basement detritus of Kelogi type as far east as VEK (loc. 86).

Most of the thicker sandstones are vertically graded from coarse, often conglomeratic, at the base upward into medium- or fine-grained. Festoon cross-bedding is common in the middle of the bed, and delicate, small-scale cross-bedding and thin lamination characterize the upper part. This assemblage of features typifies sandstones deposited in the channels of meandering streams. These sandstones are particularly common and well developed to the southeast of the Fifth Fault, and excellent examples are in localities 17, 25, and 48. The nongraded sandstones are highly variable in stratification and are commonly massive. Rootmarkings, filled fissures, and animal burrows are in some of the sandstones.

Conglomerates are composed largely of metamorphic detritus except in the southeastern part of the area where lava and welded tuff may predominate. The metamorphic detritus is from the north and west and is accompanied by clasts of the Naabi Ignimbrite. Coarse-grained quartzite is the dominant type of metamorphic rock, and pink granitic gneiss forms most of the remainder. Clast size decreases systematically in a southeastward direction. The largest clast of western derivation, at locality $63a$, is 45 cm long, and the largest at locality 13 is about 1 cm. Values for the maximum size at intermediate segments of the gorge are an exponential function of distance and can be represented by the following equation, where S is maximum diameter of clast size, in centimeters, and d is the distance southeast from locality $63a$, in kilometers:

$$\log_{10} S = 1.68 - 0.105d, \text{ or}$$
$$S = 48.4 \times 10^{-0.105d}$$

This gradual, systematic decrease in size is evidence that streams from the west decreased gradually in gradient and velocity.

Limestone clasts are common in most of the beds composed of pea-sized pebbles, and clasts of Tuff IVB are in some of the conglomerates at WK (loc. 36).

Most of the tuffs of this facies are laminated and were deposited in standing water. All of them are altered and contain zeolites, and Tuff IVB is bricklike in color and induration. Tuff IVB is exceptional in the variety of its authigenic minerals, which include analcime, chabazite, natrolite, and dawsonite. Exposures of Tuff IVA are widely scattered and of limited extent, suggesting deposition in a series of relatively small ponds. Tuff IVB was deposited in a seasonally dry saline lake, or playa, which very likely extended from JK (loc. 14) to WK (loc. 36).

Siltstones are common on the north side of the gorge between localities 4 and 14, and they are extremely rare to the west of localities $44b$ and 18. They are typically pale gray and form massive, rootmarked beds 30 cm to 1 m thick. The siltstones are weakly cemented, rather poorly sorted, and consist of detrital silt and sand, small claystone clasts (including pellets), and interstitial clay. A few samples were disaggregated and sieved, giving an average composition as follows: 44.4 percent silt, 28.2 percent claystone (both clasts and matrix), and 27.4 percent sand. The silt fraction is highly feldspathic and invariably contains more feldspar and less quartz than the sand fraction. The silt fraction also includes a small proportion of authigenic analcime. The percentage composition of the silt and sand fractions of two samples is as follows, as based on x-ray diffraction:

		Quartz	Feldspar	Hornblende	Analcime
loc. $44b$	silt	6	85	$\leqslant 5$	~ 4
	sand	25	70	$\leqslant 5$	0
loc. 14	silt	11	~ 85	$\leqslant 2$	~ 2
	sand	46	~ 52	$\leqslant 2$	0

These siltstones are probably overbank fluvial sediments, deposited either as natural levees or in the flood basin of a meandering river (see Butzer, 1971, pp. 50-51).

Claystone-pellet aggregates are generally well sorted, and the pellets are of medium sand size. They contain admixed sand and can grade into sandstone. Most of the aggregates exhibit the festoon cross-bedding of fluvial sands. They appear to be most common in the upper part of Beds III-IV (und.), and good examples are in localities 22a and 25. The content of claystone-pellet aggregates in the facies is given as 1 percent (Table 21), but the true figure may be several times this amount in view of the fact that many sections of the facies were measured before claystone-pellet aggregates had been recognized.

Channeling is widespread at the base of conglomerates and sandstones, and fifty-six channel alignments were measured. Relatively consistent orientations were found to the west of the Fifth Fault, where the three westernmost channels range from N60°W to N75°W (locs. 61, 66, 66d). Farther east, two of three channels are N40°W and N45°W (locs. 53, 77); the third is N10°W (loc. 74). Measurements in Bed IV in the Side Gorge vary rather systematically from west to east. The six westernmost channels (locs. 95-97) are consistently oriented northwest-southeast (N35°W to N65°W, averaging N58°W), whereas six channels farther to the east are dominantly east-west, and if one disparate channel is excluded, they average N83°W. The other forty measurements were made in the Main Gorge between localities 14, 36, 44b, and 83. These results are distributed more or less evenly around the compass and strongly support other evidence for meandering streams in this area.

Kleindienst (1973) inferred a northward flow direction for streams of Bed III at JK (loc. 14) on the basis of cross-bedding, clast orientation, and elongate sand-filled scours or shallow channels a meter wide and exposed for a length of 5 m. The cross-bedding and clast orientation are highly variable, as expected from meandering streams. The sand-filled channels, oriented northeast-southwest, resemble those described by Harms et al. (1963) and very likely indicate the overall flow direction rather closely.

Rootmarkings are suggestive of grass and brush, and no examples are known that point to marshland vegetation. The apparent absence of marshland rootcasts may, however, be attributable to the lack of a deposit (for example, siliceous earthy claystone) suitable for preserving their distinctive features. Dr. van Zinderen Bakker found some pollen in samples from Beds III and IV at JK (loc. 14). The Bed III pollen indicated "open grassland with the margin of open water nearby, and forest on the highlands"

(quoted in Kleindienst, 1973). Two Bed IV samples indicate open grassland. Shells of a freshwater bivalve (Unio sp.) have been found in sandstones at several localities in the southeastern part of the facies (for instance, loc. 44b). Vertebrate remains are widespread but have not been collected and studied to nearly the same extent as the faunas of Beds I and II. Most of the known fossils are from Bed IV and were collected from archeologic sites. L. S. B. Leakey has listed the known fossil vertebrates of Bed IV (1965). The faunal remains recently excavated by M. D. Leakey should add significantly to this list. The fauna of Bed IV has been characterized as follows by M. D. Leakey (in press, 3): "The faunal remains of Bed IV are usually very fragmentary, but hippos, equids, and crocodiles are common at nearly all sites. Catfish bones are also plentiful. In many respects the fauna stands close to that of upper Bed II, with the exception of Suidae and Bovidae, both of which are poorly represented and do not include the variety of species found in Bed II."

The bovid fauna, recently studied by Alan and Anthea Gentry, is indicative of savannah grassland with scrub and bush. The bovid fauna differs somewhat from that in the upper part of Bed II, but the ecological significance of this difference is not yet clear (Alan Gentry, 1974, personal communication). Diceros bicornis (black rhino) predominates in Bed IV over Ceratotherium simum (white rhino) (M. D. Leakey, 1973, personal communication).

The western fluvial facies was largely deposited by streams draining the basement terrain to the west and north. Some streams meandered, and others were of low sinuosity to the west of the Fifth Fault and in the area of the Side Gorge. Meandering streams characterized the area east of the Fifth Fault and north of the Side Gorge. The meandering streams were commonly between 1 and 2.5 m deep, as inferred from the thickness of vertically graded sandstone beds. The larger streams may have flowed more or less uniformly during the rainy seasons, in view of the fact that most of the vertically graded sandstones are unbroken by erosional hiatuses such as might be expected in a river that fluctuated greatly in depth and velocity.

The southeastern part of the facies has features which indicate that this area represents the main channelway of the Olduvai basin, with a

124

major trunk stream and a well-developed floodplain. The mixture of volcanic and metamorphic detritus shows that it drained the southeastern as well as the northwestern parts of the basin. Deep vertical cracking in claystones probably reflects intermittent desiccation of water-saturated floodplain clays. The siltstones may well be levee deposits, the small ponds documented by Tuff IVA may represent abandoned meander loops, and the playa lake represented by Tuff IVB may have been a relatively large flood basin. The relatively high content of zeolites and severe etching of quartz point to evaporative concentration of sodium carbonate in ponds and poorly drained flood basins.

Fluvial-Lacustrine Deposits

The fluvial-lacustrine deposits of Beds III-IV (und.) are about one-half claystone, one-third sandstone, and the remainder comprises conglomerate, claystone-pellet aggregate, tuff, and siltstone. The proportion of claystone is highest in the middle locality (204), and the proportion of sandstone is highest in the western locality (205). Conglomerate is restricted to locality 205. These lithologic comparisons are of dubious value, however, as the stratigraphic sections at localities 204 and 205 cannot presently be correlated.

Claystones are pale brown to mottled with reddish-brown and generally contain a small amount of silt and sand. Flakes of muscovite are conspicuous in the silt-sand fraction, which comprises both metamorphic and volcanic detritus. Quartz is generally etched, and authigenic zeolites are invariable present.

Sandstones are dominantly medium grained and well sorted, and most of the remainder are coarse, poorly sorted, and commonly conglomeratic. The medium-grained sandstones are composed of metamorphic and volcanic detritus, commonly accompanied by a substantial proportion of claystone pellets, abraded calcite crystals, and nephelinite tephra. Claystone pellets, calcite crystals, and tephra are relatively rare in the westernmost sandstones (loc. 205). The coarser sandstones contain limestone in addition to metamorphic and volcanic detritus. The volcanic detritus includes Engelosin phonolite, Naabi Ignimbrite, Sadiman phonolite, materials from Ngorongoro, and undiagnostic lava. Engelosin phonolite and Naabi Ignimbrite are most abundant in locality 205 and are relatively uncommon at localities 203 and 204, where the other types of volcanic detritus are most abundant. Quartz is generally etched, and volcanic glass and nepheline are wholly altered. Cementing materials in the sandstones are zeolites, calcite, and rare dolomite.

Conglomerates consist of pebbles and small cobbles, chiefly of metamorphic rock and Naabi Ignimbrite. They also include small subangular pebbles of Engelosin phonolite.

The siltstone and claystone-pellet aggregate are pale brown to reddish-brown, micaceous, and commonly contain nephelinite tephra. Quartz is etched and both glass and nepheline are altered, as in the associated sandstones. Analcime is abundant in the claystone-pellet aggregates and constitutes 26 percent of one sample analyzed by x-ray diffraction.

The small amount of tuff is reworked and contaminated either by detrital sand or claystone pellets. The tuffs are of nephelinite composition (Table 20), and zeolites are alteration products of the glass and nepheline.

Vertebrate bones a few centimeters or more in length are in some of the sandstones at locality 204, and small bones and bone fragments are in some of the sandstones and the sand fraction of claystones and siltstones. Ostracods occur throughout the section at locality 204 and are particularly abundant in some of the claystone-pellet aggregates and claystones.

This facies contains sediments deposited in streams and intermittent standing water. The conglomerates and most or all of the sandstones are stream-channel deposits. Abundant ostracods suggest fresh or brackish standing water, and the calcite crystals indicate saline, alkaline water. Claystone-pellet aggregates are indicative of nearby mudflats. The environment was saline and alkaline as judged from etching of quartz and the high content of zeolites.

ENVIRONMENTAL SYNTHESIS AND GEOLOGIC HISTORY

A widespread episode of faulting affected the Olduvai basin about 1.15 m.y.a., causing erosion

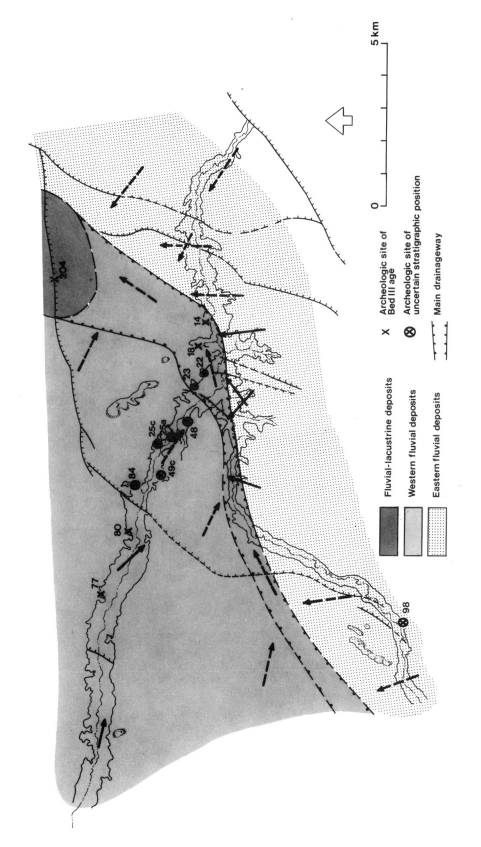

Figure 43. Paleogeography and lithologic facies of Bed III in the area of the gorge. Solid arrows indicate direction of stream flow as based on channel measurements, and dashed arrows represent flow direction inferred on other grounds.

126

of Bed II and drastically changing the paleogeography. The graben between the Fifth and FLK faults subsided, and several faults became active for the first time in the Olduvai basin (Fig. 42). The lowest point in the basin of Bed III was shifted from its Bed II position in the graben to a location about 7 km to the northeast (Fig. 43; see also Fig. 45). Nearly all of the sediment was transported and deposited by streams, and for the first time in the Olduvai region, water was supplied principally from metamorphic terrain to the west rather than from the volcanic highlands. Renewed faulting about 0.80 m.y.a. shifted the drainage axis of the basin southward about half a kilometer and caused widespread, generally shallow, erosion of Bed III in the main drainageway (Fig. 44; see also Fig. 46). The alluvial-plain environments of Bed III age continue relatively unchanged through the deposition of Bed IV, and the deposition of Bed IV was terminated about 0.6 m.y.a. by fault movements. Mozonik, a nephelinite volcano with reversed polarity in the basin of Lake Natron, is a possible source of nephelinite tephra in Beds III and IV.

The eastern fluvial facies of Beds III and IV accumulated on an alluvial plain sloping gently to the north and west from the foot of the volcanic highlands. Sediment was transported mostly by braided streams that flowed only intermittently throughout the year. The western fluvial facies accumulated on a well-watered broad alluvial plain that sloped southeastward. Most of the sediment was deposited by meandering streams as much as 2.5 m deep. The *Unio* and remains of hippo and crocodile suggest that water was present throughout the year, at least in pools within the largest channel(s). Metamorphic detritus of the western fluvial deposits are entirely from the Mozambique belt, suggesting that the headwaters of these streams did not extend appreciably west of the present headwaters of Olduvai Gorge (see Fig. 1, Chap. 1). The southeastern part of the facies represents a main drainageway with a meandering trunk stream and a well-developed floodplain. The paleogeography of Beds III and IV differs primarily in the location of the main drainageway (Figs. 45, 46). The drainageway for Bed III and equivalent sediments of Beds III-IV (und.) flowed from west to east, then changed its course to the northeast. The Bed IV drainageway lay about 0.5 km south of the Bed III drainageway, and it flowed in an easterly direction to the eastern limit of exposures.

The fluvial-lacustrine facies probably represents the Bed III drainage sump for the Olduvai basin. The size and location of this semilacustrine area can be inferred only to a limited extent. Locality 205 obviously lies near its western margin, as judged from the small proportions of lacustrine sediments and of volcanic detritus from the southeast. The eastern margin of the axial sump lay to the east of locality 203, and the northern limits of the Bed III drainage sump can be only conjectured.

Vegetation in the Olduvai basin was chiefly bush, scrub, and open grassland. Trees and perhaps strips of forest fringed the meandering rivers of the western alluvial plain and the main drainageway. Vegetation was relatively sparse on the alluvial plain to the south of the drainageway.

The climate was semiarid while Beds III and IV were being deposited. Mineralogic alterations in both fluvial and lacustrine sediments are evidence of a relatively dry climate, and a fluctuating and generally low water table is suggested by the reddish-brown color and caliche limestones of the eastern fluvial deposits. These features are common to Bed II above the widespread disconformity, particularly in the eastern fluvial deposits, and there is presently no basis for inferring a difference in climate either between Bed II and Beds III-IV, or between Bed III and Bed IV.

L. S. B. Leakey has suggested that the climate of Bed III was significantly drier than that for Bed IV (1951, 1965). This conclusion was based on differences in lithology and faunal content of Beds III and IV in the Side Gorge and eastern part of the Main Gorge. Much and perhaps all of the apparent faunal difference was a result of confusion between lithologic facies and vertical stratigraphic subdivisions. Faunal remains found in the western fluvial deposits of Beds III-IV (und.) were assigned to Bed IV, regardless of their stratigraphic level in Beds III-IV (und.). With this stratigraphic confusion resolved at JK (Kleindienst, in press), it can be seen that the Bed III faunal assemblage at JK is ecologically similar to the fauna of Bed IV. Additional fossils from Bed III and equivalent deposits must, however, be collected and studied for a fully satisfactory comparison.

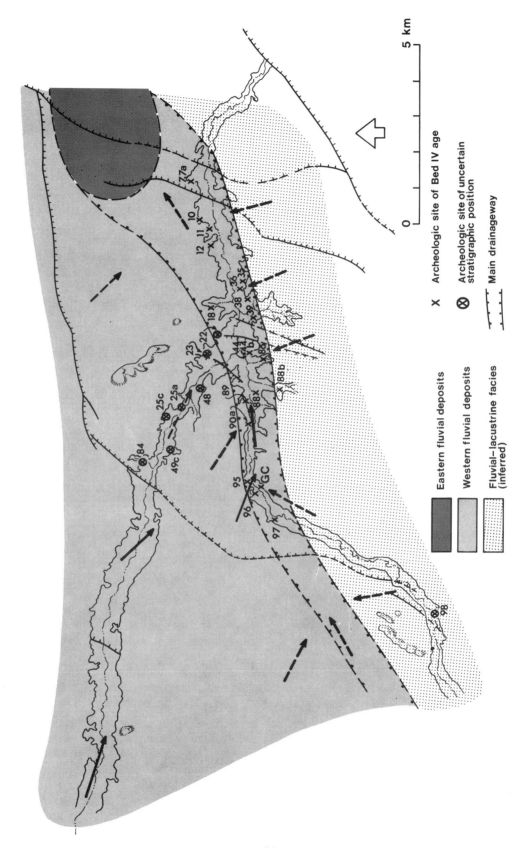

Eastern fluvial deposits

Western fluvial deposits

Fluvial–lacustrine facies (inferred)

X   Archeologic site of Bed IV age

⊗   Archeologic site of uncertain stratigraphic position

–·–·–   Main drainageway

Figure 44. Paleogeography and lithologic facies of Bed IV in the area of the gorge. See Figure 43 for meaning of arrows.

128

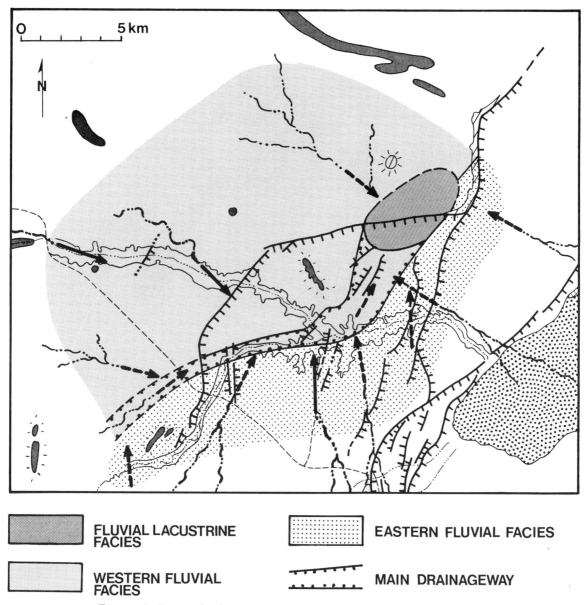

O 5 km

FLUVIAL LACUSTRINE FACIES

EASTERN FLUVIAL FACIES

WESTERN FLUVIAL FACIES

MAIN DRAINAGEWAY

Figure 45. Regional paleogeography of Bed III with inferred drainage pattern.
See Figure 43 for meaning of arrows.

HOMINID REMAINS AND ARCHEOLOGIC MATERIALS

Seven hominid fossils have been found in Beds III-IV. These are listed by M. D. Leakey (1971a) as H-2, 12, 22, 25, 28, 29, and 34. All are assigned either to *Homo erectus* or *Homo sp.* (see especially Day, 1971). Two fossils are from Bed III horizons, three are from Bed IV, and the others from undetermined levels of Beds III-IV (und.). All are from the western fluvial deposits with exception of H-22, a surface find with a reddish-brown sandstone matrix characteristic of the eastern fluvial deposits. H-25, from locality 54, is the only find relatively distant from the main drainageway.

Forty-three archeologic sites are recorded for Beds III-IV. Thirty-eight of these are listed by M. D. Leakey (1971a), and the other five are given in Table 22. The vast majority (twenty-eight) are in Bed IV, eight are in deposits correlative with Bed III, and the remaining seven are

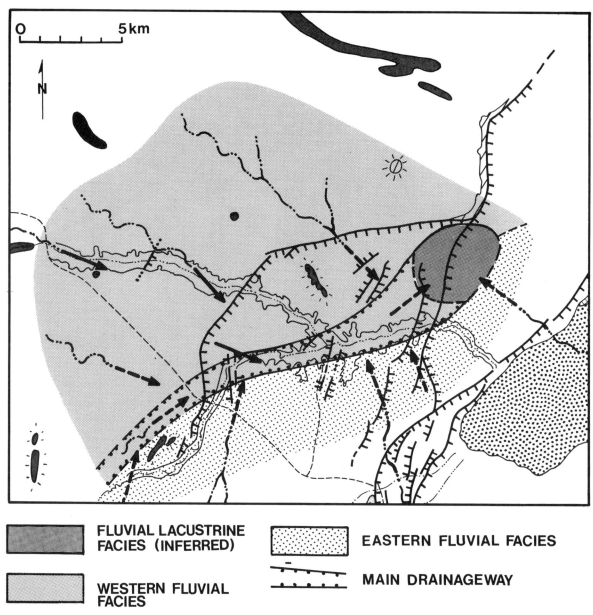

FLUVIAL LACUSTRINE FACIES (INFERRED)

WESTERN FLUVIAL FACIES

EASTERN FLUVIAL FACIES

MAIN DRAINAGEWAY

Figure 46. Regional paleogeography of Bed IV with inferred drainage pattern.
See Figure 43 for meaning of arrows.

from Beds III-IV (und.), either where the site horizon is uncertain or where the horizon is known but cannot be correlated with reasonable assurance either with Bed III or Bed IV. Sites are at various levels in both Bed III (and equivalent deposits) and Bed IV (Figs. 40, 41; Table 23).

About a dozen sites have been excavated and studied in detail, but only preliminary descriptions have been published thus far. Two of the excavated sites are in Bed III (Kleindienst, 1973), and the others are in Bed IV (M. D. Leakey, 1971b; in press, 1). The artifacts of most sites belong to an Acheulian industry, but an industry within the Developed Oldowan tradition, termed the Developed Oldowan C, is represented at a few sites (M. D. Leakey, in press, 1).

Sites are concentrated in the main drainageway and the western alluvial plain. Forty of the forty-three sites are in the western fluvial depos-

Table 22. *Previously Unrecorded Archeologic Sites in Beds III-IV*

Location and stratigraphic position	Nature of site
3 m above base of Bed IV, northwestern part of CK (loc. 7a)	Artifacts and faunal remains on claystone; buried by siltstone
West side PLK (loc. 23), 18 m above base of Beds III-IV (und.)	Artifacts and faunal remains in sandstone 2 m thick
Kar K (loc. 77) at base of Beds III-IV (und.)	Bifaces, etc., in sandstone filling channel at 3.3 m into Bed II
Southeast of Kelogi (loc. 98), near middle of 7.6-meter section of Beds III-IV (und.)	Bifaces, etc., in conglomerate and sandstone 1 m thick
Near middle of 8.5-meter thickness of fluvial-lacustrine deposits of Beds III-IV (und.) at locality 204; probably correlative with Bed III	Artifacts, principally *débitage*, in fine- and medium-grained sandstones 45 cm thick

Table 23. *Stratigraphic Context of Archeologic Sites in Beds III-IV*

Stratigraphic position	Lithofacies	Number of sites	Locality and site[a]
Bed III	Eastern fluvial	0	
Bed III	Western fluvial	7	14 (20a, 20b, 20c), 18 (18a, 18b), 77, 80
Bed III	Fluvial-lacustrine	1	204
Bed IV	Eastern fluvial	1	88c (PEK, 70b)
Bed IV	Western fluvial	26	7 (CK, 27b^1, 27c^3), 7a^2, 10 (LK, 25b^1), 11 (MK, 24c), 12 (EF-HR, 23b), 18 (TK, 18d^2, 18e^3), 35 (PDK, 53b^3, 53c^1), 36 (WK, 52$3^*$), 38-39 (51b), 44 (46d^1), 44b (HEB, 47$2^*$; includes 4 levels), 86 (VEK, 45c^1). 88 (MNK, 71d^1), 89 (FC, 63b^1), near 90a (GRC, 61), 95 (GTC, 58^1), south of 95 (GC, 65), 96 (CMK, 57a^1, 57b^3), 97 (NGC, 56)
Beds III-IV (und.); level or correlation unknown	Eastern fluvial	1	98
Same as above	Western fluvial	7	22 (16), 23 (PLK), 25a (11), 25c (10), 48 (37), 49c (Bos K, 36), 84 (9)

[a]Archeologic site designations of M. D. Leakey (1971a) are in parentheses. An asterisk designates sites described by M. D. Leakey (in press, 1). The stratigraphic position of Bed IV sites, where known or reasonably inferred, is indicated by the numbers 1, 2, and 3 used as superscripts to the site designation (e.g., CK, 27b^1). The number 1 refers to sites below the level of Tuff IVA; 2 refers to sites between Tuffs IVA and IVB, and 3 refers to sites above Tuff IVB.

its (Table 23), and of these, thirty-four are in or adjacent to the main drainageway. Two other sites are in the eastern fluvial deposits, and one of these (PEK, loc. 88b) is near the main drainageway. The other site is in the fluvial-lacustrine deposits and may be a lake-margin site. Sites in the main drainageway represent both the Developed Oldowan C and Acheulian industries; however, only Acheulian sites have been found outside the drainageway, as based on a high percentage of bifaces among artifacts lying on the surface. It is not clear whether this apparent paleogeographic difference in industries is significant,

as it is in Bed II, where Acheulian sites characteristically lie inland from the lake-margin zone with Developed Oldowan sites.

Almost without exception, artifacts and associated occupational debris of Beds III-IV are in sandstones and conglomerates of stream channels. Archeologic materials are reworked, although at some sites the artifacts are uniformly sharp, indicating that transportation was minimal (M. D. Leakey, in press, 1). Kleindienst (1973) discusses the probable nature of reworking and sedimentation of artifacts in stream channels of Bed III at JK (loc. 14). The site at

CK (loc. 7a) is an atypical one where artifacts lie on a clay surface and are buried by siltstone. Another unusual occurrence is at locality 204, to the north of the gorge, where artifacts are in sandstones that exhibit some lacustrine features and can perhaps be classed as lake-margin rather than fluvial. The association of artifacts with stream channels led M. D. Leakey to suggest that man chose to live in riverbeds during the dry season.

7

MASEK BEDS

STRATIGRAPHY AND DISTRIBUTION

The Masek Beds are the latest deposits prior to erosion of the gorge. Where exposed in the gorge, these deposits comprise roughly equal amounts of eolian tuff and detrital sediments. They have a maximum exposed thickness of about 25 m and occur over a slightly larger area than Bed I or Bed II. They crop out over the length of the Main and Side gorges and are exposed in fault scarps along the southwestern part of Olbalbal. They also crop out in a few places 5 km north of the gorge along the northern, upthrown side of the east-west extension of the Fifth Fault. Correlative deposits can also be found in the faulted terrain to the east of the main fault scarp between lakes Natron and Manyara. The thickest exposures in this eastern area appear to be about midway between Lake Manyara and Kerimasi.

Main and Side Gorges

The Masek Beds are found along the rim of the gorge, and they underlie the plain at shallow depth. The thickness of the Masek Beds varies strikingly in relation to faults, with 12 to 15 m common on the downthrown and 1.5 to 4.5 m on the upthrown side of faults (Figs. 40, 47). The Masek Beds disconformably overlie Beds III-IV in most places, although the contact can be difficult to locate where claystones of the Masek Beds overlie claystones of Beds III-IV. In a few places the Masek Beds fill channels 5 to 6 m deep cut into Bed IV, or else through Bed IV into Bed III. The Masek Beds overlie Bed II near the mouth of the gorge and along the southwestern part of Olbalbal (for example, loc. 157).

The contact between the Masek Beds and Ndutu Beds is in most places sharp and easy to locate. The uppermost 60 to 90 cm of the Masek Beds generally consist of hard, brown eolian tuff that is highly cemented by calcite and coated by dense, laminated calcrete 1 to 2 cm thick (Plate 7). The upper surface of the Masek Beds can either be smooth or knobby, with a relief of about 15 cm, and suggestive of erosion by wind. This knobby surface is locally (for instance, loc. 47) littered with irregularly shaped clasts of Masek tuff 10 to 20 cm long. The clasts, like the underlying surface, are coated with calcrete. Where the uppermost Masek tuffs lack a calcite cement and overlying calcrete (locs. 9, 84), the contact with the Ndutu Beds can be difficult to locate without mineralogic examination. In a few places the Masek and Ndutu beds are separated by an angular unconformity.

The Masek Beds are subdivided into two units, the lower of which is the thicker. The upper unit is named the Norkilili Member. The type locality for both units is Kestrel K (loc. 84). Eolian tuff and claystone in roughly equal mounts constitute 85 percent of the lower unit, and the remainder comprises sandstone and conglomerate. Tuffs are dominantly pale yellowish-brown and may be stained with red. The tephra is of nephelinite and melilitite composition, characterized by sodic augite, perovskite, melanite, melilite, and so forth (Fig. 48; Table 24). Chemically analyzed tuff is rich in iron and represents mafic tephra (see Table 26, no. 7).

Despite reworking and contamination, two tuff units can be recognized widely and are used as stratigraphic markers within the lower unit. The lower widespread tuff unit is brownish-yellow, rich in pumice, and found in many sections of the Masek Beds in the Main Gorge to the west of locality 4a. It was noted in the Side Gorge only at locality 96. This tuff generally contains 5 to 10 percent crystals, principally sodic augite and altered nepheline. The upper widespread tuff unit is pale yellowish-brown and

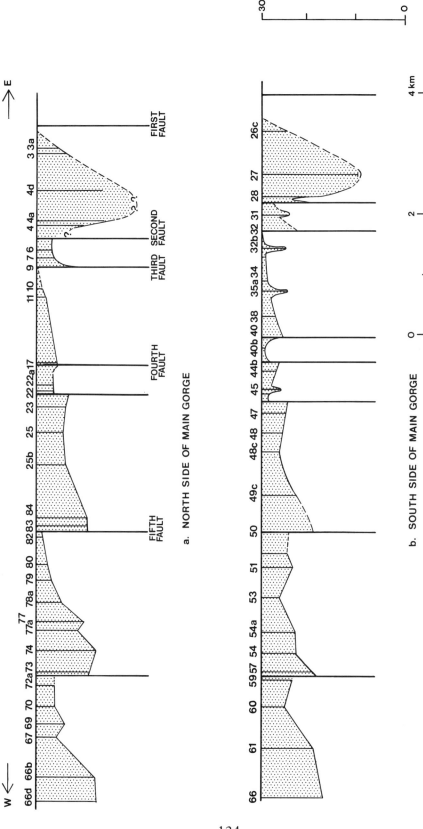

Figure 47. Cross-sections showing the thickness of the Masek Beds as exposed along the Main Gorge. Section is constructed with the top of the Masek Beds taken as horizontal. Vertical lines represent measured stratigraphic sections. Channeling is represented by the sections at localities 31, 32b, 35a, and 45.

134

Table 24. *Allogenic Materials in Tuffs of the Masek, Ndutu, and Naisiusiu Beds*

Stratigraphic unit	Lithology	Percent clay	Pumice and rock fragments		Percent nepheline	Biotite abundance	Other diagnostic mineralogic features[a]
			Percent	Composition			
Naisiusiu Beds	Eolian tuff	40-60	25-40	Nepheline phonolite	1-5	Rare	
Naisiusiu Beds (marker tuff)	Vitric tuff	0	90	Nepheline phonolite	2-5	Absent	Sphene, hornblende
Ndutu Beds, upper unit	Eolian tuff	15-30	25-60	Dominantly nephelinite; also phonolite, melilitite	10-20	Abundant or common	Perovskite > sphene; ilmenite < melanite
Ndutu Beds, upper unit (marker tuff)	Vitric tuff	0	75-90	Nepheline phonolite	2-5	Rare	Sphene, anorthoclase
Ndutu Beds, lower unit (in gorge)	Eolian tuff	5-20	10-20	Melilitite, nephelinite	5-10	Gen. rare to common	Perovskite > sphene; ilmenite < melanite
Ndutu Beds, lower unit (marker tuff)	Vitric tuff	0	90	Trachyte	0	Common	Anorthoclase
Ndutu Beds, lower unit (rim of gorge)	Clayey eolian tuff	50-65	10-40	Varied but dominantly melilitite	1-5	Rare	Perovskite > sphene
Masek Beds, Norkilili Member	Eolian tuff	20-35	25-50	Dom. nephelinite	10-20	Rare	Melanite >> ilmenite
Masek Beds, lower unit	Eolian tuff	30-50	25-50	Melilitite and nephelinite	1-10	Trace	Altered melilite common; sodalite(?) rare; ilmenite > melanite

[a]Augite is relatively common in most tuff samples and is not listed.

distinguished by brown, rounded rock fragments. This unit may be represented by one or more beds and is found throughout the length of the Main Gorge. It was noted in the Side Gorge only at CMK (loc. 96). The tuff generally contains 10 to 15 percent crystals, which commonly include augite coarser than that found in the lower tuff. Pumice fragments are generally brown, and a few are as much as 5 mm long.

The lower marker tuff lies at or near the base of the Masek Beds in most places. However, a 5- to 11-meter thickness of detrital sediments, principally claystone, underlies the lower unit in the western part of the gorge (locs. 61-66, 66b-d). These detrital sediments may be correlative with channel fillings of the lower unit in the eastern part of the Main Gorge, two of which (locs. 32b, 35a) are overlain by pumice-rich eolian tuff which probably represents the lower unit. The upper marker tuff varies greatly in stratigraphic position. It is most commonly at or near the top of the lower unit (for example, locs. 23, 25, 47), but it may lie near the middle of the lower unit of the Masek Beds (locs. 57, 83). In the syncline to the east of the Second Fault, these upper tuffs lie at or near the base of the Masek Beds, and they are separated from the Norkilili Member by as much as 17 m of nontuffaceous sediments (Fig. 49).

The Norkilili Member is generally between 1 and 4.5 m thick and extends for the length of the Main and Side gorges. It consists chiefly of reworked tuff, both eolian and fluvial, and includes a substantial proportion of sandstone and conglomerate. The uppermost 60 to 90 cm are in most places brown and highly cemented by calcite. Red spotting characterizes most of the tuffs, which range from yellowish-brown to pinkish-brown. Tuffs of the Norkilili Member differ from those of the lower unit of the Masek Beds in a wide variety of mineralogic respects (Fig. 48; Tables 24, 25). The red spotting and white, cottony dawsonite generally seem to identify tuffs of the Norkilili Member in the field.

The Norkilili Member conformably overlies the lower unit in most places. However, in the excavation at FLK (loc. 45), tuffaceous sandstone of the Norkilili Member fills a channel 4 m deep cut into the lower unit, which fills an older channel cut into Bed IV. An angular unconformity locally separates the two units on the anticlinal crest to the east of the Second Fault.

North of the Gorge

Only the Norkilili Member of the Masek Beds has been found to the north of the gorge. A

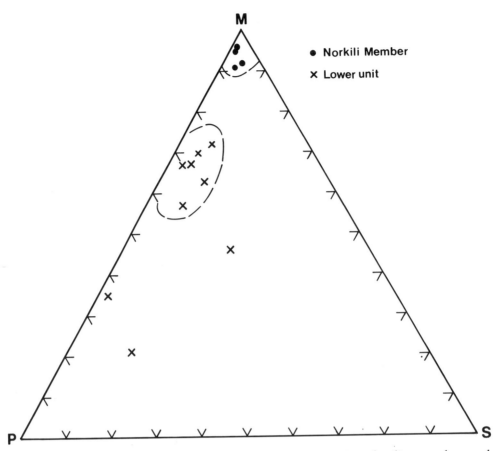

Figure 48. Composition of tuff samples from the Masek Beds plotted in a triangular diagram whose end members are melanite (M), perovskite (P), and sphene (S). Plotted points are based on about 100 grains of melanite + perovskite + sphene counted in thin sections.

1.5-meter thickness unconformably overlies Beds III-IV (und.) at locality 203. This section comprises a conglomerate of eolian tuff clasts overlain by pale reddish-brown water-worked tuff. About a kilometer west of locality 205, a hard bed of calcareous tuff typical of the Norki-lili Member lies above Beds III-IV (und.) Exposures are poor in this area, and lower Masek tuffs might possibly be present.

Southwestern Margin of Olbalbal

The lower unit of the Masek Beds is represented by as much as 19 m of deposits near the southwestern margin of Olbalbal (locs. 156-158). The lower part of the sequence consists of reddish-brown claystones with interbedded sandstones, conglomerates, and tuffs. The upper is composed of eolian tuffs and tuffaceous siltstones as much as 7 m thick. Most of the eolian tuffs are fine-grained and highly contaminated with older

volcanic detritus; consequently, they cannot be correlated with any specific tuff unit in the gorge. The base of the Masek Beds was found exposed only along the scarp of the First Fault (loc. 157). Here, 7 to 9 m of reddish-brown deposits overlie an irregular surface eroded over Bed II 4 to 6 m above the Lemuta Member. The overlying Masek tuffs are 3 to 4 m thick at this locality. The maximum thickness of 19 m was measured at and near locality 156. This figure is a minimum, as the base of the Masek Beds is not exposed. The Masek tuffs are overlain by tuffs of the Ndutu Beds.

AGE OF THE MASEK BEDS

The age of the Masek Beds can be estimated from several lines of evidence. One line is the dating of Kerimasi, source of the tuffs in the Masek Beds. That Kerimasi was the source is shown by the mineralogic similarity of its tephra deposits to the Masek tuffs. As noted earlier, the

136

Table 25. *Content of Authigenic Minerals in Tuffs of the Masek, Ndutu, and Naisiusiu Beds*

| Stratigraphic unit | Lithology | No. of samples | Percentage zeolites[a] | | | | | Percentage calcite[b] | No. of samples with dawsonite | Other minerals |
			Total	Ph	Ch	An	Nt			
Naisiusiu Beds	Eolian tuff	14	25.0	15.2	8.8	1.0	–	2.3	3/14	
Naisiusiu Beds (marker tuff)	Vitric tuff	2	50	40	–	10	–	1.6	–	
Ndutu Beds, upper unit	Eolian tuff	22	36.0	26.1	0.6	3.9	5.3	4.4	12/22	Dolomite rare
Ndutu Beds, upper unit (marker tuff)	Vitric tuff	3	53	49	–	4	–	–	–	
Ndutu Beds, lower unit (in gorge)	Eolian tuff	6	20.8	15.9	–	4.9	tr.	2.5	–	
Ndutu Beds, lower unit (marker tuff)	Vitric tuff	3	53	49	–	4	–	–	–	
Ndutu Beds, lower unit (in rim)	Clayey eolian tuff	4	20	6	11	2	1	~6	–	Dolomite in 3 samples
Masek Beds, Norkilili Member	Eolian tuff	8	32.6	17.7	5.1	2.9	6.9	5.0	6/8	
Masek Beds, lower unit	Eolian tuff	14	23.4	16.0	–	7.0	0.5	1.7	2/14	Authigenic clay is common; rare dolomite

NOTE: Percentages are averages estimated from x-ray diffractograms; presence or absence of dawsonite was determined microscopically.

[a] Zeolite abbreviations are Ph = phillipsite, Ch = chabazite, An = analcime, and Nt = natrolite.

[b] Calcite is both allogenic and authigenic.

tuffs of the lower unit of the Masek Beds compare closely in mineral composition with the lower Kerimasi tephra deposits exposed in Swallow Crater, whereas the Norkilili tuffs compare in their abundant melanite with the single sample of tephra collected from the upper Kerimasi deposits. Moreover, Kerimasi is situated to the east, upwind from Olduvai, in a location ideal for supplying the Masek tephra.

The age of Kerimasi has been bracketed within limits of 1.1 to 0.4 m.y.a. by K-Ar dating of younger and older volcanic rocks in the vicinity of Kerimasi (Macintyre et al., 1974). The most crucial dating in this regard is 0.37 m.y.a. for tephra deposits of the eruptions that produced Swallow Crater and a nearby, related tuff cone. These deposits directly overlie the Kerimasi tephra.

Magnetic polarity suggests that the Masek Beds lie entirely within the Brunhes Normal Epoch, which extends from the present to 0.7 m.y.a. All six samples measured in the laboratory by A. Brock have normal polarity. These samples are from various levels in the Masek Beds and include both calcrete and zeolitic eolian tuff. The same rock types in the Lemuta

Member gave reliable results. Thus Kerimasi appears to fall within dates of 0.7 and 0.4 m.y.a.

The amount of time represented by the Masek Beds can be estimated by comparing the volume of its detrital sediment with that in Beds III-IV, which span a period of roughly 0.55 million years. Where exposed in the gorge, the Masek Beds contain about one-third the volume of detrital sediment found in Beds III-IV. This comparison probably errs in including the eastern fluvial facies of Beds III-IV, inasmuch as the comparable facies of the Masek Beds is not adequately represented in the gorge. If the comparison is restricted to the western fluvial facies of both units, the ratio of Masek sediments to sediments of Beds III-IV is between 0.35 and 0.40, suggesting a time span on the order of 0.20 million years.

In summary, the Masek Beds fall within limits of 0.7 and 0.4 m.y.a. on magnetic polarity and K-Ar dating. If the estimate of 0.6 m.y.a. for the top of Bed IV is correct, the limits are further narrowed to 0.6 to 0.4 m.y.a. The time span estimated from the volume of sediment neatly fills the entire interval of 0.6 to 0.4 m.y.a. (see Fig. 12, Chap. 4).

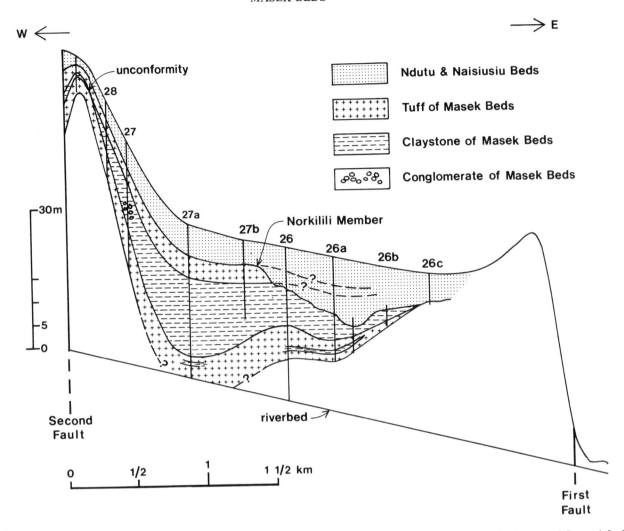

Figure 49. Cross-section of the Masek Beds as exposed on the south side of the Main Gorge between the First and Second faults. Locality 28 is near crest of anticline and locality 27a is in trough of syncline.

LITHOFACIES AND ENVIRONMENTS

The lower unit of the Masek Beds comprises two fluvial facies and an eolian facies. One of the two fluvial facies is named the eastern fluvial facies, and it has many features in common with the eastern fluvial facies of Bed II and Beds III-IV. The other is the western fluvial facies, which is similar in many respects to the western fluvial facies of Beds III-IV. The eastern fluvial deposits are found along the southwestern margin of Olbalbal (locs. 156-158), and near Kelogi (locs. 98-101a). A pebble conglomerate of this facies is also at the base of the Masek Beds at locality 26. The western fluvial deposits are found throughout the length of the Main Gorge and in the Side Gorge at least as far south as

NGC (loc. 97). Deposits of the eolian and western fluvial facies are widely interbedded, commonly on a small scale, and are not areally limited such as the facies of Beds I through IV. The Norkilili Member is regarded as a single facies.

Eastern Fluvial Deposits along the Margin of Olbalbal

The eastern fluvial deposits are here chiefly claystone and sandstone; claystone-pellet aggregate and conglomerate constitute the remainder (see Table 21, Chap. 6). Claystones are massive, root-marked, and are mottled brown and reddish-brown. Most of them are porous and have an

138

earthy luster. They contain a substantial proportion of claystone pellets and detrital volcanic sand and silt. Many of the sand grains are rounded and coated with clay, presumably as a result of wind transport. Volcanic ash is in most claystones and can be a major constituent, particularly in the uppermost claystones. The claystones have been extensively modified by chemical processes, and authigenic minerals, principally zeolites, are abundant. Dawsonite is common in some of the claystones in Olbalbal (loc. 156). Claystone-pellet aggregates are similar to the claystones in their color, mineralogic alterations, and authigenic minerals.

Sandstones occur generally as poorly defined beds that are clayey, mottled reddish-brown, and difficult to distinguish at a distance from claystones. They consist largely of volcanic detritus but may contain claystone pellets. Conglomerates form lenticular beds composed largely of pebbles and small cobbles of volcanic rock from Lemagrut, Sadiman, and Ngorongoro. Minor constituents are limestones, eolian tuff, and laminated calcrete.

Channel alignments were measured at locality 156 and range from N35°E to N35°W, averaging N10°W. All but one of the five measurements are to the northwest, and the northwesterly group by itself averages N21°W.

Faunal remains have not been found in these deposits, and the rootmarkings are suggestive of grass.

These fluvial deposits were deposited largely by streams on an alluvial plain similar in most respects to that of the eastern fluvial deposits of Bed II and Beds III-IV. However, sediment was redeposited more extensively by wind in these Masek deposits than in the older deposits. An alkaline soil environment is indicated by the zeolitic alteration.

Eastern Fluvial Deposits Near Kelogi

Sandstone is the most abundant type of sediment near Kelogi, and claystone and conglomerate form the remainder. Claystones are brown, rootmarked, and resemble those of the western fluvial facies in appearance. Sandstones are medium- to coarse-grained and commonly conglomeratic. Volcanic detritus is overall the dominant constituent, but claystone pellets can be abundant, and basement detritus is invariably present. The volcanic detritus is almost entirely from Lemagrut and the Laetolil Beds. Conglomerate clasts are chiefly of lava but include Kelogi-type gneiss, limestone, Laetolil tuff, and trachytic welded tuff. The lava is chiefly dark gray and typical of Lemagrut but includes 5 to 10 percent of green nephelinite and phonolite similar to that of Sadiman. This green lava may have been derived from the Laetolil Beds, and the welded tuff is from an unknown source.

Channeling is common, particularly at the base of the Masek Beds. The six measured alignments range from N65°W to N70°E and seem to indicate sinuous streams. As a measure of flow direction, their significance is doubtful.

Western Fluvial Facies

The western fluvial facies is about three-quarters claystone, and nearly all of the remainder is sandstone and conglomerate (Table 21). A main drainageway is represented by sediments along the southeastern margin of the facies, as in Beds III-IV.

Claystones are typically massive, brown, and rootmarked. Most have a waxy luster, but porous claystone with an earthy luster is found in many places and is particularly common at the base of the Masek Beds in its westernmost sections. They resemble claystones of the western fluvial facies of Beds III-IV in their content of detrital sand and silt, and claystone pellets and volcanic ash are common additional constituents. A substantial proportion of the sand in many claystones exhibits evidence of wind transport (for example, rounding and clay coating of rock fragments). Quartz and augite are conspicuously etched in most of the samples from the main drainageway. Authigenic analcime averages 9 percent in twelve analyzed claystones. The clay mineral is illite, with or without a small amount of interlayered montmorillonite.

Sandstones form lenticular beds of fluvial type. As in Beds III-IV, volcanic and metamorphic detritus are intermixed in varying proportions in sandstones of the main drainageway. The proportion of metamorphic detritus is highest to the northwest. Most of the sandstones contain at least a small proportion of reworked ash similar to that found in Masek tuffs. Claystone pellets are abundant in most of the medium-grained sandstones.

Conglomerates form lenticular beds as much as 3 m thick and ranging in clast size from small pebbles to boulders. Only rarely does a conglomerate form the basal layer of a fining-upward sandstone sequence. The smaller clasts to the west of the main drainageway are chiefly limestone, laminated calcrete, and metamorphic detritus; most of the larger clasts are blocks of eolian tuff. Within the main drainageway are also clasts of lava and of varied materials from the underlying units of the Olduvai Beds. Blocks of Tuff IVA, for example, are in the channel filling at FLK, and conglomerates farther to the east contain clasts of reddish-brown zeolite claystone very likely from Bed III. Eolian tuff clasts can be wholly coated with laminated calcrete 1 to 2 cm thick and having an unbroken surface (loc. 31). This calcrete was formed after the conglomerate had been deposited. Curiously, clasts of lava in the same conglomerate are not coated by calcrete.

Most of the conglomerates and many of the thicker sandstones fill steep-sided channels, the deepest of which are 5 to 6 m deep. The full width of these deeper channels is nowhere exposed. The shallower, well-exposed channels are commonly as much as 1.5 m deep and three to four times as wide as they are deep. Four of the five channels measured to the west of the Fifth Fault are aligned between N60°W and N75°W, indicating streams of low sinuosity flowing to the southeast. Channeling at the base of the thick channel fill at FLK is east-west, parallel to the main drainageway. Five well-defined channels in the vicinity of the Second Fault record flow directions ranging from N60°E to S30°E. This variation reflects either sinuous streams or a widening of the main drainageway, which permitted greater variation in overall flow direction (see Fig. 50).

Vertebrate remains are relatively scarce, and the only systematic collection is from the excavated channel at FLK (loc. 45). The identified forms are as follows (M. D. Leakey, 1974, personal communication):

Clariids (catfish)
Crocodylus sp.
Chelonia, sp. indet.
Aves, indet.
Giraffa, sp.

Primates, indet.
Rodentia, indet.
Carnivora, indet.
Suidae, sp. indet.
Ceratotherium simum
Equus, sp.
Hipparion, sp.
Hippopotamus gorgops
Tragelaphus strepsiceros sub sp.
gazelle (probably *Antidorcas recki*)
Cephalophini indet. (duiker)
Reduncini indet. (bushbuck, etc.)
Neotragini indet. (pigmy antelope)

The channel also yielded land snails identified by B. Verdcourt as *Limicolaria,* "rather like *L. martensiana catheria* Dall, a common shell on the Athi plains, etc." (1969, personal communication to M. D. Leakey).

Rodent teeth were obtained from claystone 3 m above the base of the Masek Beds at locality 4. J. J. Jaeger kindly provided the following information: "The rodent teeth from the Masek Beds belong to a fat-mouse, *Steatomys* sp. which is a Cricetidae, Dendromurinae, present in Bed I and living today in East Africa. It is *usually* found in dry areas (dry savanna). It is a nocturnal animal who burrows to a depth of 0.9 to 1.2 meters and a distance of 1.5 to 1.7 meters."

The western fluvial facies comprises sediments deposited in two types of fluvial settings. The bulk of sediments were deposited in shallow stream channels and on adjacent floodplains of an alluvial plain sloping gently to the east and southeast. The other sediments, found with certainty only below the lower marker tuff, were deposited in relatively deep channels eroded in older deposits. These channel-filling deposits are found only between FLK and the Second Fault. Both types of sediments differ generally from the western fluvial facies of Beds III-IV in a higher content of wind-borne detritus.

Eolian Facies

The eolian facies is at least 90 percent eolian tuff, and the remainder includes waterlaid tuff, laminated calcrete, limestone nodules, tuffaceous claystone, and rubble.

Eolian tuffs are generally massive, sparsely rootmarked, and form beds 60 cm to 2 m thick. Individual beds locally pinch out, and elsewhere they can bifurcate into two tuff beds. A few beds exhibit the steep, large-scale cross-bedding

140

characteristic of barchan dunes. Tuffs are characteristically rather well cemented. Component particles in the massive beds are mostly between 0.05 and 0.3 mm in diameter and compare closely in texture with the present blanket of sediment over the Serengeti Plain adjacent to the gorge.

The tuffs are generally a mixture of tephra with varied other materials: volcanic and metamorphic detritus, limestone, broken crystals of calcite, claystone pellets, and so forth. The tephra comprises vitroclasts (principally pumice), rock fragments, and crystals (principally sodic augite, nepheline, and melilite). Found rarely are lava fragments as much as 2.5 cm long. The broken calcite crystals may be of eruptive origin, as Kerimasi is a "carbonatite" volcano much like Oldoinyo Lengai. Allogenic clay forms a significant amount of most samples, and it occurs as claystone pellets, pelletoid coatings, and less commonly as a primary matrix. With increase in clay matrix, tuffs grade into tuffaceous claystones.

Waterlaid tuffs were deposited either by streams or in standing water. The stream-laid tuffs are composed of eolian sediment that has been sorted and stratified, generally to a minor extent, by running water. A few beds of small extent are laminated and consist of uncontaminated primary tephra with no evidence of wind transport. These tuffs resulted from showers of ash into rather small ponds.

Glass and melilite in all the tuffs are wholly altered, and many other materials (nepheline, plagioclase, quartz, augite) are altered less extensively. The tuffs are cemented by zeolites, with or without calcite (Table 25). Many of the tuffs up to and including the lower marker tuff also contain an authigenic clay ("illite") formed from alteration of mafic glass. Chemically analyzed authigenic clay of one tuff sample is extraordinarily rich in iron and quite unlike detrital clay in composition (see Table 26, no. 8, Chap. 8).

Laminated calcretes occur widely, both within and over the top of eolian tuff beds. They range in thickness from a few millimeters to a few centimeters. Limestone nodules are common in clayey tuff and tuffaceous claystone beneath the Norkilili Member. Rubble layers are composed of angular clasts of eolian tuff and calcrete derived from the Masek Beds. Most of the rubble fills stream channels. Rubble layers are found at various levels in the facies.

A few directional measurements were obtained from the facies. The dip, or maximum inclination, of cross-bedding ranges from S20°W to S40°W in a bed at locality 31; the cross-bedding at one place in Kestrel K (loc. 84) dips due south. A small rubble-filled channel at WK (loc. 36) is oriented north-south, perpendicular to the main drainageway.

Shells of a land snail, probably *Limicolaria* sp., are found widely but in small numbers in the eolian tuffs. The rootmarkings are suggestive of grass. The only other fossils noted are rare, unidentified bone fragments of sand size in some of the tuffs.

The eolian facies is chiefly an accumulation of wind-worked volcanic ash on a land surface of low relief. Dunes of ash are indicated by cross-bedding, but the bulk of the ash accumulated as a blanketlike deposit, much like the surface layer of sediment on the plain at Olduvai Gorge. The ash was altered and cemented to form zeolitic tuff penecontemporaneous with deposition, as shown by zeolitic tuff clasts in conglomerates at many levels. Similarly, the laminated calcrete was formed at shallow depths, as on the Serengeti Plain today.

Norkilili Member

The Norkilili Member is roughly 90 to 95 percent tuff and 5 to 10 percent sandstone and conglomerate. Most of the tuff is typically eolian, but a substantial amount (10 to 20 percent?) was deposited by streams. The Norkilili Member is regarded as a single facies because it is difficult to subdivide into genetically distinct units.

Eolian tuffs are massive and form beds 30 to 90 cm thick. Sorting and grain size are roughly the same as in eolian tuffs of the lower unit. Tephra is dominantly of nephelinite composition (Table 24). Claystone pellets, older detritus, broken calcite crystals, and so forth, are additional materials. Water-worked tuffs are stratified, moderately well sorted, and generally contain considerable detrital sand. Festoon cross-bedding of fluvial type is found widely.

Primary materials are altered as in the lower unit, and the authigenic minerals are roughly the

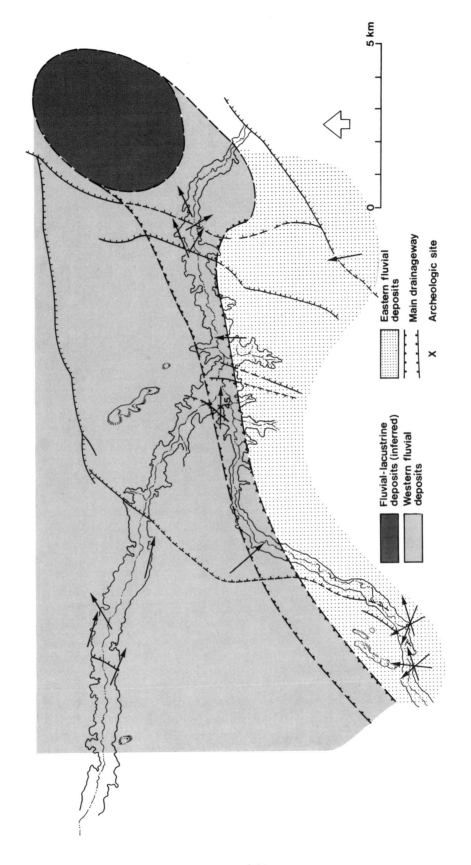

5 km

Fluvial-lacustrine deposits (inferred)

Western fluvial deposits

Eastern fluvial deposits

Main drainageway

X   Archeologic site

Figure 50. Paleogeography and lithologic facies of the Masek Beds in the area of the gorge. Arrows indicate flow direction as based on stream-channel measurements.

142

same except for the much more common occurrence of dawsonite in the Norkilili Member. A notable feature is the high degree to which the uppermost bed is in most places cemented by calcite.

Sandstones are commonly laminated or cross-bedded. Most of them are tuffaceous, and many are conglomeratic. Conglomerates are composed principally of pebbles, and the coarsest conglomerates are found to the east of the Second Fault. Pebbles to the west of the main drainageway are chiefly limestone and metamorphic detritus. To the east of the Second Fault, pebbles also include lava, red claystone of Bed-III type, and eolian tuff and laminated calcrete from the lower unit of the Masek Beds.

Channeling was found only at FLK, where a steep-sided channel 4 m deep and filled with tuffaceous sandstone is cut into the lower unit. Sandstone filling the channel exhibits smaller-scale channeling, and two of the smaller channels are oriented N65°W and N85°E, suggesting easterly stream flow.

A few land snails (*Limicolaria* sp.?) are the only faunal remains noted. The rootmarkings resemble those of grass.

The bulk of the Norkilili Member is an accumulation of wind-worked ash on a relatively flat surface, traversed by streams. This environment differed from that of the lower unit principally in the absence of clay-covered floodplains.

ENVIRONMENTAL SYNTHESIS AND GEOLOGIC HISTORY

The Masek Beds were deposited during a lengthy eruptive episode of Kerimasi and during a period of faulting in the Olduvai region. Displacements are documented for all faults along which the thickness of the Masek Beds is known (Fig. 47). Folding between the First and Second faults was also contemporaneous with deposition (Fig. 49). Here, a narrow anticline developed adjacent to the Second Fault, and a broad syncline formed to the east. This syncline contains the maximum known thickness of the Masek Beds. During folding, Beds III and IV were probably eroded away on the eastern flank of the syncline where the Masek Beds unconformably overlie Bed II. The angular unconformity within the Masek Beds on the crest of the anticline also shows that folding was contemporaneous with deposition.

Stratigraphic evidence is insufficient to show the extent of displacement, if any, along the First Fault, and whether or not the faults in the southwestern part of Olbalbal were active at this time. The First Fault may well have been, but displacement (if any) along the fault was insufficient to make Olbalbal the principal locus of deposition, as it was in later times. As evidence, streams from Ngorongoro and Lemagrut flowed to the northwest across the southwestern part of Olbalbal, whereas the present drainage in this area is northeasterly, into the lowest part. Moreover, pebbles of Sadiman lava are common in conglomerates of the Masek Beds on the west, or upthrown, side of the First Fault, whereas clasts of Sadiman lava presently are carried into Olbalbal, to the east of the First Fault.

The Olduvai basin was largely an alluvial plain of low relief while the Masek Beds were being deposited. The paleogeography was similar in many respects to that of Beds III-IV. The northwestern part of the basin sloped gently to the east and southeast joining the main drainageway which flowed eastward into the drainage sump (Figs. 50, 51). To the south of the drainageway, an alluvial surface rose toward Lemagrut and Ngorongoro. However, the drainage sump had now shifted eastward to the syncline between the First and Second faults. Another difference is in the main drainageway between the FLK fault and the Second Fault. Early in Masek sedimentation, streams in this area flowed in relatively narrow, deep valleys cut in rising fault blocks.

About half of the lower unit was deposited by streams, and about half by wind. Except for the fillings of deep channels in the main drainageway, fluvial sedimentation was generally similar to that of Beds III-IV. As evidence, the proportion of claystone, sandstone, and conglomerate is about the same in equivalent facies of both units (see Table 21, Chap. 6). Eolian tuffs of the lower unit were deposited partly in barchan dunes and partly in a thin layer over the ground surface. The direction of dune cross-bedding suggests southwesterly winds, which differ somewhat from the dominantly westerly winds of the present time. Regrettably, the directional measurements are too few for a reliable comparison of wind direction.

The overall drainage pattern of the Norkilili Member seems to have been essentially the same as for the lower unit, but the lack of floodplain

143

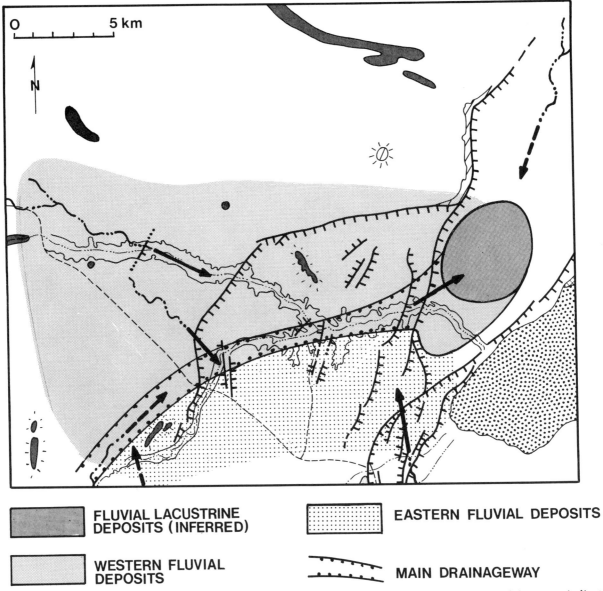

FLUVIAL LACUSTRINE DEPOSITS (INFERRED)

EASTERN FLUVIAL DEPOSITS

WESTERN FLUVIAL DEPOSITS

MAIN DRAINAGEWAY

Figure 51. Regional paleogeography of the Masek Beds showing inferred drainage pattern. Solid arrows indicate direction of stream flow as based on channel measurements, and dashed arrows represent flow direction inferred on other grounds.

clay shows that the nature of the fluvial environment has changed. Evidently the detrital clay was carried to the drainage sump, which may reflect an increase in stream gradient and valleys lacking floodplains. Earth movements presumably caused this change in fluvial regime.

The climate must have been semiarid, as judged from the large amount of eolian sediment, the calcrete and mineralogic alterations, and the paucity and nature of faunal remains. Very likely the climate was drier than that of

Beds III-IV and rather like that of the present. Eolian tuffs do not by themselves show that the Masek climate was drier than that of Beds III-IV, for their presence or absence might simply record whether or not large volumes of tephra were available. However, the high content of wind-worked *detrital* sediment in floodplain claystones, and so forth, shows that eolian processes were relatively more active in Masek time. Both the calcretes and eolian tuffs are strikingly similar to those of late Pleistocene and Holocene

144

time in the Olduvai region, and the *Limicolaria* and *Steatomys* are suggestive of dry savannah. However, the hippo, crocodile, and catfish show that water was present throughout the year, at least while the lowermost part of the Masek Beds were deposited. Three of the bovid groups (Cephalophini, Reduncini, and Neotragini) point to areas of vegetation more luxuriant than that found in the same area today. The drainage sump presumably contained a saline, alkaline perennial pond or lake, which may have been an acceptable habitat for hippo, and at least locally or intermittently for crocodiles.

HOMINID REMAINS AND ARCHEOLOGIC MATERIALS

Hominid remains and artifacts are relatively rare in the Masek Beds, as might be expected from the drier climate and diminished stream flow. The sole hominid fossil found thus far is part of a hominid mandible (H-23) provisionally attributed to *Homo erectus* (M. D. Leakey, in press, 2). This site in the main drainageway was later excavated, yielding an Acheulian assemblage near the base of a lower Masek channel cut deeply into Bed IV (M. D. Leakey, in press, 1). Isolated, sharp-edged artifacts, including bifaces, lie scattered on the surface of the lower unit in its westernmost exposures, and further investigation may well reveal worthwhile archeologic sites in this area. Isolated artifacts are also found in conglomeratic sandstones of the lower member to the east (for example, locs. 6, 7), but some of the artifacts are abraded, and all may have been derived from Bed IV, which is here channeled by the lower unit.

8

NDUTU BEDS

STRATIGRAPHY AND DISTRIBUTION

The Ndutu Beds are widespread sediments, principally tuffs, which were deposited in the Olduvai region over a lengthy period of intermittent faulting, erosion, and partial filling of the gorge. The Ndutu Beds in the gorge are subdivided into two units that record different stages in its evolution. The lower unit is largely conglomerate, sandstone, and tuff. To the west of the Second Fault these sediments were deposited locally during a more or less continuous excavation of the gorge, and there are only a few patches left today. East of the Second Fault, sediments of the lower unit filled and overflowed a proto-gorge, and extensive exposures still remain. The upper unit of the Ndutu Beds is principally tuffs that accumulated to a considerable depth over the sides and in the bottom of the gorge, after it had been eroded, on the average, to roughly three-quarters of its present depth. Tuffs of the upper unit also form a thin layer over the plain, which is exposed along the rim of the gorge. The Ndutu Beds are also exposed extensively in the tributary branches of the Side Gorge 8 to 10 km south of Kelogi, at the western foot of Lemagrut. These sediments south of Kelogi, as much as 12 m thick, are chiefly sandstone, claystone, and conglomerate. The small proportion of tuff suggests that they represent the lower unit of the Ndutu Beds. The type section of the Ndutu Beds is taken 1.4 km to the east of the Second Fault on the south side of the gorge (loc. 26, Fig. 52). Here both units are present.

Within the gorge the Ndutu Beds rest unconformably on all older units. On the rim of the gorge they are commonly separated from the Norkilili Member with an uneven disconformity littered by blocks of Norkilili tuff. Both the disconformity and the blocks are characteristically coated with laminated calcrete (Plate 7). The disconformity is a rubble-littered eolian deflation surface like that visible over parts of the plain in the vicinity of the gorge.

Lower Unit of the Ndutu Beds

The lower unit to the west of the Second Fault is represented by a thin, discontinuous deposit of eolian tuff along the rim of the gorge and by isolated patches of sediment at various levels between the rim and about 4 m above the riverbed. The tuffs exposed along the rim are generally represented by a single bed 15 to 50 cm thick. Its upper contact as well as its lower is in most places an eolian deflation surface. At Kestrel K (loc. 84) is a 2-meter thickness, which includes interbedded calcrete and two rubble-littered erosion surfaces. Most of the tuffs along the rim are rather highly contaminated with older detritus. The tephra is dominantly melilitite in composition.

The patches of sediment within the confines of the gorge are between 60 cm and 7.6 m in thickness and are principally sandstone and conglomerate. From the small size and scarcity of these exposures, it seems reasonably certain that these sediments were deposited during minor, in part localized, episodes of valley fill during the overall cutting of the gorge. By this reasoning, the highest patches of streamlaid materials should be the oldest.

The highest and presumably oldest beds in the Main Gorge are at HK (loc. 24), where a 3.4-meter thickness lies only about 6 m below the level of the plain and only slightly below the Norkilili Member. The most extensive exposures, 3.5 to 4 m thick, were found between JK (loc. 14) and DK (loc. 13) about 11 m below the plain. The thickest single exposure, 7.6 m thick, lies about midway in the sides of the gorge at locality 25a. This sequence includes the single marker tuff identified in the lower unit. It is a cream-colored, fine- to medium-grained vitric tuff 15 to 50 cm thick and of trachyte composition (Tables 25, 26). Here the marker tuff lies about 22 m above the riverbed. Of archeological interest are 60 cm to 1 m of sandstone with artifacts probably eroded from Bed IV and lying

146

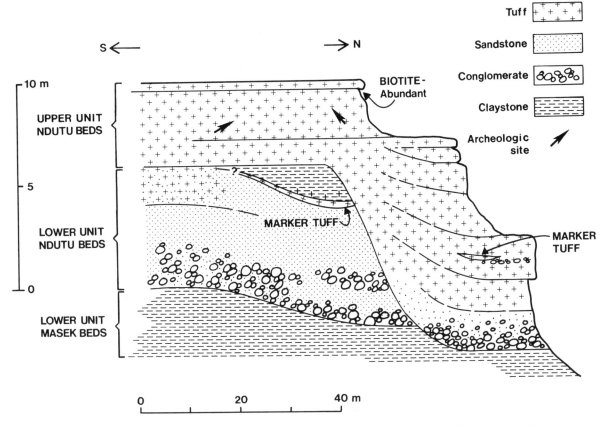

Figure 52. North-south section of the Ndutu Beds as exposed at their type locality (loc. 26).

about midway in the sides of the gorge at HK (loc. 24). A 2.6-meter thickness of conglomerate and tuff lies only 6 m above the riverbed on the south side of the gorge near the west edge of Long K. Algal limestone coating basalt 3.7 m above the riverbed near FLK-NN (loc. 45b) is the lowermost exposure of the lower unit found to the west of the Second Fault.

Scattered exposures of the lower unit are also in the Side Gorge. To the north of BK (loc. 94), about 5 m below the rim of the gorge, is a 1-meter thickness of conglomerate and sandstone that yielded a well-preserved mandible of Theropithecus jonathani. At MNK (loc. 88) is a 4-meter thickness, the base of which lies about 7 m above the riverbed and 26 m below the plain. The cream-colored marker tuff is in the middle of this sequence.

East of the Second Fault are excellent exposures of a tuffaceous alluvial sequence 6 to 14 m thick (Figs. 52, 53). The base of the sequence drops from about 62 m above the riverbed near the Second Fault to 22 m above the riverbed

near the mouth of the gorge. The marker tuff lies in the upper half of this sequence (locs. 4b, 4c, 26). Except in the axis of the syncline, these sediments are confined within a pre-existing valley, or protogorge. This sequence comprises sandstone, conglomerate, claystone, tuff, and algal limestone. The sequence is highly variable, but in general the percentage of tuff increases eastward, and tuffs and tuffaceous sandstones form between 75 and 95 percent of the sediments near the mouth of the gorge. Most of the tuffs between the First and Second faults are gray or brown and contain biotite and melilite. However, at locality 3a they are dominantly cream-colored and phonolitic. Both the lateral difference in tephra composition and an angular discordance within the sequence at locality 3a suggest that the alluvial sequence did not accumulate continuously, but included episodes of cut and fill.

The marker tuff is inclined 6 degrees to the east at locality 4c, where it is delicately laminated, rippled, and was almost certainly depos-

147

Table 26. *Chemical Analyses of Zeolite-Bearing Tuffaceous Rocks, Authigenic Mica, and Chabazite from Olduvai Gorge*

	1	2	3	4	5	6	7	8	9
SiO_2	40.71	46.04	36.32	46.03	34.79	48.68	43.89	46.5	51.81
TiO_2	2.00	1.02	2.15	0.88	2.09	0.32	2.85	2.5	0.05
Al_2O_3	14.72	19.50	13.56	17.20	13.58	16.16	11.97	7.8	17.16
Fe_2O_3	7.37	2.27	7.05	3.16	6.86	5.66	13.54	17.2	0.23
FeO	1.65	0.76	1.92	2.21	2.21	0.29	0.25		0.02
MnO	0.20	0.16	0.21	0.12	0.20	0.17	0.49	0.4	0.27
MgO	1.92	0.54	3.06	1.39	3.01	1.03	3.78	4.1	0.16
CaO	8.66	2.79	12.06	3.56	14.01	0.54	1.73	1.9	0.52
Na_2O	5.58	6.43	4.82	7.61	5.84	7.40	1.62	2.3	8.84
K_2O	3.51	4.24	3.83	4.65	3.08	5.87	6.53	6.4	1.37
H_2O^+	7.42	8.66	6.04	3.80	7.75	8.26	5.65⎫	10.9	13.03
H_2O^-	4.43	5.98	4.46	6.20	2.44	4.93	6.05⎭		6.19
P_2O_5	0.20	0.33	0.57	0.22	0.38	0.22	0.56		0.03
CO_2	1.59	0.88	3.99	3.15	3.65	0.37	1.05		
Total	99.96	99.60	100.04	100.18	99.89	99.90	99.96	100.0	99.68

1. Clayey eolian tuff of Naisiusiu Beds at the type locality. Authigenic minerals are chabazite (23%), phillipsite (8%), dawsonite (<1%), and calcite. Univ. Calif. Cat. No. 478-34.

2. Phonolitic marker tuff of Naisiusiu Beds from between locs. 4a and 4b. Tuff is almost entirely glass altered to phillipsite; minor authigenic calcite is present. U.C. Cat. no. 478-33.

3. Eolian tuff of upper unit of Ndutu Beds from north side of gorge to north of FLK. Authigenic minerals are phillipsite (36%), natrolite (6%), analcime (1-2%?), dawsonite (≤1%), and calcite. U.C. Cat. no. 478-38.

4. Phonolitic marker tuff from the upper unit of the Ndutu Beds at loc. 4. Tuff is about 75% vitric materials altered to phillipsite and analcime, which form 60% and 10% respectively of the tuff. Minor calcite is present. U.C. Cat. no. 478-39.

5. Tuffaceous claystone of upper unit of the Ndutu Beds at loc. 49b. Authigenic minerals are natrolite (29%), chabazite (8%), analcime (4%), calcite, and dolomite. U.C. Cat. no. 478-40.

6. Trachyte marker tuff of the lower unit of the Ndutu Beds at MNK (loc. 88). Tuff is chiefly vitric material altered to phillipsite (60%) and analcime (10%). Minor authigenic clay and illite are present. U.C. Cat. no. 478-42.

7. Vitric tuff probably of original melilitite composition from the lower unit of the Masek Beds at HEB (loc. 44b). Vitric material is altered to "illite" cemented by phillipsite (20%). U.C. Cat. no. 478-43.

8. Clay-mineral fraction <2μm from sample no. 7. All iron is calculated as Fe_2O_3.

9. Chabazite of authigenic origin from claystone of Bed II 3 m above the Lemuta Member. Chabazite occurs as small paper-thin veinlets. The composition of the unit cell as based on 72 oxygen atoms is as follows:

Si: 25.84, Al: 10.09, Fe: 0.12, Mg: 0.12, Ca: 0.28, Na: 8.54, K: 0.87, and H_2O: 31.97 (R. A. sheppard, 1974, personal communication).

NOTE: Analyses 1-7 are by H. Onuki. Analysis 8 is by R. N. Jack using x-ray fluorescence. Analysis 9, by E. E. Engleman, was provided by R. A. Sheppard (U.S. Geological Survey).

ited in ponded water. This site lies on the steeper, western flank of the syncline, and the 6-degree dip is clearly a result of folding.

Another, younger unit of conglomerates and sandstones with algal limestone lies topographically lower in the sides of the gorge between the First and Second faults. Near the mouth of the gorge, they are found between 4.6 and 9 m above the riverbed, and probably correlative conglomerate near the Second Fault lies 11.5 m above the riverbed. In conglomerates near the mouth of the gorge are blocks of tuff from the alluvial sequence on the slopes above, proving that the conglomerates are younger than the alluvial sequence.

Upper Unit of the Ndutu Beds

The upper unit along the rim of the gorge is a single massive horizontal bed of eolian tuff. The tuff is 30 cm to 2 m thick to the west of the Second Fault, whereas it is 3 to 4.5 m thick in the syncline to the east. In many places the horizontal bed at the rim of the gorge slopes abruptly downward as a mantle over the slopes of the ancestral gorge. Eolian tuffs of the slopes are commonly between 3 and 7.5 m thick. The valley-slope deposits interfinger in the bottom of the gorge with dominantly waterlaid deposits as much as 24 m thick (Fig. 54). A longitudinal profile of the waterlaid deposits shows that they

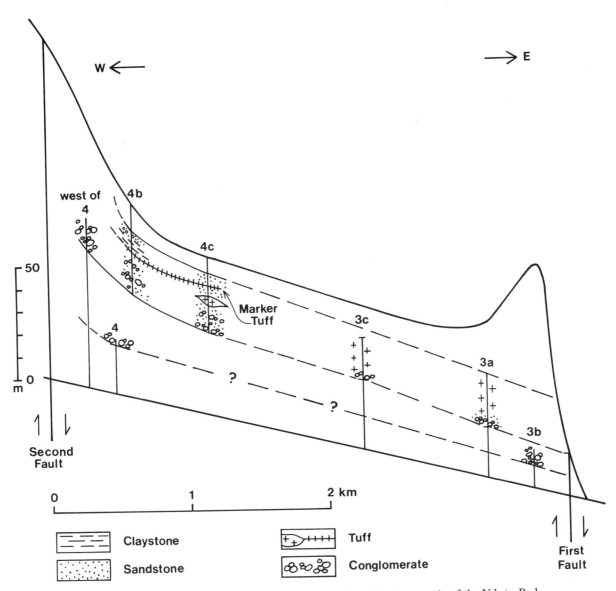

Figure 53. Longitudinal profile showing deposits of the lower unit of the Ndutu Beds in the Main Gorge between the First and Second faults.

have about the same overall gradient as the present riverbed (Fig. 55). The valley-bottom sequence usually comprises a basal bed of conglomerate and sandstone 60 cm to 3 m thick overlain by water-worked tuffs. Colluvial deposits with detritus from the Olduvai Beds are found on both the bottom and sides of the ancestral gorge.

Tuffs of the upper unit are dominantly of nephelinite composition and contain much nepheline and sodic pyroxene and smaller amounts of sphene, ilmenite, melanite, perov-

skite, biotite, and melilite. Biotite is extremely abundant in some beds, and individual crystals are as much as 2 cm in diameter. A biotite-rich bed is widespread near the top of the upper unit in the valley-bottom sequence to the east of the Second Fault (for example, Fig. 52). Phonolite and melilitite ejecta are not rare, and a melilite-rich tuff is the topmost bed of the upper unit in a few places (for instance, loc. 9).

A yellow, medium-grained vitric tuff is the only marker bed found useful in correlating over large distances in the gorge. It ranges from

149

Figure 55. Longitudinal profile of the Main Gorge showing water-worked deposits of the upper unit of the Ndutu Beds.

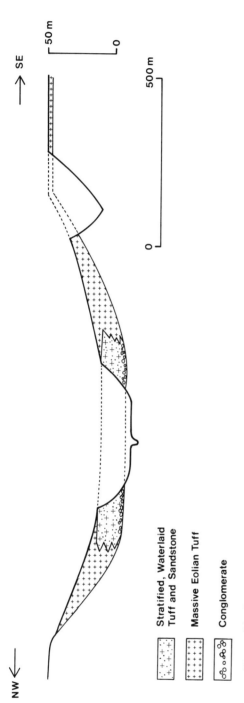

Figure 54. Cross-section of the Main Gorge showing the upper unit of the Ndutu Beds between the east edge of Long K (near loc. 37) and a point on the north side of the gorge between localities 19 and 20. Dotted lines show the inferred extent of the upper unit prior to erosion.

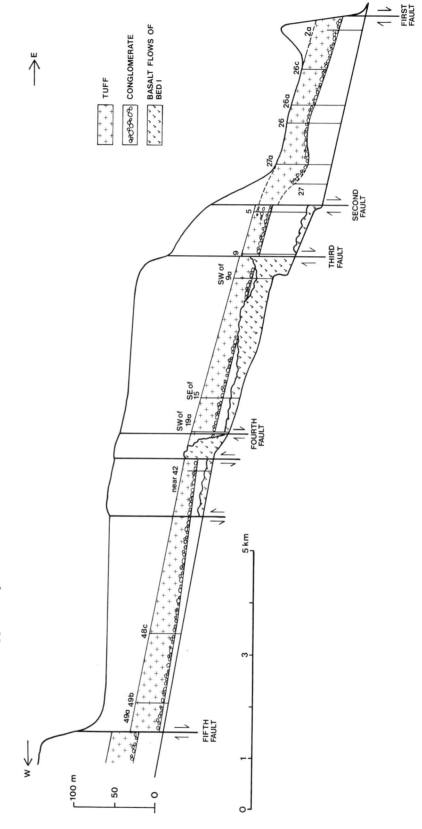

150

30 cm to 2.4 m in thickness and is commonly the basal bed of the valley-slope deposits; in the valley-bottom section it lies near the base of the tuffs (Fig. 52). This tuff is 75 to 90 percent altered glass and rock fragments, 2 to 5 percent nepheline, and less than a percent of biotite. The pumice contains crystals of nepheline, anorthoclase, and sphene. A phonolite composition is indicated both by the mineral content and the chemical composition (Table 26, no. 4).

The upper unit is faulted in many places. The largest offset in the gorge is along the First Fault, where the displacement is at least 18 m, and very probably 23 m or more. Waterlaid deposits are offset 13.4 m by the Fifth Fault on the south side of the gorge (loc. 49a). Most or all of this displacement along the Fifth Fault took place while the upper unit was being deposited, for the uppermost 8 to 10 m of Ndutu tuffs, some with steep initial dips, are banked up against the fault scarp. Valley-slope tuffs overlapping the Third Fault are offset about 1 m. East of the Second Fault, the upper unit appears to be thickest in the synclinal axis (Fig. 55), suggesting that folding continued during deposition of the upper member.

Criteria for Distinguishing Upper and Lower Units

The upper and lower units of the Ndutu Beds can be distinguished on the basis of several criteria. Waterlaid deposits of the lower unit are at various levels in the gorge, whereas waterlaid deposits of the upper unit are confined to a particular interval in any one segment of the gorge (Fig. 55). Sandstones of the lower unit generally are better cemented than those of the upper unit, and sandstones of the lower unit are commonly brown or pale reddish-brown, whereas those of the upper unit are almost invariably gray. Conglomerates in the lower unit are commonly much coarser than those of the upper unit, and they generally have a much higher proportion of clasts from the Olduvai Beds. Most conglomerates of the upper unit contain a high proportion of quartz and quartzite pebbles. Algal limestone is widespread in the lower unit but has not yet been identified in the upper unit.

Tuffs differ, on the average, in mineral com-

position. Melilitite tuffs predominate in the lower unit, whereas nephelinite tuffs predominate in the upper (Table 24). The ratio of perovskite to sphene is generally higher in the lower unit than it is in the upper (Fig. 56). Water-worked tuffs of the lower unit are generally better sorted and contain a larger percentage of quartz and feldspar than water-worked tuffs of the upper unit. Authigenic natrolite and dawsonite are more common in the upper unit than in the lower (see Table 25, Chap. 7). On the rim of the gorge, eolian tuffs of the lower unit generally contain much more clay and are more highly burrowed and rootmarked than those of the upper unit. Finally, the two units differ in their alteration. Quartz and nepheline are generally more altered in the lower unit (Table 27), and shells of the land snail *Limicolaria* are white and partly decomposed in the lower unit, whereas they generally appear fresh and unaltered in the upper unit. Notwithstanding these criteria, some individual exposures are difficult to classify accurately. Probably the greatest difficulties were encountered in distinguishing eolian tuffs of the two units along the rim of the gorge.

Ndutu Beds outside Olduvai Gorge

Eolian tuffs of the Ndutu Beds blanket the Serengeti Plain for a considerable distance to the north and west of the gorge and are exposed discontinuously over a sizable area to the east and northeast. From Pickering's mapping and my own scattered observations, it seems likely that eolian tuffs of the Ndutu Beds and associated calcretes originally covered an area of at least 15,000 km^2.

Both units of the Ndutu Beds are found along the southwestern margin of Olbalbal, where they form the uppermost deposits on the easternmost fault scarp. This fault is a major branch of the First Fault, and it extends southward about 7 km from near the mouth of the gorge. The lower unit is principally sandy reworked tuff and conglomerate, The best-exposed and thickest section (4.3 m) is at locality 158, where it overlies eolian tuffs of the Masek Beds. The upper unit, 1.2 to 2.4 m thick, is eolian tuff, some of which exhibits the large-scale cross-bedding of sand dunes.

Both the upper and lower units of the Ndutu

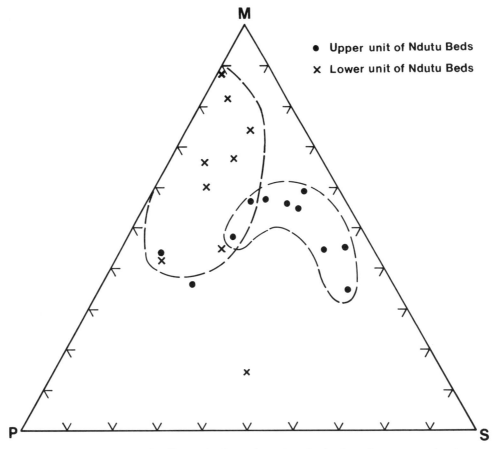

● Upper unit of Ndutu Beds
✕ Lower unit of Ndutu Beds

Figure 56. Compositions of tuff samples from the Ndutu Beds plotted in a triangular diagram whose end members are melanite (M), perovskite (P), and sphene (S).

Beds are widespread but thin along the western margin of Olbalbal at the level of the plain both north and south of the gorge. In thickness and lithology they compare closely with exposures in the rim of the gorge to the west of the Second Fault. Eolian tuffs of the upper member are widely exposed over the plain to the north of the gorge. A maximum thickness of about 6 m was noted in a small valley about 5 km north of the gorge (loc. 205). Several thin calcretes are within this sequence.

AGE OF THE NDUTU BEDS

The upper unit of the Ndutu Beds very likely covers the period from roughly 60,000 to about 32,000 yr. B.P. The lower unit presumably spans most of the interval between the upper unit and the Masek Beds, or from about 60,000 to 400,000 yr. B.P. Radiometric and geochemical methods have thus far been applied only to the

upper unit. Two samples analyzed by D. L. Thurber are beyond the limit of radiocarbon dating (in M. D. Leakey et al., 1972). One of these was calcrete from the base of the upper

Table 27. *Solution of Quartz in Eolian Tuffs of Olduvai Gorge*

Stratigraphic unit	Most common degree of solution[a]	Maximum solution
Namorod Ash	None visible	None visible
Naisiusiu Beds	Frosted	Slightly etched with some etch pits 2-3 μm deep
Ndutu Beds, upper unit	Etched, frosted (av. 1-3 μm?)	10 μm
Ndutu Beds, lower unit	3-10 μm	30 μm
Masek Beds, lower unit	Variable; generally 10-15 μm?	30 μm

[a]Measurements refer to the average thickness of quartz dissolved from a grain in which the original outline is preserved by zeolite, clay, or calcite.

Table 28. *Solution of Nepheline in Upper Pleistocene Eolian Tuffs of Olduvai Gorge*

Stratigraphic unit	Nature of samples	Solution in least-altered grains	Solution in most-altered grains
Namorod Ash	All samples	0	Slight etching; maximum of 5 μm?
Naisiusiu Beds	Less-altered 7 samples	0-5 μm	Gen. 5-15 μm (av. ~10 μm); max. ~20 μm
	More-altered 5 samples	5-10 μm	Gen. 20-30 μm; (av. ~25 μm); max. ~40 μm
Upper unit of Ndutu Beds	Less-altered 4 samples	10-20 μm	Gen. 30-70 μm; (av. ~45 μm); max. ~80 μm
	More-altered 8 samples	50-60 μm	$\geqslant$75-100 μm

unit at locality 6 ($\geqslant$30,000 yr. B.P.); the other was a collection of *Limicolaria* shells from the upper part of the upper unit to the east of the Second Fault ($\geqslant$29,000 yr. B.P.). Another date, of 3,340 ± 800 yr. B.P., was obtained from the collagen of bones associated with artifacts near the top of the upper unit at locality 26 (Berger and Libby, in press). This date is clearly too young to be accepted as valid and very likely signifies contamination by modern C-14, especially radiocarbon from weapons-testing deposited via ground water into the sample (R. Berger, 1974, personal communication). It should be noted that the bones were collected from the uppermost 15 cm of sediment beneath the ground surface, and the dated sample of collagen was very small. Both factors enhance the possibility of serious contamination.

Dates of 32,000 and 56,000 yr. B.P. have been obtained by J. L. Bada from bones of the upper unit using aspartic acid measurements. One bone was obtained from the lower part of the upper unit along the rim of the gorge at FLK. The upper unit is represented here by a 1.5-meter thickness of eolian tuff. This sample gave a date of 56,000 ± 3,000 yr. B.P. (Bada and Protsch, 1973). The other bone sample was obtained from a site with artifacts in eolian tuffs near the top of the Ndutu Beds at locality 26, where the upper unit is as much as 13 m thick (Fig. 52). This sample gave a date of 32,000 yr. B.P. (J. L. Bada, 1974, personal communication). The racemization rate used for these dates was calibrated from bones of the Naisiusiu Beds yielding a C-14 date of 17,550 ± 1,000 yr. B.P.

The alteration of nepheline was earlier used to obtain an age for the upper unit (M. D. Leakey et al., 1972). This estimate was based on the difference in degree of nepheline alteration be-

tween eolian tuffs of the Naisiusiu Beds and the upper unit of the Ndutu Beds. Twelve samples were taken from both the Naisiusiu Beds and the upper unit of the Ndutu Beds and were divided into two groups differing in the average amount of nepheline dissolved. The Naisiusiu samples could be readily subdivided into a less altered group of seven and a more altered group of five; the Ndutu samples comprised a less altered group of four and a more altered group of eight. The amount of solution was then measured on the least and most altered grains of each sample. These measurements require both the original outline of the nepheline grain and relict nepheline in the center of the cavity or pseudomorph. Because many of the larger nepheline grains are entirely altered in most of the Ndutu samples, there is no way to determine the maximum amount that would have been dissolved if larger grains had been present. Exact comparison is impossible, as the average thickness of dissolved nepheline varies not only within the same stratigraphic units but within the same sample. Comparative data are probably best for the most altered grains in the less altered samples and the least altered grains in the more altered samples (Table 28). Both comparisons suggest that the thickness of nepheline dissolved in the upper unit is roughly five times that dissolved in the Naisiusiu Beds. A radiocarbon date of 10,400 yr. B.P. on the Naisiusiu Beds was used to calibrate the alteration rate. Assuming a constant rate of solution, the upper unit of the Ndutu Beds was calculated to have an average age of 50,000 ± 20,000 years. The date of 10,400 yr. B.P. is now suspect, as will be discussed later, and an age of 15,000 to 17,000 yr. B.P. seems more likely for eolian tuffs of the Naisiusiu Beds. Using these figures, the upper unit should average 75,000 to

87,500 yr. B.P., which is clearly too old if the aspartic-acid dates are correct. The dates based on nepheline alteration are the less reliable, inasmuch as the rate of alteration is highly sensitive to the chemical environment. Assuming the aspartic-acid dates to be correct, nepheline was altered more rapidly prior to 17,500 yr. B.P. than over the past 17,500 years. Differences of salinity and pH in the soil environment can account for differences of this magnitude, with the higher rates reflecting a higher pH and salinity. Because nepheline varies in chemical composition, one must also consider the effect of compositional differences on rate of solution. In the case of plagioclase feldspar, for example, the rate of alteration is highly sensitive to the content of calcium, with anorthite ($CaAl_2Si_2O_8$) the most rapidly altered. To check this possibility, J. Stormer (University of California, Berkeley) chemically analyzed with the electron microprobe twenty-two nepheline crystals altered to varying degrees. The nephelines were mostly from the Ndutu Beds but included some from the Masek and Naisiusiu beds. The nepheline crystals vary considerably in composition, but they exhibit no correlation between composition and degree of alteration.

Dating of the lower unit is rather indirect and unsatisfactory. Macintyre et al. (1974) have suggested that an episode of faulting to the east of the main Manyara-Natron fault was contemporaneous with the eruption of Swallow Crater and Kisetey (~370,000 yr. B.P.). This faulting, although relatively minor compared to that 1.1 to 1.2 m.y.a., is possibly related to and contemporaneous with displacements at Olduvai which resulted in subsidence of the Olbalbal graben, erosion of the gorge, and deposition of the lower unit of the Ndutu Beds. Using this line of evidence, the oldest sediments of the lower unit could be as much as 370,000 years old. The oldest of the known tuffs were deposited when the gorge had been cut only 2 to 3 m below the top of the Norkilili Member. These tuffs are mineralogically rather similar to tuffs of the Norkilili Member and may represent the terminal eruptions of Kerimasi roughly 400,000 years ago. Faunal remains of the lower unit include middle Pleistocene forms (M. D. Leakey, in press, 2), which fits with time span suggested by the lines of evidence given here.

Measurements of the amount of dissolved quartz can be used to get a rough (minimum?) estimate of the age of the lower unit in the rim of the gorge. The solution of quartz grains in the lower unit is compared to that of the upper unit, whose age is known. This comparison is restricted to eolian tuffs on the rim of the gorge, where deposits of the two units are generally rather similar, and where the topographic setting and chemical environment would be expected to remain most nearly the same. Unfortunately the amount of solution is difficult to measure accurately in the upper unit. Nevertheless it is clear that quartz has dissolved to a greater extent in the lower unit than in the upper (Table 27). In most stratigraphic sections the degree of solution decreases abruptly in going from the lower to the upper unit. This fact points to a hiatus between the two units. Using either the most common or the maximum solution measured, deposits of the lower unit appear to have an average age about three times that of the upper, or on the order of 150,000 years. This figure has a large degree of uncertainty, in view of the wide range in values within both the upper and lower units.

LITHOFACIES AND ENVIRONMENTS

The lower unit of the Ndutu Beds comprises eolian and fluvial deposits. Eolian deposits are found on the rim of the gorge, and they presumably extend widely over the plain. Some are also found within the gorge. Fluvial sediments are chiefly confined to the gorge and its tributaries and to Olbalbal. The upper unit comprises eolian, fluvial, and colluvial deposits. Eolian tuffs are found in the rim and on the sides and bottom of the ancestral gorge. Fluvial sediments are confined to the valley bottom, and colluvial deposits are both on the sides and in the bottom of the valley. Originally the largest volume of sediment was probably in the fluvial deposits, but selectively greater erosion of the valley-bottom sediments has reduced their volume to a quarter or less of the present exposures.

Lower Unit of the Ndutu Beds

Eolian deposits of the lower unit consist of eolian tuffs and associated calcretes and rubble layers. Calcretes and rubble deposits are found only along the rim of the gorge in the thicker exposures. Eolian tuffs are widespread along the rim of the gorge and are interbedded with fluvial deposits within the gorge and are most common in the alluvial sequence to the east of the Second Fault. Eolian tuffs of the rim are rootmarked, massive, and yellowish-brown or yellowish-gray. Most beds are poorly sorted and have a high content of claystone pellets and pelletoid coatings. Eolian tuffs within the gorge are massive or slightly stratified and are gray, brown, or pale reddish-brown. They are typically rather well sorted and have been redeposited to some extent by running water.

Fluvial deposits within the gorge are sandstone and conglomerate with smaller amounts of tuff, claystone, and algal limestone. Conglomerates of the Main Gorge contain clasts from the walls of the gorge together with older detritus. Pebbles of quartz and quartzite, derived from the west, are common in many conglomerates and are the dominant clasts in a very few. All of the Olduvai Beds up to and including the lower unit of the Ndutu Beds are represented in conglomerate clasts, which are as much as 1 m long in the coarser conglomerates. Particularly common as clasts are the more resistant rock types such as basalt of Bed I, zeolitic tuff of Beds I and II, and tuff of the Norkilili Member. Some blocks of Norkilili tuff are densely cemented with calcite and coated on one side with laminated calcrete, showing that the cement and calcrete were present before the blocks were eroded.

The sandstones are a varied mixture of materials from the Olduvai Beds together with older detritus. Many if not most of the sandstones also contain volcanic ash. Fluvial tuffs are generally much reworked and highly contaminated with older detritus. An exception is the cream-colored marker tuff, which is reworked only slightly and was locally deposited in ponded water. Claystones are found principally in the sequence of sediments conformably overlying the marker tuff (for example, locs. 4b, 25b, 26, 88). South of Kelogi, claystone is interbedded at varying levels with sandstone and conglomerate.

Algal limestone is found principally as a coating 1 to 4 cm thick on clasts in the coarser conglomerates. West of the Second Fault, it is restricted to conglomerates lower than midway in the sides of the gorge. To the east, it is both in the alluvial sequence near the rim of the gorge and in the conglomerates at lower elevations. The limestone is somewhat porous, laminated, and can be botryoidal.

The lower unit in the southwestern part of Olbalbal is dominantly fluvial deposits but includes eolian tuffs. The fluvial deposits are chiefly conglomerate and tuffaceous sandstone. The conglomerates can be coarse, and they consist mostly of lava, limestone, and clasts of eolian tuff from the lower unit. The lava is entirely from Lemagrut except for a very few pebbles of green porphyritic lava from Sadiman or the Laetolil Beds.

Both eolian and fluvial deposits are altered, and authigenic minerals are in nearly all beds. Zeolites and calcite typically cement the sandstones and tuffs (Table 25), and analcime is in the claystones. Dawsonite is in a few of the tuffs.

Rare shells of *Limicolaria*(?) and isolated bones are the only faunal remains noted in the eolian tuffs. Rootmarkings in the tuffs are suggestive of grass. Vertebrate remains also have been found in sandstones and conglomerates. Some of these are probably from older beds, but others represent the contemporaneous fauna. Remains of catfish are locally common (for instance, loc. 26b), and the *Theropithecus* mandible, though delicate, is not abraded, indicating that it belonged to the contemporary fauna.

Upper Unit of the Ndutu Beds

The eolian deposits are tuffs, the lithology of which differs in the different topographic settings. Eolian tuffs of the plain and rim of the gorge are massive, rootmarked, and pale yellowish-brown. They can be somewhat clayey and are contaminated to a variable extent with older detritus. Calcareous horizons represent cemented, weakly developed paleosols, and minor erosion surfaces are in some of the thicker exposures. Eolian tuffs mantling the sides of the gorge can either be massive or highly cross-bedded, and they lack calcareous horizons. Where present, the original upper surface of the tuffs mantling the valley sides slopes smoothly downward toward the axis of the gorge, generally at an angle of 4 to 8 degrees (Fig. 54). The base of the valley-slope deposits is irregular, and eolian tuffs

155

may be banked against overhanging cliffs. Eolian tuffs of the valley bottom can be massive, evenly stratified, or exhibit large-scale eolian cross-bedding. They intergrade with tuffs classed as fluvial, and the two categories are arbitrarily subdivided on the basis of whether the primary texture and structure of the eolian sediment has been modified to a greater or lesser extent by running water.

Fluvial deposits comprise tuff, sandstone, claystone, and conglomerate. A basal conglomerate is widespread, and lenticular conglomerates are at higher levels. In most places the basal conglomerate contains abundant pebbles of quartzite that are subrounded to well rounded. Clasts are also of materials derived from the Olduvai Beds, including eolian tuffs of the upper unit. Locally the conglomerates consist wholly of detritus from the Olduvai Beds (for example, near locs. 27, 42). Boulders and large blocks are present locally, as for example adjacent to overhanging cliffs (loc. 27).

Sandstones are found with the basal conglomerate and are composed of varying mixtures of older detritus with contemporaneous tephra. Claystones were found only in the lower part of the fluvial sequence on the north side of the gorge (locs. 49a-49c). Claystone beds are generally thin and are interbedded with tuff or sandstone. Claystone pellets are abundant in some of the claystone layers.

Fluvial tuffs are generally rather well sorted and are either thinly bedded or delicately cross-bedded. Augite is characteristically concentrated in thin laminae. Older detritus is commonly present but varies greatly in amount. The tephra in all these tuffs is the same as that in the eolian tuffs except for its higher degree of sorting. A small proportion of tuffs in the fluvial facies is thinly laminated, poorly sorted, and was very likely deposited in ponded water. The biotite-rich tuff to the east of the Second Fault appears to represent a primary shower of ash into water.

Colluvial deposits are found principally within a 2-kilometer segment of the gorge to the west of the Third Fault. These exposures lie both on the slopes and bottom of the pre-existing valley. The most common kind of colluvial deposit is a poorly sorted brown clayey tuff that contains as much as 50 percent of detritus from the sides of the gorge above the level of the exposures. Initial dips are 5 degrees in the deposit at DK

(loc. 13). These sediments resulted from the mixture of wind-borne tephra with older detritus by sheetwash on the slopes of the valley. A conspicuous reddish-brown deposit of clayey colluvium 9 m thick is exposed just to the west of the Third Fault on the south side of the gorge. This deposit has steep initial dips and represents a clayey talus fan of detritus from Beds II and III. Large slumped blocks, generally embedded in eolian tuff, are another type of colluvial deposit.

Tuffaceous sediments are altered and cemented by zeolites, calcite, and so forth (Table 25), and dawsonite is common in at least half of the eolian tuffs. Fluvial tuffs are altered to about the same extent as eolian tuffs, and both of them contain the same wide variety of authigenic minerals. Chemical analysis of altered tuff and tuffaceous claystone are given in Table 26 (nos. 3-5). Claystones are rich in zeolites, the most common of which is analcime.

Isolated vertebrate remains are in both eolian tuffs and fluvial deposits, but they have not been studied from an ecologic standpoint. Shells of land snails, found principally in eolian tuffs, are the most common fossils in the upper unit. A small collection, identified by A. Gautier and T. Pain, contained two species of *Limicolaria, L. caillaudi* and *L. martensiana* (A. Gautier, 1972, personal communication).

ENVIRONMENTAL SYNTHESIS AND GEOLOGIC HISTORY

The Ndutu Beds were deposited during a period of extensive faulting which resulted in subsidence of the Olbalbal graben to form the present drainage sump. Major displacements were along the First, Second, and Fifth faults, and the syncline between the First and Second faults was formed largely at this time. Evidence for the time of folding is found on the north flank of the syncline, where the waterlaid marker tuff of the lower unit dips 6 degrees east, and a tuff of the Masek Beds dips 10 degrees east, showing that most of the folding took place following deposition of the marker tuff in the Ndutu Beds. The upper unit of the Ndutu Beds also appears to be gently folded here, and the major folding may have accompanied the latest phase of subsidence of the Olbalbal graben, which resulted in erosion of the upper unit of the Ndutu Beds. The complex geologic history of the Ndutu Beds is summarized as follows:

1. The Olbalbal graben subsided intermittently following deposition of the Masek Beds and resulted in stages of erosion and partial filling of the gorge by sediments of the lower unit of the Ndutu Beds. The alluvial sequence that includes the marker tuff represents a period when sediment accumulated widely in the gorge. To the west of the Second Fault, the northern margin of the early-stage valley coincided rather closely with the present margin of the gorge, as shown by patches of sediment in the sides of the gorge only slightly below the rim. The southern margin of the early-stage valley has not been located but may have been considerably north of the present margin on the assumption the valley became wider as it was eroded deeper. To the east of the Second Fault, the alluvial sequence filled a protogorge whose margins corresponded more or less to those of the present gorge (Fig. 57). The stream gradient varied considerably at different times, as inferred from the range in sediment type from coarse conglomerate to claystone at different levels. The main stream flowed more continuously than at the present time, as evidenced by catfish remains and widespread algal limestones, neither of which is found presently.

Ash showers fell at intervals on the Serengeti Plain, and the ash was extensively redeposited by wind. Some of it was transported into the gorge, and the remainder accumulated on the plain to form eolian tuffs. The earliest tuffs, lacking in biotite, are rather similar in composition to tuffs of the Norkilili Member, and they may also have come from Kerimasi. The later, biotitic tuffs are from Oldoinyo Lengai, with or without contributions from the smaller vents in the vicinity of Kerimasi (for example, Swallow Crater). The source of the trachytic marker tuff is unknown.

2. Shortly after the marker tuff of the lower unit was deposited, the graben subsided further, resulting in erosion of a narrow gorge to very near the level of the present gorge. Conglomerates, sandstones, and tuffs of the lower unit were deposited in this protogorge, and to the east of the Second Fault they accumulated almost to the base of the alluvial sequence.

3. The upper unit of the Ndutu Beds was deposited after the gorge had been filled to a considerable depth by alluvium, and the valley bottom was relatively wide. The valley at this stage has been termed mature by L. S. B. Leakey (1965), but this is an oversimplification, as the valley sides are in some places smooth and gently sloping, but are elsewhere steep and rugged. Overhanging cliffs are preserved beneath eolian tuffs of the upper unit in at least several places. Initially, quartzose sands and gravels were the principal sediments deposited in this broad valley. Shortly thereafter, most of the sediment was volcanic ash transported at least partly in the form of dunes by winds from the northeast. In view of the large volume of volcanic ash deposited in a short time, the river was very likely periodically choked with sediment, and its channel was braided. The lack of algal limestone and fish remains may reflect either the high content of tephra in the stream or a decrease in stream flow. The uneven valley sides were mantled to a considerable depth with wind-borne tephra, resulting in smooth and gentle slopes (Fig. 54). The volcanic ash of the upper unit was erupted from Oldoinyo Lengai, and its large volume shows that it represents a major eruptive episode.

4. A final, severe phase of faulting resulted in folding of the syncline to the east of the Second Fault and in subsidence of the Olbalbal graben by at least 23 m.

The climate was semiarid and perhaps like the present while the Ndutu Beds were deposited. Ubiquitous evidence for a relatively dry climate is in the zeolitic eolian tuffs and calcretes, which are found both within and at the base of the lower unit and at the base of the upper unit. Calcareous horizons within the upper unit represent incompletely developed calcretes. Regarding the *Limicolaria* species of the upper unit, Gautier states that, "As it stands, the fossils of the [upper unit of the] Ndutu Beds indicate no substantial differences in biotope and climate between fossil and present-day environment" (1972, personal communication).

HOMINID REMAINS AND ARCHEOLOGIC MATERIALS

The only hominid fossil referable to the Ndutu Beds is H-11, a palate and maxillary arch of *Homo* sp. (M. D. Leakey, 1971a). It was found

157

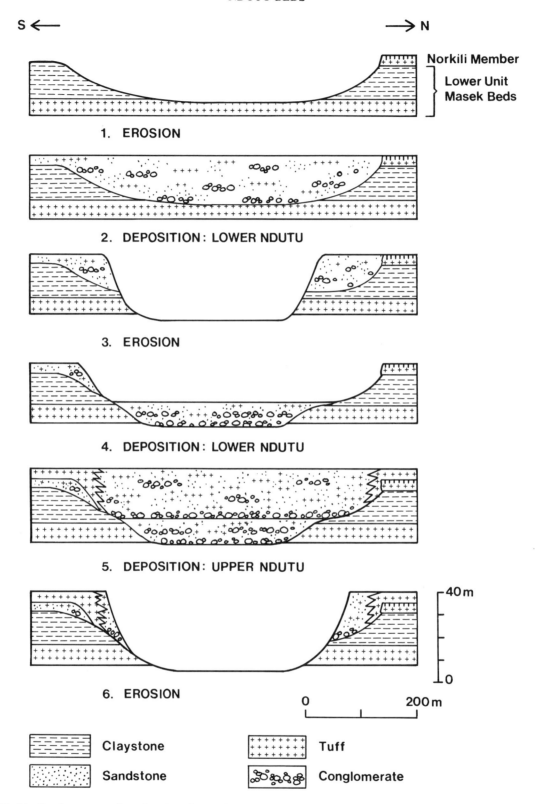

Figure 57. Idealized cross-section showing the evolution of the Main Gorge between the First and Second faults from the end of Masek deposition to the beginning of Naisiusiu deposition. Section is based principally on exposures at localities 4d and 26.

on the surface of sediments of the lower unit about 10 m below the rim of the gorge slightly to the west of DK (loc. 13).

Archeologic sites were found in two places, both of which were situated close to the main stream below the Second Fault. One site is a small concentration of artifacts in claystone about 2 m below the top of the lower unit in locality 4b; the other comprises two small concentrations of artifacts and bones in eolian tuff near the top of the upper unit at locality 26 (Fig. 52). Artifacts from the two sites are archeologically indistinguishable, and the assemblage "appears to represent a Middle Stone Age indus-try of somewhat indeterminate character" (M. D. Leakey et al., 1972, p. 332).

Isolated artifacts of Middle Stone Age affinity occur widely in eolian tuffs of the upper unit. They are most common in valley-bottom deposits of the Main Gorge (for example, locs. 49b, 49c) or in sediments deposited adjacent to the main stream east of the Second Fault. Judging by their widespread occurrence, the total number of isolated artifacts in the upper unit must be quite large. A few artifacts of the same type were also noted in the conglomerates, probably of the lower unit, that are exposed 8 to 10 km south of Kelogi.

9

NAISIUSIU BEDS

STRATIGRAPHY AND DISTRIBUTION

The Naisiusiu Beds are chiefly eolian tuff that occurs as a thin layer over the Serengeti Plain and is represented in the gorge and its tributary gullies by remnants of a thicker blanket of sediment deposited after the Ndutu Beds had been largely stripped away and the gorge had been eroded to nearly its present level. In the rim of the gorge the Naisiusiu Beds are a discontinuous deposit most commonly between 30 cm and 1 m, and at most 3 m thick. They are generally underlain and overlain by laminated calcrete, and the uppermost calcrete is either at the surface or covered by unconsolidated wind-worked volcanic ash. A calcrete or thin calcareous horizon is typically found near the middle of the Naisiusiu Beds in the thicker exposures. Erosional remnants of the Naisiusiu Beds cover probably 5 to 10 percent of the sides of the gorge. Most exposures are between 50 cm and 2.5 m thick, but thicknesses of 4.5 m are not rare, and the thickest section is 10 m (loc. 4d). From the field relationships and thickness, it seems likely that the Naisiusiu Beds originally mantled half or less of the valley slopes. The upper 15 to 30 cm of the valley-slope deposits are commonly hard, calcareous, and thinly coated with calcrete. A yellow marker tuff is widespread either within or at the base of the valley-slope deposits. The Naisiusiu Beds in the bottom of the gorge commonly have a basal bed of conglomerate and sandstone overlain by eolian tuff. The basal conglomerate is as much as 2.5 m thick, and its base ranges from the level of the present riverbed to as much as 5.5 m above it. The conglomerate lies highest either where the riverbed is cut into basalt, or just upstream (for example, loc. 41). The gorge in these areas of resistant basalt generally has been eroded on the order of 4 m since the Naisiusiu conglomerates were deposited. To the east of the Second Fault, eolian tuffs of the Naisiusiu Beds are exposed in the present riverbed, showing that the gorge was at least slightly lower than at present when the tuffs were being deposited.

The type section is taken at a hillock of Naisiusiu Beds that lies 110 m west of the Second Fault on the north side of the gorge (Fig. 58). The section is 7.3 m thick and comprises a basal unit 2.2 m thick of sands and gravels, a middle unit 1.5 to 1.8 m thick of poorly sorted tuffaceous sandstones, and an upper unit 3.1 to 3.3 m thick of eolian tuffs (M. D. Leakey et al., 1972). The yellow marker tuff is near the top of the tuffaceous sands, about a meter above the excavated concentration of bones and artifacts.

The eolian tuffs consist of nepheline phonolite tephra contaminated by claystone pellets and other detritus. Cream-colored phonolitic pumice fragments are in most of the tuffs, and the dominant minerals are augite and nepheline. Accessory minerals are sphene, perovskite, melanite, biotite, and rare altered sodalite(?) (Fig. 59; see also Table 24, Chap. 7). The marker tuff is a fine- to medium-grained phonolitic pumice tuff that is yellow to pale orange (see Table 26, no. 2, Chap. 8). It is discontinuous and has an average thickness of 10 cm where it is present. Its crystals include augite, nepheline, anorthoclase, sphene, and hornblende.

Eolian tuffs of the Naisiusiu Beds differ in several respects from eolian tuffs of the upper unit of the Ndutu Beds. On average, the Naisiusiu tuffs are softer, finer-grained, and contain more clay. Nepheline and biotite are much less common, and chabazite is much more common than in the Ndutu tuffs (see Table 25, Chap. 7). Finally, primary minerals are altered less in the Naisiusiu tuffs (see Table 27, Chap. 8), and in a few places, the Naisiusiu and Ndutu beds could be distinguished only on the basis of nepheline alteration.

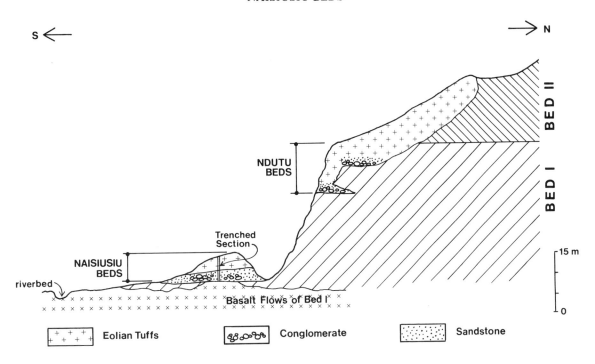

Figure 58. Cross-section showing the Naisiusiu Beds and the upper unit of the Ndutu Beds at the type locality of the Naisiusiu Beds 110 m west of the Second Fault. Vertical exaggeration is about 2x.

Eolian tuffs of the Naisiusiu Beds mantle much of the terrain adjacent to the gorge, and they thicken and coarsen, overall, to the northeast. Eolian tuffs commonly lie either at the surface or beneath unconsolidated Holocene sediments along the southern margin of Olbalbal. They extensively cover the scarp of the First Fault to the north of the gorge, and they are unfaulted where they overlie the fault. A 6-meter thickness of eolian tuffs is exposed near the top of the scarp at locality 1. Here the lower tuffs are medium- to coarse-grained and locally almost free of contaminating detritus. Sphene is relatively abundant, and the ratio of sphene to melanite is higher than in tuffs of the Ndutu Beds. Eolian tuffs are exposed widely along the east-west extension of the Fifth Fault, 5 km north of the gorge. Here they are 60 cm to 2 m thick over the plain and as much as 5 m thick in valleys cut through the fault scarp (for example, loc. 205). They overlap the foot of the hills of basement rocks 7 km farther to the north and can be traced into the basin of Lake Natron directly west of Oldoinyo Lengai.

Tuffs as much as 5 m thick are on the rim of Swallow Crater, to the east of Kerimasi and southeast of Oldoinyo Lengai. This tuff is coarser and contaminated less than the Naisiusiu tuffs in the vicinity of the gorge. It is identified as belonging to the Naisiusiu rather than the Ndutu beds on mineral composition and the degree of nepheline alteration. These tuffs are overlain and underlain by laminated calcretes, as at Olduvai. Unconsolidated, wind-worked ash overlies the upper calcrete, and below the lower calcrete is tephra erupted from Swallow Crater.

AGE OF THE NAISIUSIU BEDS

Radiocarbon dates on organic materials in the Naisiusiu Beds range from $10,400 \pm 600$ to $17,550 \pm 1,000$ yr. B.P. Dates of 17,000 and $17,500 \pm 1,000$ yr. B.P. were obtained from ostrich eggshell and the collagen of bone from below the marker tuff at the type section (Appendix A; also M. D. Leakey et al., 1972). Dates of $10,400 \pm 600$, $11,000 \pm 750$, and $15,400 \pm 1,200$ yr. B.P. have been obtained from bones collected from eolian tuffs on the northern side of the gorge where it is cut by the Fifth Fault (L. S. B. Leakey et al., 1968; Berger and Libby, in press). One of the dated collagen samples (10,400 yr. B.P.) was a composite from several bones, whereas the other two were from

161

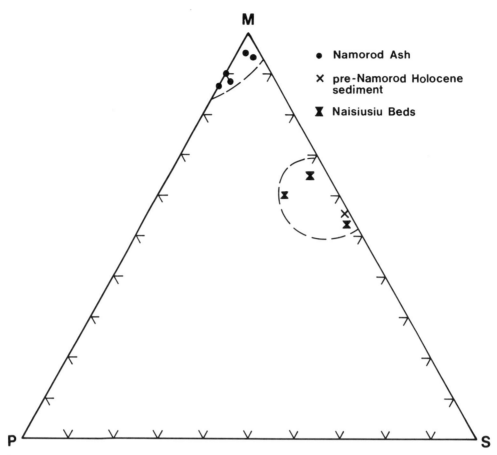

Figure 59. Composition of samples from the Namorod Ash, pre-Namorod tuffaceous sediment of Holocene age, and tuffs of the Naisiusiu Beds, plotted in a triangular diagram whose end members are melanite (M), perovskite (P), and sphene (S).

single large bones. Geologic relations suggest that the bones were deposited together over a short period of time. Aspartic acid measurements were obtained by J. L. Bada from two bones of this site, one of which yielded the C-14 date of 11,000 yr. B.P. Both measurements give dates on the order of 20,000 yr. B.P., using bones from the type section for temperature calibration (Bada, 1974, personal communication). This large discrepancy between the aspartic-acid dates and the younger of the C-14 dates cannot be explained by slightly different temperature histories of the type section and the tuffs at the Fifth Fault. Thus, either the younger (or all) of the C-14 dates from the Fifth Fault are in error, or else the geologic relations are misleading and the aspartic-acid dates are incorrect. The amount of dated collagen from each sample was very small, only 6 to 7 percent of

the amount required to fill the C-14 counter (Bada, 1974, personal communication). Consequently, slight contamination of the bones by modern C-14 could seriously affect the dates. The bones were collected no more than 60 cm below the surface and were to some extent accessible to contaminating organisms. The erroneous date of 3,340 ± 800 yr. B.P. on bones from the Ndutu Beds clearly illustrates the serious effects of contamination. Further work on bones from the Fifth Fault is needed before the C-14 dates of that site can be accepted as valid.

Laminated calcretes are widespread at the base, top, and within the Naisiusiu Beds and are potentially useful for dating the Naisiusiu Beds. The origin of the calcretes must be understood, however, before the C-14 dates can be correctly interpreted. Briefly, the calcium carbonate of the calcretes at Olduvai has been derived from "natrocarbonatite" ash, from mafic glass and melilite, and from wind-deposited particles of calcite (see Hay, 1970b). Calcium carbonate from these various sources was transported in solution downward through the permeable surface layers of sediment and deposited as calcrete over horizons of lower permeability, generally over paleosols or cemented tuff. Upward capillary transport of calcium carbonate has played either a minor role or none at all. Thus the calcretes at the base and within the Naisiusiu Beds were formed beneath eolian tuffs, and a C-14 date ideally should give the time when calcium carbonate was leached from already deposited tuff. The calcrete overlying the Naisiusiu Beds was formed, and may be presently forming, beneath the unconsolidated surface layer of sediment on the plain.

Two sources of error may affect the radiocarbon dates from the calcretes. Either young calcium carbonate might be deposited in the pore space of older calcretes to give an age too young, or older calcium carbonate might be incorporated to give an age too old. Deposition of young carbonate in older calcretes is probably negligible, for most of the calcretes are virtually impermeable. The one dated calcrete at the base of the Ndutu Beds gave a C-14 date of ≥30,000 years, showing that it did not have a measurable amount of young carbonate. Incorporation of older carbonate is potentially a more serious problem, and Williams and Polach (1971) have shown that C-14 dates from pedogenic carbonates of the arid zone in Australia are commonly 500 to 7,000 years too old.

The validity of the calcrete dates in the Olduvai region can be tested by dating calcretes in a vertical sequence and by comparing calcrete dates with the C-14 dates from organic materials. No discordant dates are among the fifteen from calcretes in the vicinity of Olduvai Gorge and Oldoinyo Lengai (Appendix A). Seven samples from the post-Naisiusiu calcrete gave dates of $2,190 \pm 105$ to $9,130 \pm 130$ yr. B.P. A calcrete from within the Naisiusiu Beds gave a date of $14,580 \pm 210$ years and that from the base gave dates of $15,850 \pm 270$ to $22,470 \pm 420$ yr. B.P. Dates of 1,300 and 2,050 yr. B.P. from calcretes near Oldoinyo Lengai are particularly significant, as they are from an alluvial fan of ejecta from the post-Naisiusiu eruptive phase of Lengai, which produced the Namorod Ash of Olduvai Gorge. These two dates are reasonable for the alluvial deposits, in view of their slight degree of alteration. The calcrete dates also compare rather well with the other C-14 dates. Three of the four calcrete samples from the base of the Naisiusiu Beds give dates older than the dated materials from below the marker tuff. The other, slightly younger calcrete date (15,850 yr. B.P.) may well reflect a real difference in the age of the lowermost Naisiusiu deposits in different places. It is also possible that the younger calcrete was formed only after a substantial part of the Naisiusiu Beds had been deposited. The date from the middle of the Naisiusiu Beds (14,580 yr. B.P.) fits reasonably well with the dates from organic materials, and the post-Naisiusiu calcrete is everywhere considerably younger than the dated organic materials except for the suspect dates of 10,400 and 11,000 yr. B.P. In conclusion, the Olduvai calcrete dates very likely err by no more than a few thousand years, and they are probably considerably more accurate than this. Age limits suggested by the calcrete dates are about 15,000 yr. B.P. or slightly less for the upper part of the Naisiusiu Beds and slightly more than 22,000 yr. B.P. for the oldest deposits. These limits are in accord with the more reliable of the C-14 dates on organic materials and are accepted as the best presently available for the Naisiusiu Beds.

LITHOFACIES AND ENVIRONMENTS

The Naisiusiu Beds consist largely of eolian tuffs, and fluvial and colluvial deposits constitute the remainder. These lithofacies are closely related to the present-day topography, and fluvial and colluvial deposits occur only in the bottom of the gorge, whereas eolian tuffs are found chiefly on the sides of the gorge and over the plain. A lacustrine or fluvial-lacustrine facies presumably lies at depth in Olbalbal. Eolian tuffs are relatively fine-grained and poorly sorted, and most of them contain at least 50 percent of clay in the form of clasts and pelletoid coatings (see Table 24, Chap. 7). They are moderately altered and cemented by zeolites and calcite (see Table 25, Chap. 7). Eolian tuffs of the valley sides have been locally redeposited by sheetwash, and they commonly contain angular clasts as much as 25 cm long of eolian tuff from the upper unit of the Ndutu Beds. Eolian tuffs of the valley bottom have been locally reworked by streams. Hillocks as much as 3 m high of eolian tuff of the Naisiusiu Beds lie on the plain above Olbalbal, about 1 km northwest of locality 1. The tuff of these hillocks is unstratified and differs from typical eolian tuffs of the Naisiusiu Beds only in a slightly coarser grain size and better degree of sorting. Bone fragments and *Limicolaria* shells occur widely in eolian tuffs but are rarely abundant. Fossil animal burrows 5 to 8 cm in diameter are widespread and locally common in the more clayey of the eolian tuffs along the rim of the gorge. Mammalian remains are concentrated in the topmost 60 cm of eolian tuff in one area along the trail at the Fifth Fault on the north side of the gorge (between locs. 82 and 83).

Tephra of the eolian tuffs originated in eruptions of Oldoinyo Lengai, and the volume of tephra is much less, perhaps only a tenth of that in eolian tuffs of the upper unit of the Ndutu Beds. In view of the fine texture and poor sorting of most eolian tuffs, it seems likely that the constituent particles of these tuffs were transported in a thin zone near the ground, perhaps supplemented by dust clouds, rather than in the form of extensive ash falls or sand dunes. The hillocks of tuff on the plain may, however, represent erosional remnants of trailing ridges left by barchan dunes.

The fluvial deposits are chiefly conglomerate and sandstone but include a small amount of water-worked tuff. The conglomerates consist of well-rounded pebbles of quartz and quartzite together with detritus obviously derived from the Olduvai Beds (for example, blocks of basalt from Bed I). These conglomerates are similar in most respects to the conglomerates at the base of the upper unit of the Ndutu Beds. Scattered bones can be found in the sands and conglomerates. Eolian tuffs have been modified in the sides of the gorge by colluvial processes, but purely colluvial deposits were noted only in the type section of the Naisiusiu Beds, where they are 1.5 to 1.8 m thick and lie between fluvial and eolian sediments. The colluvial sediments are clayey tuffaceous sandstone consisting of Naisiusiu tephra mixed with detritus of Bed I from adjacent slopes to the north. The sediments dip a few degrees to the south, and very likely they were deposited by sheetwash at the northern margin of the valley bottom. Faunal remains were exposed in an excavation to obtain materials for C-14 dating at the type section of the Naisiusiu Beds. One level consisted largely of broken bones and teeth together with some fragments of ostrich eggshell. The bones were too broken for identification, but the teeth belong almost exclusively to *Equus burchelli*, according to D. A. Hooijer (in M. D. Leakey et al., 1972, p. 335). Bone fragments were also found at a lower level in the sheetwash deposits, associated with the main concentration of artifacts.

ENVIRONMENTAL SYNTHESIS AND GEOLOGIC HISTORY

The Naisiusiu Beds were deposited over a rather short span of time after the gorge had been eroded to approximately its present level. They are not displaced where they overlie faults, showing that movements had ceased along the known faults. Eruptions of Oldoinyo Lengai were the main source of sediment, and this ash was for the most part extensively reworked by wind before it was deposited and lithified.

The overall climatic regime resembled that of the present day. The few identified mammalian forms are of species living in the same area today, and the calcretes and zeolitic alteration point to a semiarid climate. The climate was, however, somewhat moister and cooler than it is at present. Evidence for a cooler or moister climate is found in the relatively high content of claystone clasts and pelletoid coatings around sand grains. These were very likely derived from the margin or dried floor of small impermanent ponds or lakes in the area between Oldoinyo Lengai and the gorge. Rainfall is presently insufficient to create ponds or lakes in the depressions of this area. Cooler temperatures are inferred from the aspartic-acid racemization of bones from the type section of the Naisiusiu Beds. Using aspartic acid from bones dated by C-14, Schroeder and Bada (1973) obtained an average temperature of $18.3°C$ for the past 17,550 years. This temperature refers to the ground rather than to the air, and is approximately $7.3°C$ cooler than the ground temperature measured in 1974. In other tropical to temperate regions of the world, the climate was relatively cool from 20,000 to 10,000 yr. B.P.; after this it rapidly warmed to the vicinity of the present temperature. Using this model of secular temperature variation, the average temperature for the period from 20,000 to 10,000 yr. B.P. was at least 8°C cooler than it is now (J. L. Bada, 1974, personal communication). In view of the much lower temperatures of Naisiusiu time, the wetter conditions may have resulted from lower evaporation rates rather than increased rainfall.

HOMINID REMAINS AND ARCHEOLOGIC MATERIALS

The human skeleton found by H. Reck in 1913 appears to fall within the time span of the Naisiusiu Beds. This skeleton was found in the sides of the gorge, and its age was controversial until 1933, when most of the participants to the controversy agreed that it represented a relatively recent intrusive burial into Bed II. The skeleton is now identified as *Homo sapiens.* Collagen from a sample of bones yielded a C-14 date of 16,920 ± 920 yr. B.P. (Protsch, 1973, p. 157),

which is close to the other two dates from the type section of the Naisiusiu Beds.

Isolated artifacts are widely scattered through eolian tuffs of the Naisiusiu Beds, and one archeologic site is known. This site, at the type section of the Naisiusiu Beds, was first recorded in 1931, when a trial trench was dug. In 1969 it was further excavated by M. D. Leakey, and artifact concentrations were found at two levels in the sheetwash deposits (M. D. Leakey et al., 1972). Materials from one of the two levels gave the C-14 dates of 17,550 ± 1,000 and 17,000 yr. B.P., indicating that the skeleton was at least roughly contemporaneous with the archeologic site. The artifact assemblage contains a high proportion of microliths, and M. D. Leakey has commented that "no exact parallel for the industry from the Naisiusiu Beds is known at present. It may represent a local variant of the Kenya Capsian, although the geometric element is less pronounced and the industry as a whole seems to be less specialized, or alternatively, it may prove to be an early form of Wilton hitherto unrecorded in Tanzania" (1972, p. 340).

10

HOLOCENE SEDIMENTS

STRATIGRAPHY AND DISTRIBUTION

Alluvium of Holocene age occurs widely in the bottom of the gorge and forms the uppermost deposits in the floor of Olbalbal. A layer of ash is interbedded near the top of the alluvial sequence, and patches of ash and tuffaceous sand are found on the sides of the gorge. The ash stratum is the only marker bed in the Holocene deposits and is termed the Namorod Ash, its name being taken from a hill at the northern foot of Lemagrut. Pre-Namorod tuffaceous sands locally underlie the Namorod Ash in the bottom of the gorge. Calcrete overlain by wind-worked tuffaceous sediment forms the surface layer of the plain.

Sediments in Olduvai Gorge

The valley-bottom deposits are exposed in the walls of the present stream channel, which are commonly vertical and 2 to 5 m high. These sediments are thickest and exposed best between the First and Second faults. Here the lower 3 to 4 m of beds are dominantly waterlaid sands, silts, clays, and gravels. Wind-deposited tuffaceous sand or reworked ash is interbedded 3 m below the level of the Namorod Ash near locality 27. The Namorod Ash overlies the waterlaid sediments and grades up into 30 cm to 1 m of tuffaceous silt forming the topmost layer of valley-bottom section.

The Namorod Ash in the bottom of the gorge is medium to dark gray and 45 cm to 1.2 m thick. It ranges from unconsolidated to moderately indurated. In most places it is wholly reworked, but near locality 3 the basal 5 to 8 cm is a primary ash-fall stratum that preserves the impressions of plant stems buried in the showers of ash. The original thickness of air-fallen tephra is not known accurately. The thickness of 5 to 8 cm is a minimum, and a figure on the order of

20 to 25 cm is likely on the assumption that the ash layer exposed in the walls of the streambed comprises the tephra which fell into the valley bottom together with that washed from the sides into the bottom of the gorge. The tephra is medium-grained, crystal-lithic ash of melilite nephelinite composition. The principal minerals are nepheline, sodic augite, and melilite, which average about 20, 12, and 5 percent, respectively. Accessory minerals are biotite, melanite, apatite, perovskite, sphene, and rare sodalite (Table 29). Most of these mineral grains are thinly coated with finely crystalline lava or devitrified glass. A small amount of nepheline-bearing pumice is in some samples. Most of the ash particles are between 0.2 and 0.5 mm in diameter, but flakes of biotite may be as much as 1 cm across. The Namorod Ash ranges from unconsolidated to moderately indurated. Glass and melilite are partly altered in most of the deposit. Some of the melilite crystals in a given sample may be wholly altered, whereas others are fresh, and the cores are selectively altered in still other crystals. These differences in alteration presumably reflect compositional differences in the melilite. Nepheline crystals are at most lightly etched. Alteration is greatest in the indurated samples, which are cemented by calcite and zeolite (phillipsite). The zeolite forms as much as 20 percent of the most altered samples.

The wind-worked ash below the Namorod Ash (loc. 27) is of nephelinite composition and is contaminated with older detritus. It differs mineralogically from the Namorod Ash in a much higher content of augite, ilmenite, and sphene, in a lower content of nepheline and biotite, and in the absence of melilite (Table 29). Some but not all of this difference may be a result of eolian sorting and contamination by older materials.

The Namorod Ash probably belongs to the same eruptive phase as the alluvial fan of dark-

166

Table 29. *Composition of Namorod Ash and Holocene Eolian Sediment*

Sample No.	Nature and location of deposit	Brown clasts, coatings, and pellets	Nephelinite lava	Augite	Nepheline + feldspar + quartz	Apatite + melilite	Melanite	Ilmenite	Sphene	Perovskite	Biotite
1	Namorod Ash in gorge (av. of 2)	1.1	53.5	11.0	21.2	4.9	5.9	0.1	0.3	0.3	1.6
2	Pre-Namorod tuffaceous sediment in gorge (loc. 27)	24.3	1.1	37.1	21.8	3.5	5.6	2.4	3.2	0.3	0.7
3	Crest of active S barchan	2.6	1.1	71.2	3.7	5.7	5.7	3.4	5.4	0.8	0.3
4	Ridge 3.5 km NNW of loc. 200	9.7	46.6	8.2	18.6	5.5	6.1	2.1	0	0.6	2.4
5	Hummock 3.5 km N of loc. 10	21.8	15.8	31.3	23.1	2.0	2.3	0.7	2.3	0.3	0.3

NOTE: Data were obtained by S. G. Custer from point-counting of thin sections.

gray reworked tephra on the northern foot of Oldoinyo Lengai. The two sets of deposits are mineralogically similar in most respects and are altered to about the same extent. The principal mineralogic difference is in the content of biotite, which is more common in the Namorod Ash.

Sediments in Olbalbal

Olbalbal contains a wide variety of Holocene sediments. The central and lowest part of Olbalbal is a relatively flat area that measures about 12 km in length and 3.0 to 5.5 km in width (see Fig. 2, Chap. 1). It represents the Holocene drainage sump and is underlain by relatively fine-grained sediments. Alluvial fans of coarse detritus from the volcanic highlands border the central area on the south and east. Wind-deposited silt and ash cover these alluvial fans in places. A field of dunes now stabilized by vegetation borders the north edge of the low central area and extends northeastward beyond the limits of the Olbalbal fault graben.

The low, central part of Olbalbal presently has a relief of about 15 m. The highest area is adjacent to the mouth of the gorge, and the lowest is a seasonal marshland covering an area of about 1.5 km^2 in the northwest corner of the central area. Olduvai Gorge empties into a sinuous channel 1 to 3 m deep that follows the western margin of Olbalbal northward into the

area with seasonal marsh. Water flows into the marshland at times of heavy rains, and the marshland was covered by water 1 to 1.5 m deep during and after the exceptionally heavy rains of late 1961 and early 1962. A much larger adjacent area was flooded at the same time (Fig. 2). The marsh is brackish, which seems surprising in view of the large volume of dissolved salts periodically supplied to Olbalbal over the past few hundred thousand years. The loss of salts may be largely attributable to leakage of water along faults to a lower level in the hydrologic system, very likely to Lake Natron which lies about 650 m below the lowest point in Olbalbal.

Holocene deposits in the central part of Olbalbal are an assortment of fluvial, eolian, and semilacustrine sediments. Their base is not exposed, and their thickness is unknown. In the banks of the channel along the western margin are exposed fluvial sands, silts, and gravels, which are locally overlain by a layer of eolian silt. A pit 1.8 m deep was dug along the western edge of the marshland area to expose the sediments of the present drainage sump. Clay is the dominant type of sediment, and medium-grained sand forms most of the remainder. A small amount of claystone-pellet aggregate was identified. Both the clays and sands contain reworked Namorod Ash, claystone pellets, and claycoated, pelletoid grains of detrital sand. Ostracods are abundant in the surface layer of clay deposited in the flooding of 1961/62, and a few ostracods were found in clay at the base of the pit. Pits 1.5 m deep were dug in the lowest area of the southwestern part of Olbalbal 3 to 3.5 km south from the mouth of the gorge. In all of these pits, a layer of grayish-brown eolian silt overlies stratified wind-worked sand-size tephra, chiefly of the Namorod Ash.

Sediments of the Plain

Holocene deposits of the plain comprise a laminated calcrete overlain by a layer of unconsolidated sediment with tephra of the Namorod Ash. Grassland covers this surface layer of sediment. The calcrete is present over at least 90 percent of the plain and is generally between 1 and 2 cm thick. In most places it overlies the Naisiusiu Beds, but it may also coat the Ndutu Beds and metamorphic rocks. The surface layer of unconsolidated sediment adjacent to the gorge is a thin, blanketlike deposit generally between 15 and 60 cm in thickness. In places it is eroded away, and locally it thickens to a meter or more in the form of low sand hummocks. A few kilometers north of the gorge the surface layer is represented by dunes and their trailing ridges in an east-west belt that has a width of about 5 km (Fig. 60; Plate 11). The western margin of the belt lies roughly north from the intersection of the Fifth Fault and the Main Gorge. To the north and west of the dune belt, the unconsolidated surface layer is relatively thin and even, as it is to the south of the dune belt. Gullies in the northern area expose as much as 60 cm of semiconsolidated wind-worked ash beneath the unconsolidated surface layer with Namorod tephra. The semiconsolidated ash is gray to yellowish-brown, only slightly altered, and is judged to be much closer in age to the Namorod Ash than to the Naisiusiu Beds.

The belt of dunes and dune trails curves to the northeast between the First and Second faults and continues in a northeasterly direction for about 25 km, where it ends in the basin of Lake Natron to the west of Oldoinyo Lengai.

AGE OF THE HOLOCENE DEPOSITS

The Holocene deposits are clearly much younger than the Naisiusiu Beds. The most altered of the Namorod Ash is altered much less than the Naisiusiu tuffs. Moreover, Holocene sediments in the gorge were deposited only after the Naisiusiu Beds had been extensively eroded. On the basis of these facts, the Namorod Ash and other Holocene sediments in the bottom of the gorge were very likely deposited during the later part of Holocene time. In latest Holocene time, the present riverbed was cut into the valley-bottom deposits.

The most significant C-14 date is $1{,}370 \pm 40$ yr. B.P. (LJ 2979; J. L. Bada, 1974, personal communication) and was obtained from a large shell of the land snail *Achatina* sp. collected from waterlaid silts 2 m below the Namorod Ash in locality 4d. Fluvial sediments between the *Achatina* and the Namorod Ash very likely represent a rather short period of time, and the Namorod Ash is probably on the order of 1,200 or 1,300 years old. Radiocarbon dates on calcretes from the north side of Oldoinyo Lengai are compatible with this age for the Namorod Ash. These dates, 1,300 and 2,050 yr. B.P., were obtained from calcretes associated with stream-worked tephra of Namorod affinity (Appendix A).

Seven C-14 dates are available for the calcrete at the top of the Naisiusiu Beds. The dates range from $2{,}190 \pm 105$ to $9{,}130 \pm 130$ yr. B.P., and are more or less evenly distributed between 2,190 and 6,630 yr. B.P. The calcrete at all these localities shows no evidence of erosion and was almost certainly formed beneath the present surface layer of unconsolidated sediment. The surface layer contains Namorod tephra at all localities where it was sampled, indicating that it was deposited, or at least extensively reworked, more recently than about 1,250 yr. B.P. Either the calcrete dates are too old because of contamination, or else the calcretes were formed before the Namorod Ash was erupted. The latter alternative implies that the Namorod Ash was reworked into the surface layer without exposing the calcrete to erosion. In order to test the likelihood of contamination, a C-14 date was obtained from the silt-size detrital particles of calcite in a sample of the surface layer. This date is $1{,}200 \pm 100$ yr. B.P., which is far too young to account for the older of the calcrete dates by incorporation of older carbonate. Consequently, it seems most likely that the calcretes were formed, either partly or wholly, before the Namorod Ash was deposited. A fruitful line of approach would be to date different layers of

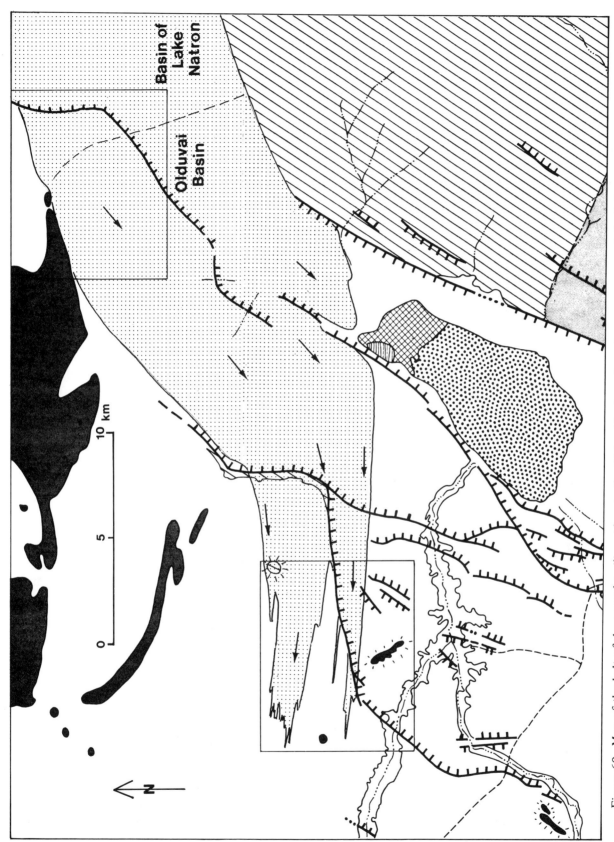

Figure 60. Map of the belt of dunes and trailing ridges in the Olduvai basin and southwestern corner of the Lake Natron basin. Dune belt is shown by stippled pattern, and arrows show direction of dune movement as indicated by dune forms and trailing ridges. Rectangular outlines show areas covered by the aerial photographs of Plates 11 and 12.

169

the same calcrete sample. This would aid immensely in interpreting the present dates from bulk samples of calcrete.

EOLIAN SEDIMENTATION ON THE PLAIN

Eolian sediment on the plain was studied in order to learn more about the nature and results of eolian processes in the Olduvai region. This study in modern sedimentation helps to understand the origin of the eolian tuffs of the Olduvai and Laetolil beds, which are similar, except in their alteration, to the Holocene sediment on the plain. Sediment samples were analyzed by sieving, heavy-mineral separation, and microscopic examination including point-counting. Grain-size data are given mostly in phi (ϕ) measures, a practice currently followed in modern sedimentation studies. A good reference to phi measures is Inman (1952). Very briefly, the phi scale is logarithmic, and the particle diameter in millimeters is represented by the negative log to the base 2. Thus decreasing grain sizes correspond to increasingly larger phi units. As an example, 1/2 mm equals 1ϕ, 1/4 mm equals 2ϕ, and 1/8 mm equals 3ϕ. The median diameter in phi units is designated by Md_ϕ. Sorting is represented by σ_ϕ (sigma phi), and the smaller values correspond to better sorting. Heavy minerals were obtained by the sink-float method, and the terms *heavy* and *light* minerals refer to the fractions which sink and float, respectively, in bromoform (S.G. 2.87). Point-counting is a method of microscopically measuring the volume percentage of constituents in a thin section (Galehouse, 1971). The data of Table 29 are based on counts of about three hundred grains.

Several criteria can be used to distinguish Namorod tephra from older tephra, which is similar in many respects to that of the Naisiusiu Beds. Most important, the Namorod Ash is mineralogically distinctive (Table 29, nos. 1 and 2), and one of the more diagnostic minerals is melilite, which occurs both as isolated crystals and as microphenocrysts in lava. The proportions of melanite, sphene, and perovskite differ greatly from those found in the older sediments, including the Naisiusiu Beds (see Fig. 59, Chap. 9). Clasts of lava form roughly half of the Namorod Ash but are relatively rare in the older sediments. Another difference is in the shape of euhedral augite crystals, which are most commonly stubby in the Namorod Ash but slender in the older deposits. Finally, much of the older detritus has pelletoid clay coatings, whereas particles of the Namorod Ash generally lack clay coatings.

The blanket of sediment unrelated to dune processes is massive, gray, and dominantly of sand size. Its median diameter is generally close to 3ϕ (= 0.125 mm), the boundary between fine and very fine sand (Table 30). Sorting is poor, and σ_ϕ averages 0.93. The fraction finer than sand (that is, <0.0625 mm) is chiefly silt and averages 15 percent in the analyzed samples. The deposit becomes increasingly finer grained to the south of the gorge and is a brown silt near the foot of Lemagrut. To the southwest, near Silal Artum, the silt grades into dark gray clayey soil.

Over most of its extent this layer is a mixture of Namorod tephra, older ash and detritus, claystone pellets, and biogenic materials. Nephelinite tephra predominates among the older ash particles, and the detrital fraction includes metamorphic debris, tuff, limestone, calcite, claystone, and so forth. Brown pelletoid coatings are common on the older tephra and detrital grains. Organic materials include shell fragments, plant fibers, opal phytoliths, and so forth. The opal phytoliths form as much as 5 percent of the silt-clay fraction.

Namorod tephra forms the bulk of the deposit (about 75 percent?) in the area of northeast-trending sand ridges, and it probably averages between 30 and 50 percent of the deposit to the west. In this western area the content of Namorod tephra decreases from north to south, and it is a minor constituent in the brown silts at the foot of Lemagrut, which consist chiefly of claystone particles.

The surface of this layer of sediment is being redeposited by winds at the present time, particularly during the dry season. Similar reworking during the past 1,250 years seems adequate to account for the mixture of Namorod tephra through the entire thickness of the sediment layers.

Within the dune belt are active barchan dunes, sand hillocks and dunes stabilized by vegetation, and linear sand ridges. The active barchan dunes are three in number and lie near the western margin of the belt. The other landforms are widely distributed, and the type of landform

Table 30. *Grain-Size and Heavy-Mineral Data for Eolian Sediments of Holocene Age in the Olduvai Region*

	Sample no.	Location and nature of deposit	Md_ϕ	σ_ϕ	% heavies	Md_ϕ of heavies	σ_ϕ of heavies	Md_ϕ of lights	σ_ϕ of lights	% silt + clay
Active dunes	1	Crest of S barchan	2.59	0.30	94	2.60	0.29	2.50	0.31	<1
	2	Low on S flank of S barchan	2.40	0.30	87	2.36	0.34	2.59	0.34	<1
	3	Crest of Middle barchan	2.76	0.26						<1
Hummocks and Ridges	4	Thin sand layer 30 cm from N flank of S barchan	2.75	0.42	55	2.82	0.43	2.59	0.37	<1
	5	N ridge 280 m E of S barchan	2.78	0.42	47	2.85	0.42	2.70	0.42	2
	6	S ridge opposite no. 5	2.81	0.48	51	2.85	0.50	2.79	0.49	4
	7	Hummock on S ridge of S barchan	2.78	0.53	40	2.90	0.42	2.68	0.50	2
	8	Ridge 5 km N of Naibor Soit			38					
	9	Hummock 4 km NE of TK	2.95	0.56	36	3.08	0.50	2.90	0.56	3
	10	Hummock 4 km E of loc. 1	2.62	0.51	34	2.78	0.46	2.53	0.54	<1
	11	Hummock 7 m high, S part of Natron basin	2.06	0.81	37	2.20	0.69	1.99	0.83	4.5
	12	Hummock 3 m high, near no. 11			35					
Thin surface layer on plain	13	Wake of S dune between nos. 5 and 6	2.91	0.51	36	3.00	0.51	2.88	0.56	15.5
	14	Near loc. 83	3.00	1.32						18.5
	15	N side of Naibor Soit	2.95	0.94						16
	16	Near locality 22	2.95	0.72						8
	17	Near head of TK	3.20	0.90						23
	18	6 km N of Naibor Soit			32					21
	19	11 km N of loc. 1	2.91	0.68						

bears a close relationship to topographic position (Plate 12). Relatively low and sloping areas are characterized by sand hillocks and dunes of various shapes that are stabilized by vegetation. The most common form of stabilized dune is classed as parabolic; that is, it has the outline of a *U* or *V* when viewed from the air (Plate 12), and the long axis of the dune form parallels the prevailing wind direction, with the open end facing the prevailing winds. These dunes have an average length of about 500 m, and the open end of the dune form is generally between 100 and 200 m in width. Most of the fixed hillocks and dunes are found in the down-faulted area which continues northward from Olbalbal into the basin of Lake Natron. Another area of fixed hillocks and dunes is on the southern, downthrown side of the east-west extension of the Fifth Fault to the north of the gorge. Linear, parallel sand ridges characterize the higher plains areas. The ridges are generally rather broad, between 30 cm and 2 m high, and spaced at intervals of 60 to 150 m. The broader of the ridges are difficult to recognize on the ground but can

be readily identified in aerial photographs (Plate 12). The ridges are relatively even in height toward the northeast and become hummocky in the western part of the dune belt. Individual ridges can be traced on aerial photographs for as much as 8 km. Sand of the dunes and dune ridges is a mixture of Namorod Ash with older tephra and other detritus.

The active dunes and their related sediment

Table 31. *Dimensions of Active Dunes near Olduvai Gorge*

	Length	Width	Height	Date measured
Southern dune	35 m	65 m	5.3 m	6/7/70
	30	60-65	5.0	9/16/73
Middle dune	25	33	1.8	6/7/70
	21	24	1.5	9/16/73
Northern dune	25	59	4.5	6/7/70
	25	34	3.9	9/16/73

NOTE: Length represents the distance between the trailing edge and the front edge of the slipface, measured parallel to the direction of movement. The width is the maximum dimension measured perpendicular to the length.

171

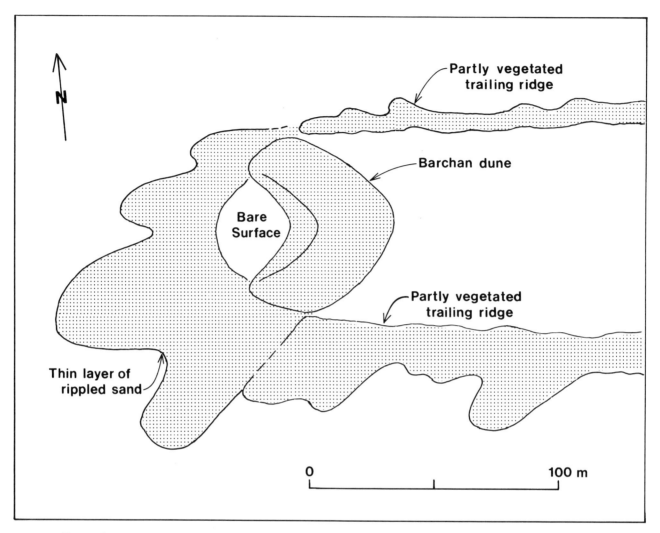

Figure 61. Map of southern barchan dune and associated sand deposits as observed in September 1973. Sand of the dune and related deposits is shown by stippled pattern.

have been studied by M. D. Leakey and myself at intervals between 1969 and 1973. Within this period the dunes have consistently maintained their barchan shape (Fig. 61) but have become reduced in size (Table 31). Low, hummocky sand ridges generally 30 cm to 1 m high and 5 to 25 m wide trail out from the dune flanks in an eastward direction. The south ridge of each dune is the higher and wider, and sand hillocks as much as 2.4 m high are on the southern trailing ridge of the southern barchan. These ridges represent the sand that was carried off the dunes by cross winds and trapped by vegetation along the flanks of the dunes in traveling westward. Judging from the relative size of the trailing ridges, cross winds from the north are either much stronger or blow more continuously than those from the south. The sand ridges from the active dunes can be traced as much as 4 km eastward. Unconsolidated sediment in the path of the dune is incorporated in the dune and its ridges, and a hard, bare surface is left in the immediate wake of the moving dune. The amount of sediment incorporated by a dune from the surface layer must roughly equal the amount lost to its trailing ridges in order for a dune to travel an appreciable distance. A dune will waste away where the incorporated sediment is less than the amount lost to the ridges. Extinct dunes are represented by the point at which the two trailing ridges come together in the form of a V (Plate 11).

A thin layer of eolian sediment is deposited in the wake of the moving dunes. In July 1972, thin patches of sediment were found in the wake of the southern dune at a distance of 100 m, and sediment formed a continuous layer 5 to 10 cm thick at a distance of 280 m. Sediment of this thin layer is indistinguishable from that of the surface layer 15 to 60 cm thick which covers the plain outside of the dune belt (see Table 30).

In order to show the rate at which the active dunes are moving, their location on aerial photographs taken in January 1958 was compared with their position in June 1970. The southern dune had traveled 190 m and the northern dune 225 m. All three dunes had decreased in size between 1958 and 1970, and a fourth and northernmost dune had become extinct in this period (Plate 11). These figures give average annual rates of 18 m for the northern dune and 15 m for the southern dune. Movement of the southern dune was also measured monthly by F. K. Lili and M. D. Leakey between September 1969 and July 1971. Measurements were of the trailing edge of the dune with reference to a benchmark. The dune moved 32.1 m over this period, averaging 17.5 m per year. It moved during all months of the year, and the rate showed no systematic seasonal differences. One can infer from these rates that the layer of sediment 5 to 10 cm thick at a distance of 280 m in the wake of the dune was deposited within the past sixteen to eighteen years. This emphasizes the rapid rate at which sediment is presently being redeposited on the plain by wind.

The three active barchan dunes are the last of a broad field of barchan dunes that traveled into the Olduvai basin for a distance of about 35 km. The older sand ridges represent the trailing ridges of former dunes, and sand hillocks on the trailing ridges resulted either from particularly strong cross winds or periods when the parent dune moved more slowly than average. Large parabolic dune forms represent short, hummocky trailing ridges of large dunes which join at the point where the dune had wasted away. It is not clear as to why these should occur principally in low areas or sloping surfaces, irrespective of whether the slope is upward or downward.

Sand of the dune belt is dominantly of fine sand size, and Md_ϕ averages about 2.60, which represents a median diameter of 0.17 mm

(Table 30). The sand is well sorted, and σ_ϕ is 0.26 to 0.30 in the samples that were sieved. Sand of the trailing ridges in the Olduvai basin is finer grained (average Md_ϕ = 2.80) and sorted more poorly (σ_ϕ = 0.42 to 0.56) than that of the active dunes. Heavy minerals form at least 35 percent of the eleven analyzed samples (Table 30), and they form about 90 percent of the active dunes.

The proportion of Namorod tephra decreases from east to west in the belt of dunes and trailing ridges. It constitutes about 75 percent of the sand in the eastern part of the belt, where the sand ridges trend northeast, and it generally forms between 25 and 50 percent, averaging perhaps 35 percent, in the western area with east-west ridges. Surprisingly, the composition changes rather abruptly in the zone where sand ridges change direction. The difference in composition is illustrated by two samples from different parts of the belt (Table 29, nos. 4, 5). The westward decrease in content of Namorod tephra implies that the western part of the plain was already covered by a layer of unconsolidated tuffaceous sediment when Namorod Ash was erupted. The sediment layer must have been thicker than the Namorod Ash in this area, for Namorod tephra is subordinate to other materials in the thin surface layer adjacent to the dune belt.

The high content of heavy minerals in the active dunes is clearly a result of selective sorting by wind. As evidence, heavy minerals constitute about 90 percent of the dunes and half or less of the trailing ridges. Data were obtained from the southern active dune and its trailing ridges to document this difference (Fig. 61; Table 30). The highest proportion of heavy minerals (94 percent) was found at the crest of the dune, and the heavy-mineral content dropped abruptly from about 85 percent at the dune margin to 55 percent in a thin layer of sand only 1 m distant. Another striking difference between dunes and their trailing ridges is in the shape of the augite grains. These are dominantly euhedral and slender in the dune, and the ratio of length to width is generally three to five. Augite grains of the trailing ridges are dominantly either subequant or tabular. Augite is slightly coarser in the dunes than in the ridges. Thus the dunes are enriched in those grains which, by reason of specific gravity, shape, and size, are most difficult to transport by saltation, the process by which sand grains are carried by wind. In other words, grains typical of the trailing ridges would be moved by wind a greater distance over a given period of time. The compositional difference between the dunes and their trailing ridges is attributed to selective sorting by cross winds, which should deplete the dunes in the fraction most readily transported. This latter fraction is found in the trailing ridges.

HOLOCENE GEOLOGIC HISTORY

The Naisiusiu Beds were eroded and the riverbed further entrenched in the early part of Holocene time. Only late in Holocene time ($\sim$ 1,500 to 2,000 yr. B.P.?) did fluvial sediment accumulate in the bottom of the gorge. These sediments reached a thickness of about 4 m, after which the Namorod Ash was deposited over the entire basin and was reworked by streams in the bottom of the gorge. The latest sediments in the valley bottom are silts that are at least partly eolian in origin. The stream had begun to cut the present narrow valley either during or after the deposition of the silt, and it has continued to erode the valley-bottom sediments up to the present time. The two Holocene episodes of erosion may have resulted either from a change in climate or further subsidence in Olbalbal. Although Holocene faulting has not been demonstrated, the past history of faulting in the Olduvai basin makes it seem a reasonable likelihood.

On the plain, tephra from relatively minor eruptions together with older detritus was redeposited by wind to form a surface layer of sediment. Beneath this layer a widespread calcrete was formed. Radiocarbon dates on the calcretes suggest that these sediments accumulated slowly from about the end of the Pleistocene 10,000 years ago to about 1,250 yr. B.P., when the Namorod Ash was discharged. The volume of fallen Namorod tephra was probably greatest in the southern part of the Natron basin. Here it was redeposited by wind to form dunes which were driven first in a southwesterly direction and then along a westerly course into the central part of the Olduvai basin, 30 to 35 km distant from the source area. The three active dunes at the western margin of the belt of dunes and trailing ridges are the last of the field of dunes that developed initially from Namorod tephra. These dunes are presently traveling at a rate of about 17 m per year, or 59 years per kilometer. This rate gives 1,900 years as the time required to travel from the source area of the dune field. The Namorod Ash is appreciably younger than this ($\sim$ 1,250 yr. B.P.?), which suggests that the dunes traveled more rapidly in the earlier, northeast-to-southwest leg of their journey than in the later, westerly course.

PLATES

Plate 1. South side of the Main Gorge at the Second Fault. Here the gorge is about 75 m deep.

Plate 2. North side of the Main Gorge at JK (loc. 14). Here Bed II is 26 m thick, and Bed III is 14 m thick. Excavations by M. Kleindienst are out of sight in background.

Plate 3. North side of the Main Gorge at RHC (loc. 80) viewed in a northwest direction. Beds II and III-IV (und.) are 23 and 16 m thick, respectively. Numbers (1), (2), and (3) represent the lower, middle, and upper disconformities of Bed II. *Cong.* designates a conglomerate overlying the uppermost disconformity and formed largely of blocks from Tuff IID.

Plate 4. Southern part of RHC (loc. 80) viewed from the east. This area is seen at the left side of Plate 3. *Cong* designates the conglomerate with blocks of Tuff IID.

Plate 5. Western side of CK(loc. 7*a*) viewed from the east. Bed III is 12.5 m thick.

Naisiusiu Beds
Ndutu Beds

Ndutu Beds
Norkilili Mbr.

Norkilili Mbr.
Lower Unit

Lower Unit, Masek Beds
Beds III-IV (und.)

Plate 6. View looking southeast in Kestrel K (loc. 84), the type locality of the Masek Beds. The lower unit of the Masek Beds is 10 m thick, and the Norkilili Member is 4.5 m thick.

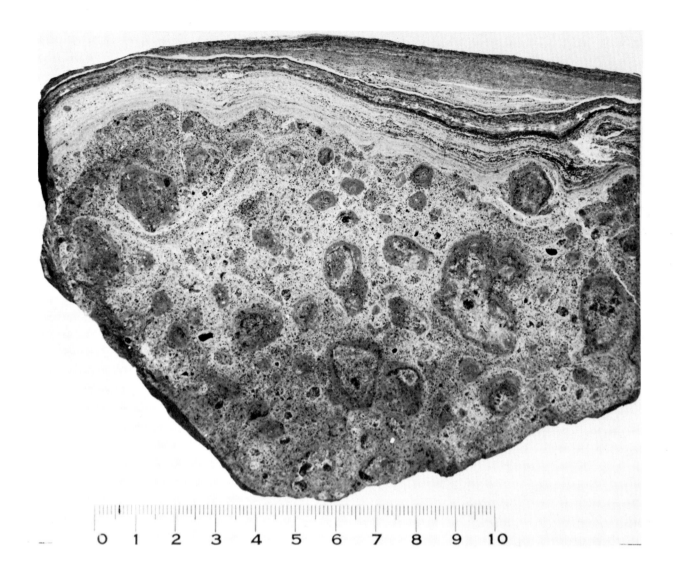

Plate 7. Sample of laminated calcrete overlying calcareous tuff of the Norkilili Member. Sample was collected from the east side of CK (loc. 6), where the calcrete is overlain by the Ndutu Beds. The calcrete yielded a C-14 date of $\geqslant$ 30,000 yr. B.P. Scale is in centimeters.

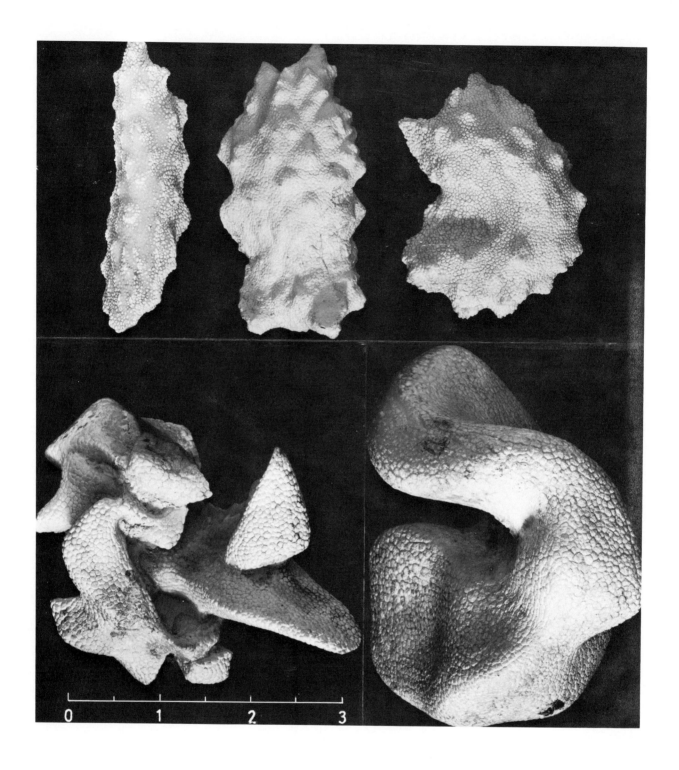

Plate 8. Chert nodules from lacustrine deposits of Bed I at RHC (loc. 80). Photograph is from Hay (1968). Scale is in inches.

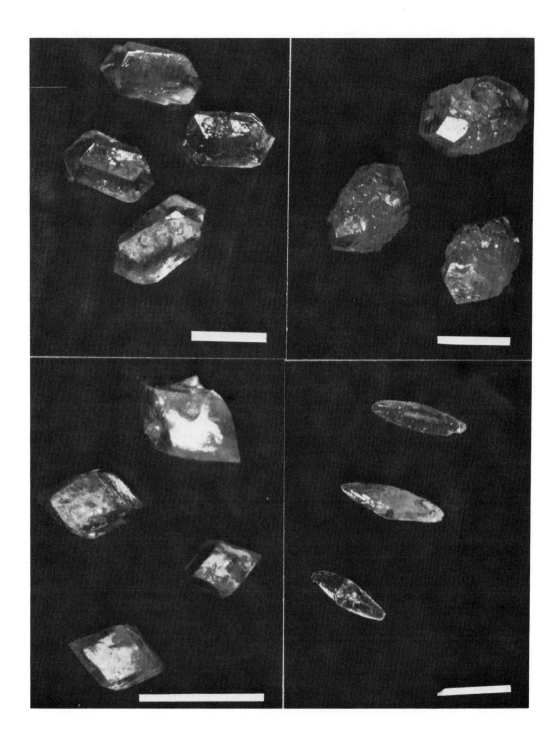

Plate 9. Calcite crystals from lacustrine deposits of Beds I and II at RHC (loc. 80). Length of bar is 1.0 mm. Crystals in (a) lower left were collected 6.1 m below Tuff IF; (b) upper left are from 9.5 m below Tuff IF; (c) upper right are from 14.6 m below Tuff IF; (d) lower right are from 25 cm above the base of Bed II. Crystals of (a) have rhombohedral faces of the form [0221]. Crystals of (b) have scalenohedral side faces of the form [8.4.12.1], and the end faces are unit rhombohedra of the form [1011]. Crystals of (d) have scalenohedral side faces of the form [1341].

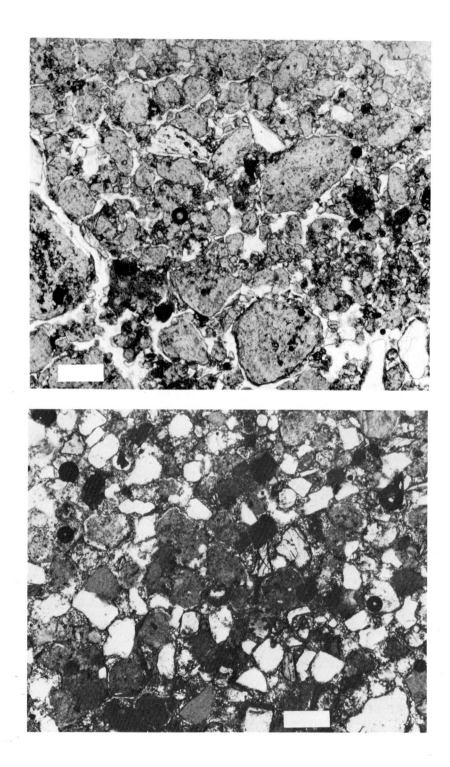

Plate 10. Photomicrographs of samples with claystone pellets as seen in thin sections (plain light). Upper photograph is of claystone-pellet aggregate collected 2 m above Tuff IID in locality 20. Lower photograph is of sandstone with about 30 percent claystone pellets (dark grains), which was collected from Beds III-IV (und.) at locality 203. Length of bar is 0.25 mm.

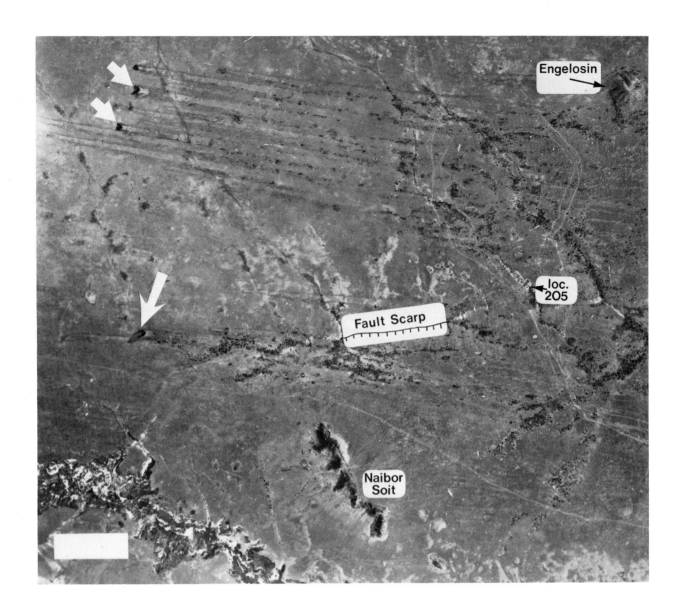

Engelosin

loc. 205

Fault Scarp

Naibor Soit

Plate 11. Aerial photograph of part of the Serengeti Plain to the north of the gorge showing trails left by barchan dunes in travelling westward (to left). Large arrow indicates dune studied in detail and depicted in Figure 61. Three barchan dunes, which appear black, are in the upper left-hand corner. The northmost dune was extinct in June 1970; two dunes presently active are marked by smaller arrows. Scale is indicated by white strip in lower left, which represents 1 km. Area of photograph is outlined in Figure 60. Photograph was taken on Jan. 21, 1958, and is no. 57,35TN5 (Surveys and Mapping Division, Dar es Salaam, Tanzania).

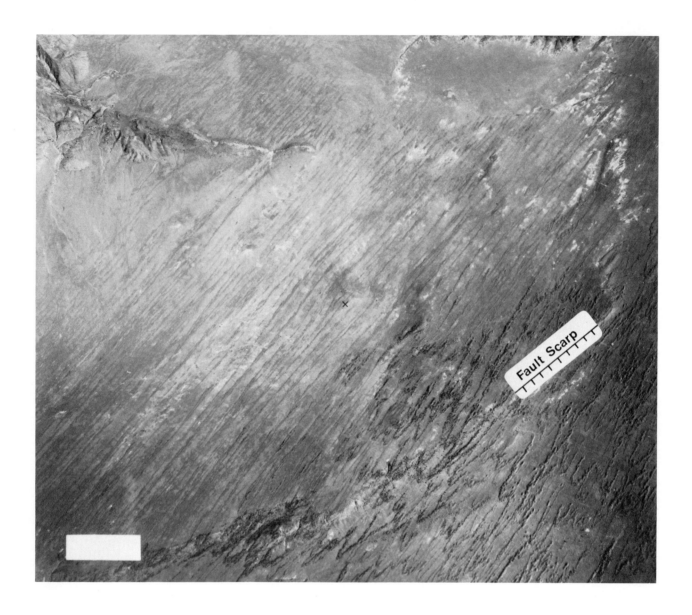

Plate 12. Aerial photograph of a part of the plain at the northeastern edge of the Olduvai basin (see Fig. 60). In center and left are linear dune trails on the plain; in lower right are U- and V-shaped dune forms fixed by vegetation and lying in a lower area in the northern end of the Olbalbal fault graben. White strip represents 1 km. Photograph was taken on Feb. 28, 1958, and is no. 72 of 35TN15 (Surveys and Mapping Division, Dar es Salaam, Tanzania).

11

EVOLUTION OF THE OLDUVAI BASIN

STRUCTURAL DEVELOPMENT

The Olduvai basin has been deformed more or less continually from the time of Bed I to the present. From Bed II onward, displacements can be demonstrated along faults within the basin, and in general the number of active faults increases up through deposition of the Masek Beds, after which the southeastern part of the basin becomes a locus of intense faulting. The nature of the faulting has been determined primarily by differences in thickness of equivalent stratal units across faults. These are a sensitive measure of fault displacements, and they allow a reconstruction of the pattern of faulting through time.

The eastern part of the Bed I basin was downwarped, as shown by paleogeographic reconstruction (see Fig. 14, Chap. 4), and it may have been faulted locally. Fault displacement on the order of 40 m is suggested by stratigraphic relationships between the mouth of the gorge and locality 157, 2 km to the south (see Fig. 10, Chap. 4). Slight movement on the fault near FLK is suggested by the pattern of thickness and paleogeography of Tuff IF to the southeast of the fault (see Fig. 16, Chap. 4). This pattern is nearly the same as in sediments of Bed II (for example, the bird-print tuff) in the same area, where folding and faulting can readily be demonstrated.

Evidence of deformation is lacking in the lower part of Bed II up through the Lemuta Member. The widespread disconformity above the Lemuta Member marks the time, about 1.6 m.y.a., when a graben began to subside between the Fifth and FLK faults. Subsidence continued throughout the deposition of Bed II, and adjacent parts of the basin were deformed.

Beds III-IV contain a record of movements along new faults in addition to those of the graben. The earliest movements of this phase of faulting are tentatively correlated with major Rift-Valley movements along the main fault between lakes Manyara and Natron. Movements along all of the old and several new faults are documented by the Masek Beds, and the fault block between the First and Second faults was folded into a broad syncline. Evidence is not clear as to whether the Olbalbal graben began to subside at this time. Subsidence, if any, must have been minor in amount in view of the fact that the drainage sump still lay to the west of Olbalbal.

The Ndutu Beds record the most severe phase of deformation, which lasted from about 0.4 to 0.03 m.y.a. Dominant features of this phase are subsidence of the Olbalbal graben and southeastward tilting of that part of the basin which lay west of the graben. Within the gorge, the major displacements were along the easternmost faults. Numerous faults were active in the southwestern part of Olbalbal and on the lower slopes of Ngorongoro and Olmoti. The Naisiusiu Beds (~15,000 to 22,000 yr. B.P.) postdate the latest known fault movements. However, the two post-Naisiusiu episodes of erosion in the gorge are suggestive of minor subsidence in Olbalbal along faults not now exposed.

The pattern of faulting in the Olduvai basin is depicted in Figure 62, which is an idealized cross-section showing deformation of the basin for periods represented by different stratigraphic units. In this reconstruction the tops of all stratigraphic units are taken as horizontal. The amount of graben subsidence for Bed II is conjectural; however, displacements for Beds III-IV and the Masek Beds are based on stratigraphic measurements. Displacements shown for the upper unit of the Ndutu Beds are measured offsets, which include both intra- and post-Ndutu movements. Fault displacements were determined for the lower unit of the Ndutu Beds by subtracting displacements of the upper unit from the total

175

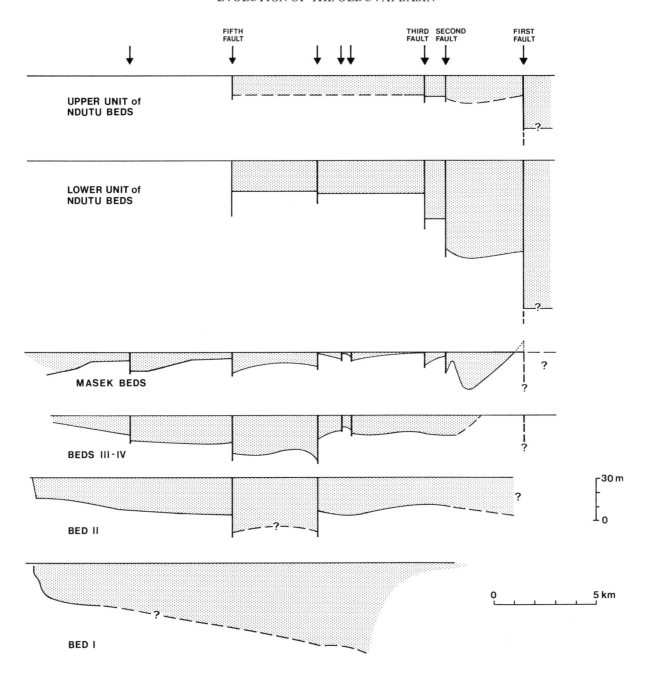

Figure 62. Diagram showing deformation penecontemporaneous with deposition for each of the major stratigraphic units in Olduvai Gorge. Diagram is based on exposures in the Main Gorge between the First Fault and the western margin of the basin. Stippled pattern for Bed I through the Masek Beds represents the observed thickness of these deposits; for the Ndutu Beds it represents the amount of displacement along known faults. The cross-section for the lower unit of the Ndutu Beds represents all displacement between the deposition of the Norkilili Member and the deposition of the upper unit of the Ndutu Beds. The cross-section for the upper unit of the Ndutu Beds represents all displacements between the deposition of the lower unit and the Naisiusiu Beds. Arrows indicate fault locations.

offset of the top of the Masek Beds. Amounts of offset for the First Fault are conjectural in all of the cross-sections.

Faulting in the Olduvai basin seems to have been more continuous than inferred by Macintyre et al. (1974) for the major Rift Valley to the east. This seeming contrast of episodic and relatively continuous movement may not be wholly real, as the duration of fault episodes in the Rift Valley can rarely be determined accurately, and faulting may be relatively continuous over long periods. Isaac (1968), for example, was able to document faulting contemporaneous with deposition through a considerable part of the Olorgesaile Formation, a mid-Pleistocene deposit in the main part of the Rift Valley about 170 km north-northeast of the Olduvai basin.

PALEOGEOGRAPHIC DEVELOPMENT

The drainage sump, or lowest part of the basin, has changed in nature and location through Pleistocene time (Fig. 63). A perennial saline lake about 10 km long and 5 km wide occupied the basin during the deposition of Bed I and the lower part of Bed II. About 1.6 m.y.a. the lake was abruptly reduced to about one-third of this size and restricted to the graben between the Fifth Fault and the fault near FLK. The lake disappeared about 1.2 m.y.a., near the end of Bed II deposition, and the graben was occupied by an alluvial plain with intermittent, localized areas of ponded water. The drainage sump for Bed III (~1.15 to 0.80 m.y.a.) was a fluvial-lacustrine area, locally a playa lake, which had shifted about 7 km northeast from its latest Bed II position. Sumps for Bed IV (~0.80 to 0.60 m.y.a.) and the Masek Beds (~0.60 to 0.40 m.y.a.) shifted progressively southeastward toward Olbalbal, which has been the lowest part of the basin during roughly the past 400,000 years. The progressive eastward shift in the sump reflects the eastward tilting of the Olduvai basin. Changes in the nature and size of the drainage sump may be wholly a result of deformation, although climatic change cannot be readily ruled out as a contributing factor. Leakage of water along fault zones may account, at least partly, for the abrupt disappearance of the perennial lake of Bed II in the graben.

The nature of the stream drainage within the

Olduvai basin has changed considerably through Pleistocene time, largely as a result of fault-induced changes in paleogeography. Numerous relatively small streams or rivers seem to have supplied the large lake of Bed I and the lower part of Bed II. When the lake was greatly reduced in size by faulting, drainage from the volcanic highlands became integrated into a main stream, which flowed westward into the east end of the lake (see Fig. 39, Chap. 5). With the deposition of Bed III, drainage from the west had become integrated into a main stream which flowed eastward into the drainage sump (see Fig. 45, Chap. 6). An integrated, eastward-flowing system of this type has existed from the time of Bed III to the present.

A paleogeographic problem not yet resolved is the extent to which the western drainage limits of the Olduvai basin have changed through time. The eastern boundary in the volcanic highlands has remained about the same, but the western limit is defined only to the extent of never having reached westward beyond the present drainage divide separating Olduvai drainage from that to Lake Victoria (see Fig. 1, Chap. 1). This latter inference is based on the fact that detritus from the Tanganyika shield is absent in the Olduvai Beds, and the present drainage divide to the west of Olduvai Gorge corresponds, at least approximately, to the boundary between the Mozambique belt and the Tanganyika shield. The divide may, however, have moved westward as much as 10 or 20 km to its present position through Pleistocene time. A westward increase in the drainage area could help explain the seemingly increased stream flow in the western part of the basin as recorded in Beds III-IV and possibly in the upper part of Bed II.

ENVIRONMENTAL CHANGE

The Olduvai Beds contain evidence both of gradual progressive change and of rapid, commonly short-term changes in environment during Quaternary (= Pleistocene + Holocene) time. The major overall change is in the direction of increased aridity. The lowest of the fossiliferous beds, between the lavas of Bed I and Tuff ID, contains faunal elements indicating a climate moister than that documented for any later deposits of the Olduvai sequence. Particularly sig-

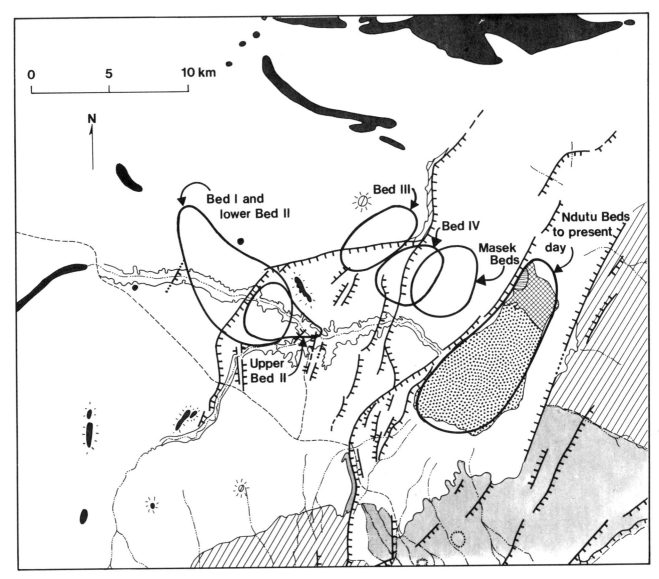

Figure 63. Map showing inferred location of the drainage sump
for different stratigraphic units in the Olduvai basin.

nificant are urocyclid slugs, which require damp evergreen forest. Data are presently insufficient to show whether the climate became progressively drier from Bed II to Bed IV, but eolian sediments in the Masek Beds provide evidence of a climate drier than that of Beds III and IV. Faunal remains in the Masek Beds indicate that fresh or brackish water was available throughout the year, thus showing that the climate of Masek time was moister than at present. Both the lower and upper units of the Ndutu Beds represent an overall climate and environment rather like the present but intermittently moister. Water flowed in the Main Gorge throughout the year, at least during part of the time represented by the lower unit, and lacustrine tuffs correlative with the upper unit of the Ndutu Beds are widespread in the floor of Ngorongoro crater, documenting a lake level higher than at present.

Relatively rapid environmental changes are exhibited at many levels in the Olduvai Beds. Much of the evidence for rapid change may reflect fault-induced paleogeographic changes, but some of the evidence points to fluctuations in climate. Several of these rapid environmental fluctuations are listed here.

1. The large ecological difference between the lower and upper faunal horizons of Bed I clearly shows that a drier environment replaced a moister one about 1.70 m.y.a. This faunal change is more likely climatic than tectonic in origin in view of the fact that the "moister" fauna corresponds to a time when the Bed I lake was large (and high), and the "drier" fauna corresponds to a time when the Bed I lake was small (and low). The lake rapidly increased in size (and level) following deposition of the beds with the "drier" fauna, thus showing that the dry episode was followed as well as preceded by a moister episode.

2. The Lemuta Member represents a relatively dry episode about 1.65 m.y.a. that lasted at least 20,000 years. The best evidence for a drier climate is provided by the lower tongue of the Lemuta Member, which is widespread and consists of detritus that was eroded as well as transported by wind.

3. Eolian and coarse alluvial sediments of Tuff IIB ($\leqslant$1.60 m.y.a.) are both underlain and overlain by lacustrine and lake-margin sediments. Tuff IIB was deposited at a time of deformation and is found over a relatively small area. Consequently, the drier environment represented by the tuff may well reflect a temporary paleogeographic change caused by earth movements. Comparable environmental changes of ambiguous origin are recorded at higher levels in Bed II (see Chapter 5).

4. Upper Pleistocene lacustrine sediments in the floor of Ngorongoro crater record fluctuation in salinity and lake level. Ostracodal marls yielding C-14 dates of 24,400 ± 690 and 27,900 ± 500 yr. B.P. record relatively fresh water and a lake level at least 2.5 m above the present mud flats. The ostracodal sediments are underlain by clays with evidence of highly saline, alkaline lake water. The lake level and salinity of this stage were probably about the same as those of the present lake. The high lake level may well be correlative with the high level in Lake Nakuru ($\geqslant$21,000 yr. B.P.), in the Rift Valley to the north, and with the level of Lake Chad ($\geqslant$22,000 yr. B.P.), in the Saharan region (Butzer et al., 1972). Curiously, no evidence was found of a high level between 7,000 and 10,000 yr. B.P. in Ngorongoro, correlative with the widespread high level in Rift-Valley lakes to the north (Butzer et al., 1972).

5. The Naisiusiu Beds ($\sim$15,000 to 22,000 yr. B.P.) were deposited at a time of low levels in Rift-Valley lakes to the north. Presumably, therefore, the climate was relatively dry, which accords with the eolian origin of the Naisiusiu Beds. On the basis of amino-acid studies, the climate was considerably cooler ($\geqslant$8°C) when the Naisiusiu Beds were deposited than it is now (Bada, 1974, personal communication). Thus the climate in the Olduvai region appears to have been cool and relatively dry during the last major worldwide glacial stage, which reached a maximum about 19,000 yr. B.P.

Lest the evidence for climatic change be over-emphasized, it should be noted that the overall climatic regime has fluctuated within moderately narrow limits at Olduvai during the past two million years. The climate has remained semi-arid, and all of the available evidence points to prevailing winds from the east or northeast, as at the present time. Easterly winds are indicated for Bed I by the distribution of tuffs, which extend to the west and possibly the southwest from Ngorongoro and Olmoti, the eruptive sources. Similar wind directions are suggested by the air-fallen tuffs of Beds II, III, and IV, inasmuch as the most likely eruptive sources lay to the east and northeast. Easterly or northeasterly winds in the lower part of Bed II are more clearly indicated by the northeastward coarsening of tephra in the Lemuta Member, and by the westward interfingering of eolian tuffs in the Lemuta Member with lake-margin claystones. Finally, both eolian and air-fall tuffs of the Masek, Ndutu, and Naisiusiu beds have originated in eruptions to the east and northeast.

12

PATTERNS OF HOMINID ACTIVITIES

Olduvai Gorge is a source of information about several different aspects of hominid activities over the past 1.8 million years. The paleogeography of hominid occupation can be determined from the distribution of artifacts, occupation sites, and hominid remains in different lithofacies. Hominid dietary and hunting practices can be inferred to some extent from the butchery sites and faunal debris on occupation sites. The size of communal groups can be estimated to some extent from the size of occupation sites, at least in Beds I and II (M. D. Leakey, 1971a). Artifacts provide a picture of man's technology with regard to the utilization of stone, and, finally, a knowledge of the sources of raw materials used for artifacts can show the hominid territorial and/or trading range. Some of these matters are archeological, but others are primarily geological and are considered in this chapter. One aspect, already discussed with regard to individual stratal units, is the paleogeography of hominid occupation. The other concerns the sources of stone and the selectivity of different rock types for different types of tools.

ENVIRONMENTS OF HOMINID OCCUPATION

Evidence of hominid activities correlates closely with lithofacies and hence with paleogeography. Within each stratal unit, archeologic sites and hominid remains are found principally in lithofacies with evidence of a perennial supply of fresh water and relatively abundant game and vegetation. Very little evidence of hominid occupation is found in eolian tuffs, which were deposited in a dry savannah, or in reddish-brown zeolite-rich claystones, deposited in relatively barren alluvial floodplains. The size of the hominid population was probably greatest during the time of Beds I through IV, and only sparse,

scattered evidence of hominid occupation is found in the younger deposits.

Evidence of hominid activities in Bed I is restricted to lake-margin deposits (Table 32). Hominid activities seem to have been concentrated along the eastern margin of the lake, and only two archeologic sites have been found in sediments deposited near the western margin. An extraordinary feature is the undisturbed nature of several of the eastern lake-margin occupation sites. The "Zinjanthropus" floor was preserved intact by a shower of ash, and the other undisturbed sites were preserved beneath lacustrine clays.

Bed II records an increase in the geographic distribution of hominid activities. Archeologic sites below the lowermost disconformity are all or nearly all in lake-margin deposits, but they are found over a much wider area than in Bed I (see Fig. 31, Chap. 5). Artifacts seem to be most common in sediments deposited along the southeastern margin of the lake, but the relative degree of concentration is much less than in Bed I. Following erosion of the disconformity, hominid activity was again highly concentrated in the southeastern lake-margin terrain (see Figs. 27a, b, Chap. 5), and artifacts are more abundant in the lower augitic sandstone than in any deposit of equal volume in the Olduvai sequence. The availability of chert at the lake margin is one reason for the abundance of artifacts in this area.

Hominid occupation is recorded over a broad zone around the southeastern margin of the lake for the interval between the lower augitic sandstone and Tuff IID. Evidence of hominid activity is particularly abundant in the area exposed by the Side Gorge. Most sites of this interval were situated within a kilometer of the lake margin, but others were in the main drainageway more distant from the lake (see Figs. 32-35, Chap. 5). Terrain to the west of the lake was

Table 32. *Relative Concentration of Archeologic Sites in Different Lithofacies of Bed I through the Masek Beds*

Stratigraphic unit and lithofacies		Total thickness measured (m)	No. of archeologic sites	Meters of section/site
Bed I	Lake	133	0	∞
	Lake-margin (eastern)	139	18	7.7
	Lake-margin (western)	123	2	61.5
	Alluvial-fan	316	0	∞
	Alluvial-plain	97	0	∞
Bed II	Lake	164	1	164
	Lake-margin (eastern)	102	4	20.5
	Lake-margin (western)	67	2	33.5
	Lake-margin (Kelogi)	19	2	9.5
	Alluvial-fan	106	0	∞
	Eolian	90	0	∞
	Eastern fluvial-lacustrine (all)	518	46	11.3
	Eastern fluvial-lacustrine (Main Gorge)	406	33	12.3
	Eastern fluvial-lacustrine (Side Gorge)	112	13	8.6
	Western fluvial-lacustrine	268	8	33.5
	Eastern fluvial	91	0	∞
Beds III-IV	Eastern fluvial	292	2	146
	Western fluvial (all)	530	40	13.2
	Western fluvial (main drainageway)	217	34	6.4
	Western fluvial (outside main drainageway	316	6	52.7
	Fluvial-lacustrine	19	1	19
Masek Beds	Western fluvial	214	1	214
	All other facies	256	0	∞

occupied sparsely, judging from the scarcity of artifacts in this area. Archeologic sites of this interval are principally stream-channel concentrations resulting from reworking of campsites which were situated either in the streambed or along the banks. A few relatively undisturbed sites are known, and one of these (loc. 12, EF-HR) is an Acheulian site that "may represent a small temporary camp on either side of a shallow water course" (M. D. Leakey, 1971a, p. 124).

Above the level of Tuff IID, hominid occupation is recorded widely over the basin. Several concentrations of artifacts, particularly bifaces, are in channels associated with the uppermost disconformity to the west of the Fifth Fault. Sites are in varied environmental settings in the area to the east. Some of these sites are concentrations of artifacts in stream channels (BK, loc. 94; TK, loc. 16, site 19a), but others are undisturbed occupation sites on paleosols of uncertain paleogeographic context (TK, loc. 16, sites 19b, d). Artifacts are scattered through marshland deposits at JK (loc. 14) and SC (loc. 90a, site 60).

As noted earlier, the known Acheulian and Developed Oldowan B sites appear to differ in their paleogeographic distribution. Sites of the Acheulian industry, as defined by the percentage of bifaces, lie more than 1 km inland from the margin of the lake, whereas contemporaneous Developed Oldowan B sites tend to lie within 1 km of the lake margin. If the Acheulian industry is defined by the type rather than the percentage of bifaces, one atypical Acheulian site (loc. 88, MNK, site 71c) lies within 1 km of the lake. This correlation between paleogeography and type of industry does not resolve the question as to whether the Acheulian and Developed Oldowan B industries reflect different cultural traditions of two groups of hominids (M. D. Leakey, 1971a, p. 272) or two "tool kits" used for different purposes by the same group of hominids.

Evidence of hominid activities in Beds III and IV is almost entirely restricted to deposits of the main drainageway and tributary streams from the west (see Figs. 43, 44, Chap. 6). Only two sites (locs. 88b and 98) are in deposits of a stream draining the volcanic highlands, and another site is in sandstones with a mixture of lacustrine and fluvial features (loc. 204). This distribution presumably reflects the more continual availability of water in streams draining the metamorphic terrain to the west. The relative scarcity of Bed III sites compared to those in Bed IV, reflects, at least partly, the fact that Bed III consists largely of stream-deposited sediment from the volcanic highlands, whereas Bed IV comprises chiefly sediments of the main drainageway.

Hominid occupation was greatly reduced at the time the Masek Beds were deposited. Only a single hominid bone and archeologic site have been found in the Masek Beds, and both of these were in sediments of the main drainageway. This reduction in evidence of hominid activities presumably reflects the climatic desiccation and large area of dry savannah. A more or less similar dry climate has continued to the present day. Artifacts are sparse but widely scattered in the Ndutu Beds, suggesting a hominid population not greatly different from that recorded by the Masek Beds. The two known archeologic sites (locs. 4b, 26) were situated close to the stream in the Main Gorge. The Naisiusiu Beds contain evidence of sparse occupation in and near the gorge. The one known archeologic site and a skeleton of *Homo sapiens* have yielded C-14 dates of about 17,000 yr. B.P., suggesting a single main period of occupancy.

I have attempted to show the concentration of archeologic sites in different lithofacies of Bed I through the Masek Beds by dividing the total thickness of measured sections by the number of archeologic sites in the facies, thus obtaining the average thickness of measured section per site (Table 32). This is a more accurate measure than the number of sites per lithofacies, for the lithofacies vary greatly in thickness and extent. It is also very likely more accurate than the number of sites per unit of cross-sectional area, inasmuch as the measured sections include nearly all of the *known* data covering both the lithofacies and its archeologic content. This is because sections were measured to include nearly all archeologic sites, and additional sites were encountered in the measuring of sections.

From the data of Table 32, archeologic sites are relatively most abundant in the eastern fluvial-lacustrine deposits of Bed II within the Side Gorge (7.4 m/site), in lake-margin deposits of

Bed I (7.7 m/site), and in deposits of the main drainageway of Beds III-IV (6.4 m/site), which represent a subfacies of the western fluvial deposits. These facies, or subfacies, with the highest concentration of sites presumably represent the environments most suited to hominid occupation at different periods of time in the Olduvai basin. These data suggest, although they do not prove, that the intensity of hominid activity may have been comparable in these environments of Beds I, II, and III-IV.

UTILIZATION AND TRANSPORT OF STONE BY HOMINIDS

The many archeologic sites in the Olduvai Beds afford an excellent opportunity to examine the utilization and transport of rock in the form of artifacts over a long time span and through several paleolithic industries. The present analysis is a preliminary one and relies to a large extent on M. D. Leakey's published data (1971a) on the composition of artifacts from the excavated sites in Beds I and II, and on unpublished data regarding the artifacts of Bed IV. She has summarized the general pattern for Beds I and II as follows:

Except in the lower part of Bed II, when chert became available for a time, the artefacts are made almost exclusively from fine-grained lavas, quartz and to a lesser extent from quartzite (see Table 7). In Bed I, with the exception of the 'Zinjanthropus' level at FLK I, 80-94 per cent of all the heavy-duty tools such as choppers, polyhedrons, discoids, spheroids, etc., are made from lava, while in Middle and Upper Bed II quartz and quartzite are the most common material. . . .

Quartz and quartzite (and chert when available) are the most usual materials for the light-duty tools, light-duty utilised flakes and *dèbitage*. DK, however, is an exception, since lava is the predominant material for all the artefacts with the exception of the light-duty utilised flakes. This is also the only known site where the amount of lava *dèbitage* is sufficient to suggest that the heavy-duty tools may have been made on the spot. At other sites where the majority of heavy-duty tools is of lava the scarcity of lava *dèbitage* indicates that the tools were made elsewhere, presumably at the sources of the raw material [1971a, pp. 262-263].

Three factors have clearly determined the choice of raw material for artifacts: size and shape of available pieces, mechanical properties of the rock, and proximity of the source. The role of size and shape is illustrated by the preference for rounded cobbles about 8 cm in average diameter for the manufacture of choppers in the

Oldowan industry. By contrast, large cobbles were chosen to supply the large flakes from which most of the Acheulian handaxes and cleavers were made. Cobbles of dense, homogeneous lava were preferred over vesicular or foliated lava for choppers, and lavas with a weak foliation seem to have been preferred for bifacial tools of the Acheulian sites. The distinctive properties of chert account for its extensive use as flakes and light-duty tools. The majority of artifacts at all excavated sites in Beds I and II are made of materials obtainable within a distance of 4 km, and at most sites the majority of artifacts are less than 2 km from possible sources. However, sites with a large number of artifacts almost invariably contain a small proportion that are of materials available no closer than 8 to 10 km. As an example, lava from the volcanic highlands is present at most of the archeologic sites in the western part of the Main Gorge. The proportion of these materials from more distant sources in and near the basin appears to increase upward in the sequence, at least to the base of Bed III. A relatively few artifacts at various levels in the sequence are of material obtained from distant sources outside the basin. Only in the Naisiusiu Beds do these exotic materials constitute an appreciable proportion of the artifacts.

Hominid usage of several different rock types will serve to document in more detail the distance of transport and patterns of utilization of raw materials. These rock types are lava of Sadiman, phonolite of Engelosin, quartzite, gneiss of Kelogi type, and chert.

Lava of Sadiman was used principally for heavy-duty tools, particularly choppers. Most or all of the lava was obtained in the form of cobbles from stream channels draining the north side of Sadiman. The high proportion of Sadiman lava choppers in Bed I and the lower part of Bed II indicates that it was selected in preference to other lavas. Moreover, as M. D. Leakey observed, "It is abundantly clear that the best choppers were made of the phonolite-nephelinite rocks [of Sadiman]" (1966, personal communication). The dense, homogeneous nature of the Sadiman lava was presumably the principal reason for its selection, but its attractive green color may have been an additional factor. The percentage of Sadiman lava among choppers at excavated sites is markedly lower above Tuff IIB

Table 33. *Percentage of Choppers of Sadiman Lava in Selected Sites of Beds I and II*

1. Bed I, FLK-N (loc. 45*a*)	44.7% of 76 choppers
2. Bed II, lower augitic sandstone, HWK-E (loc. 43, site 48*c*)	50% of 62 choppers
3. Bed II, beneath upper augitic sandstone at FC (loc. 89, site 62*a*)	11.7% of 51 choppers
4. Bed II, beneath Tuff IIC at SHK (loc. 91, sites 68*a-c*)	7.6% of 106 choppers
5. Bed II, floor above Tuff IID (loc. 16, site 19*d*)	12.5% of 32 choppers

NOTE: Data are from M. D. Leakey, 1966, personal communication.

than it is below (Table 33), despite the fact that lava cobbles from Sadiman were very likely available in stream channels within a distance of 2 km at all these sites with the possible exception of SHK. Those sites with the higher percentage of Sadiman lava choppers are Oldowan and Developed Oldowan A, whereas the sites with a lower percentage are Developed Oldowan B. Hence the stratigraphic change in proportion of Sadiman lava choppers may be attributable to the decreased utilization of Sadiman lava in the Developed Oldowan B.

The use of Sadiman lava decreases even further in younger deposits, and only 6 heavy-duty tools of Sadiman phonolite were found among the 511 heavy-duty tools excavated in Acheulian sites of Bed IV (M. D. Leakey, 1974, personal communication). Sadiman lava is represented in the excavated Acheulian site in the Masek Beds by only 2 out of 2,288 pieces of *débitage*. This site also yielded 238 tools, none of which is of Sadiman lava.

The phonolite of Engelosin has not been reported to occur on any Bed I site, but it is found near the base of Bed II, and it becomes progressively more abundant upward in the sequence as high as Beds III-IV. This phonolite was very likely available only in outcroppings and in talus at the foot of Engelosin, where it occurs as tabular pieces. The phonolite is dense, has a weak foliation, and is particularly well suited for the manufacture of bifacial tools.

The lowest known occurrence is a battered lump 1.4 m above the base of Bed II at FLK-N (loc. 45*a*, site 40*g*). Two artifacts were found by

M. D. Leakey in her recent excavation in the lower augitic sandstone at HWK-EE (loc. 42b). Artifacts made of this lava were obtained from stratigraphically higher excavations in the Side Gorge: FC (loc. 89, site 62a) and SHK (loc. 91, sites 68a-c). An isolated flaked piece of phonolite was found in sandstone above Tuff IID at FC. Engelosin phonolite is a common raw material for bifaces in Acheulian sites of Bed II above the level of Tuff IID to the west of the Fifth Fault. All of the known Bed II sites with Engelosin phonolite artifacts lie between 9 and 11 km from Engelosin. No artifacts of this phonolite have been found in the archeologic sites of Bed II that lie in the eastern part of the Main Gorge above the lower augitic sandstone. The availability of lava cobbles in nearby stream channels may account for the fact that Engelosin phonolite was utilized only rarely in the eastern area.

Beds III-IV record a sudden increase both in the proportion and distribution of Engelosin phonolite. It has been found throughout both the Main and Side gorges, and its most distant occurrence is at Kelogi (loc. 98), 18 km from the source. It was used particularly as a raw material for bifaces, and almost all of the handaxes and cleavers at one excavated site at HEB (loc. 44b) are made of this rock (M. D. Leakey, in press, 1). The most highly finished handaxes and cleavers in Bed IV are made of this phonolite, and it is tempting to attribute its selection to its attractive green color as well as to functional properties.

Quartz and coarse quartzite of the type found on Naibor Soit are the principal quartzose rocks used for artifacts. Specific varietal types of Naibor Soit quartzite (for example, green and brown) can be found in most large collections of quartzite artifacts. Naibor Soit is located centrally, and most archeological sites lie within a radius of 5 km. However, a substantial proportion of the artifacts are of coarse quartzite in the lower Bed II site at Kelogi, 13 km from Naibor Soit. Noteworthy is the 26.8-kilogram anvil of coarse quartzite at SHK (site 68c), 3 km from Naibor Soit.

Quartz and coarse quartzite were to a considerable extent selected in preference to lava for light-duty tools, light-duty utilized flakes, and débitage (M. D. Leakey, 1971a). The hardness of quartz and the sharp edges of fractured pieces

account for its selection for tools of this type. Although lava seems to have been preferred generally for bifaces, coarse quartzite was the chief raw material for bifaces at all of the Acheulian sites in Bed II and Beds III-IV to the west of the main drainageway, probably because quartzite outcroppings were nearer than large cobbles of lava. More difficult to explain is the dominance of quartz bifaces in Bed II at DK-EE (loc. 12a), where lava cobbles were available nearby. This site lies only 160 m distant and at about the same level as the Acheulian site at EF-HR (loc. 12), where three-quarters of the bifaces are of lava. All of the handaxes in the excavated site in the Masek Beds are of white quartzite (M. D. Leakey, in press, 1).

A few artifacts of medium-grained quartzite are found in many sites of Bed I through Bed IV. The easternmost recorded occurrence is in Bed III at JK (Kleindienst, in press), about 15 km east of hills of similar quartzite. Clearly the medium-grained quartzite was a less desirable raw material than the coarse quartzite.

Gneiss of Kelogi type seems to represent a material, like the medium-grained quartzites, which lacked mechanical attributes making it particularly useful for any type of artifact. A few pieces are found in many of the archeologic sites of Beds I and II, as for example DK (Bed I, site 22), the "Zinjanthropus" floor of Bed I at FLK, HWK (Bed II, sites 48a, 48c), SHK (Bed II, sites 68a-c), FLK (Bed II, site 42), and TK (Bed II, sites 19a-e). The gneiss was utilized for heavy-duty tools, hammerstones, and so forth. Similar gneiss is extremely rare in the excavated Acheulian sites. Gneiss artifacts in the Main Gorge sites are probably 8 to 10 km from the nearest outcroppings of similar gneiss.

Chert was highly regarded for flakes and other light-duty tools. Not only was it extensively utilized when it became available in the basin, but it was carried into the basin, probably from distant sources, when it could not be obtained locally. Within the basin, chert was formed in lake deposits of Beds I and II. This chert occurs principally as nodules of irregular shape with reticulate surface markings (Plate 8). It is typically white and opaque or milky and translucent; a little is yellowish-brown. The chert consists of microcrystalline quartz and resembles the flints of England and the mid-continental U.S.A. in mechanical properties. Only one chert horizon is in Bed I, and this is stratigraphically lower than the lowest of the archeologic sites. Chert occurs at various horizons in Bed II between Tuff IF and Tuff IIB. Artifacts made of locally derived chert have been found only in Bed II and the Ndutu and Naisiusiu beds.

Bed II archeologic sites with locally derived chert are restricted to lake-margin deposits, and chert artifacts are abundant only within 1 km of probable source areas. Two of these sites are below the widespread disconformity, and several others are above it. One of the two lower sites is a concentration of chert artifacts in a channel cut through a nodule horizon (loc. 201). The other (HWK, loc. 43, site 48b) contains a scattering of chert artifacts and lies at least 1 km distant from lacustrine deposits that could be expected to have chert. The main chert unit and equivalent deposits contain a factory site (at MNK, loc. 88) in addition to sites with transported artifacts (loc. 38, site a; loc. 42b; loc. 43, site 48c; loc. 43a; loc. 45, site a; loc. 45a, site 40h; loc. 45d; loc. 46c, site 44a). The chert factory site is situated within a part of the main chert unit where the nodules had been eroded away. More than 30,000 specimens, mostly of chert modified by hominids, were obtained from an area of 10 m^2 (Stiles et al., 1974). Curiously, a substantial fraction and possibly more than half of the artifacts are of chert which differs from that in the main chert unit nearby and which must have been transported to the factory site. The δO^{18} values for this transported chert are significantly lower than for undisturbed nodules at MNK, FC, and PEK. The isotopic values signify that the transported chert was formed in water less saline than that of the nodules in the vicinity of MNK, and very likely the exotic chert originated to the south, farther from the center of the lake, at a distance of perhaps a few hundred meters to a kilometer. Chert artifacts are widespread and locally abundant within the lower augitic sandstone as far west as HWK-E (locs. 42b, 43, 43a), about a kilometer from the likely source area at the southeastern margin of the lake (Fig. 27a). Isolated artifacts of similar chert are found in slightly younger sandstone as much as 1,500 m farther to the east (loc. 38) and northeast (loc. 19; Fig. 27b).

Chert artifacts in the Naisiusiu and Ndutu beds were made from nodules of Beds I and II that were exposed in erosion of the gorge. Artifacts in the one excavated site in the Naisiusiu Beds are principally chert of the type presently found exposed in the Side Gorge, 7 km distant (M. D. Leakey et al., 1972).

Isolated flakes of gray to brown chert from an unknown and probably distant source have been found in Beds I and II. The "Zinjanthropus" floor of Bed I yielded two flakes, and a single flake was found at three Bed II sites: FC, site 62a; EF-HR, site 23a; and SHK, sites 68a-c (M. D. Leakey, 1971a). M. R. Kleindienst found a few flakes of chert in Bed III at JK (in press). The nature of this chert in Bed III is unknown. These scattered occurrences compare with the Acheulian sites of Olorgesaile, which contain a few small fragments and flakes of exotic chert (Isaac, 1968, part IV, p. 33).

A few other exotic rock types have been found in archeologic sites at Olduvai Gorge. One artifact of altered but unmetamorphosed gabbro was found in the lower augitic sandstone at HWK (site 48c). This gabbro was very likely derived from the Tanganyika shield to the west, and artifacts of similar gabbro are common in an archeologic site on the western shore of Lake Ndutu, about 45 km to the west of the Bed II site. Two lumps of ochre were obtained from the archeologic site in the upper part of Bed II at BK (L. S. B. Leakey, 1958). Obsidian constitutes about 5 percent of the artifacts excavated from the Naisiusiu Beds (M. D. Leakey et al., 1972). The obsidian is rhyolitic (n = 1.494), and as far as I am aware, the nearest source of rhyolitic obsidian is at the south side of Lake Naivasha, 270 km to the north.

Several conclusions emerge from this analysis of raw materials. Hominids at all levels were at least moderately selective in their choice of raw materials for particular tool types, although proximity to a source was also a major factor. There are, however, a few anomalies such as the "Zinjanthropus" floor of Bed I, where quartzite was utilized for heavy-duty tools to a much greater degree than in nearby archeologic sites of Bed I. Another example is in Bed II, where two nearby Acheulian sites at about the same level used different materials for bifaces. From Bed I through the Ndutu Beds, raw materials were obtained almost exclusively from within or near the Olduvai basin. Tools may have been made generally at the sources of raw materials, as suggested by M. D. Leakey (1971a), but the chert factory site of Bed II contains a large proportion of chert from sources very likely at least a few hundred meters distant.

Hominids ranged widely over the basin from Bed I upward, as shown by a small but significant proportion of materials from distances of at least 8 km at most sites yielding a large number of artifacts. The degree of hominid mobility seems to have increased through time, at least through the period represented by Beds I and II, as based on the overall upward increase in proportion of materials from the more distant sources. Rare artifacts of materials from distant sources outside the basin may have been obtained from hominid groups with a territorial range bordering the Olduvai basin.

NOTE ADDED IN PROOF

M. D. Leakey in 1974 found hominid fossils in the Laetolil Beds, thus considerably extending the record of human evolution in the Olduvai region. In 1975 she organized a field party, based at Laetolil, for systematic collecting and stratigraphic study. As to be expected, this field work substantially modified the conclusions about the Laetolil Beds presented in Chapter 1. The Laetolil Beds are a regional tephra deposit far thicker and more extensive than recognized by Kent (1941) or myself (see p. 17-18). The thickest section measured, about 130 m, lies 9 km northeast of Laetolil and 17 km west of Sadiman volcano, the probable source. These deposits are 15 to 20 m thick at a distance of 25 to 30 km farther to the southwest, and 10 to 15 m thick at lakes Masek and Ndutu, 30 km to the northwest. They are a conformable sequence with the possible exception of the topmost 15 to 23 m of sediments which overlie an eroded surface with a relief of about 8 m in the vicinity of Laetolil.

The surface of the Laetolil Beds is deeply eroded and widely overlain by a series of mafic lava flows, including the dated vogesite flow ($2.39 \pm .09$ m.y.a.), erupted from numerous small vents to the south and southwest of Lemagrut. These lavas are overlain by flows from Lemagrut.

Eolian tuffs are the principal type of deposit in the Laetolil Beds, and stream-worked and ash-fall tuffs constitute nearly all of the remainder. The tephra is of nephelinite and melilitite composition. Most of the fossils are in eolian tuffs of the upper 30 m of the Laetolil Beds beneath the possible disconformity. The fauna is terrestrial and comprises mammals, reptiles, amphibians, birds (including eggs), and gastropods (including urocyclid slugs). In view of the fauna, particularly the slugs, the climate must have been considerably moister than where the eolian tuffs of Olduvai Gorge were deposited. Rootmarkings are relatively coarse and occur principally as localized concentrations suggestive of scattered brush and small trees rather than grassland.

The Laetolil Beds are very likely somewhat more than 3 m.y. old. They are undoubtedly much older than the flow dated at $2.39 \pm .09$ m.y.a., and Sadiman, the probable source of Laetolil tephra, has yielded a single K-Ar date of 3.7 m.y.a. Biotite is in several of the Laetolil tuffs, and the one sample dated thus far, from the upper, fossiliferous part of the formation, has given a K-Ar date of $3.2 \pm .1$ m.y.a. (G. H. Curtis and R. E. Drake, 1975, personal communication).

GLOSSARY OF GEOLOGIC TERMS

Accretionary lapilli. Volcanic pellets, commonly exhibiting concentric structures, owing to the accretion of fine ash or dust around raindrops falling through an eruption cloud, or to accretion around a nucleus fragment which rolls along the ground.

Allogenic. Term meaning generated elsewhere, and referring especially to materials of detrital or pyroclastic origin transported into the basin of deposition. Contrasted to authigenic.

Anhedral. A term applied to those minerals that are not bounded by their own crystal faces. Contrasted to euhedral.

Anticline. A type of fold in which the sides or limbs of the fold slope away from the hinge line on either side.

Aphanitic. Rock texture in which the crystalline constituents are too small to be distinguished with the unaided eye.

Ash, volcanic. Uncemented pyroclastic material consisting of fragments under 4 mm in diameter.

Ash flow. An avalanche of volcanic ash, generally a highly heated mixture of volcanic gases and ash, traveling down the flanks of a volcano or along the surface of the ground and produced by the explosive disintegration of viscous lava in a volcanic crater or by the explosive emission of gas-charged ash from a fissure or group of fissures.

Authigenic. A term applied to minerals which originated in sediments at the time of, or after, deposition. The term emphasizes local derivation by chemical action rather than solid matter transported.

Barchan. A dune having a crescentic ground plan, with the convex side facing the wind.

Basalt. Olivine-bearing feldspathic extrusive rock having a color index of at least 40.

Bimodal. Term applied to a sediment with two grain-size maxima, as for example a mixture of fine-, medium- and coarse-grained sand in which the amount of medium-grained sand is subordinate to the other two.

Calcrete. The more heavily cemented calcareous layers, or duricrusts, within or beneath soils of arid or semiarid regions. It commonly grades downward into less-cemented calcareous deposits.

Caliche. A calcareous zone within or beneath soils of arid or semiarid regions. It commonly grades downward into less-cemented calcareous deposits.

Carbonate. A compound containing the radical $CO_3^=$. Carbonate rock is used in this report to include both limestone ($CaCO_3$) and dolomite ($CaMg(CO_3)_2$).

Chert. Cryptocrystalline varieties of silica composed mainly of microscopic chalcedony aggregates and/or quartz particles.

Clast. An individual constituent of detrital or pyroclastic sediment or sedimentary rock produced by the physical disintegration of a larger mass. Examples are a silt particle, pebble, or pumice fragment.

Clay. The term clay carries three implications: (1) a natural material with plastic properties; (2) component particles less than 1/256 mm in diameter; and (3) a composition of alumino-silicate minerals of sheetlike structure, termed *clay minerals.*

Colluvial deposits. Deposits composed chiefly of the debris from sheet erosion deposited by unconcentrated runoff or slope wash, together with talus and other mass-movement accumulations.

Color index. In petrology, the sum of the dark or colored minerals in a rock expressed as a percentage. It may be the sum of minerals actually present (that is, modal), or it may be calculated from a chemical analysis (that is, normative).

Devitrified. Pertaining to a formerly glassy (vitreous) rock which crystallized after solidification, usually at high (near-magmatic) temperatures.

Euhedral. A term applied to minerals bounded by their own crystal faces.

Facies. The "aspect" belonging to a geological unit of sedimentation, including mineral composition, type of bedding, fossil content, and so forth. Sedimentary facies are really segregated parts of differing nature belonging to any genetically related body of sedimentary rock.

Gneiss. Coarse-grained, banded metamorphic rocks in which the schistosity is poorly defined because of the preponderance of quartz and feldspar over micaceous minerals.

Graben. A block, generally long in relation to its width, that has been downthrown along faults relative to the rocks on either side.

Granule. Sedimentary particle larger than sand and

Granule. (Continued)
smaller than pebbles (2-4 mm in diameter).

Horst. A block, generally long in relation to its width, that has been uplifted along faults relative to the rocks on either side.

Ignimbrite. Deposit of a hot tephra flow (= ash-flow deposit).

Ijolite. Coarse-grained igneous rock composed chiefly of nepheline and augite. It may contain accessory melanite, sphene, and biotite.

Inselberg. Prominent steep-sided residual hill or mountain rising abruptly from plains.

Lapilli. Volcanic ejecta ranging from 4 to 32 mm in diameter.

Lapilli tuff. Consolidated fragmental volcanic rocks of primary eruptive origin having a grain size from 4 to 32 mm.

Lithofacies. The rock record of any sedimentary environment, including both physical and organic character.

Loess. A homogeneous, nonstratified, unindurated deposit consisting predominantly of silt, deposited primarily by the wind.

Mafic. Pertaining to or composed dominantly of the magnesian rock-forming silicates. Used here to refer to volcanic glass with a color index of about 40 or more.

Melilitite. An extrusive rock composed primarily of augite and melilite. Minor minerals may be nepheline, sphene, perovskite, and so forth.

Nephelinite. An extrusive rock composed primarily of pyroxene and nepheline. Minor minerals may include melilite, sphene, melanite, and so forth.

Palagonite. An alteration product of mafic glass that appears waxy or resinous and yellowish-orange to reddish-brown.

Paleosol. A buried soil.

Pebble. Sedimentary particle with an average diameter between 4 and 64 mm.

Pedogenic. Pertaining to soil-forming processes.

Phonolite. An extrusive rock consisting largely of alkali feldspar and containing at least 5 percent nepheline, either modal or normative. See Color Index.

Phyllosilicates. Silicate minerals with a sheetlike structure, as for example mica and most of the clay minerals.

Phytolith, opal. Particles of opal found in many plants and particularly abundant in grasses.

Pleochroic. The property of exhibiting different colors in different directions by transmitted polarized light.

Pyroclastic. A general term applied to detrital volcanic materials that have been explosively or aerially ejected from a volcanic vent. Also, a general term for the class of rocks made up of these materials.

Quartzite. A granular metamorphic rock consisting essentially of quartz.

Salic. A term applied to the group of standard normative minerals including quartz, the feldspars, and the feldspathoids. The corresponding term for the silicic and aluminous minerals actually present in a rock is Felsic.

Sand. Sedimentary particles between silt and granule size (1/16-2 mm).

Siliceous earth. A white or cream-colored silica-rich deposit that is friable, porous, and has an earthy luster. Silica is in the form of fine-grained particles of biogenic opal.

Silt. Sedimentary particle between clay and sand size (1/256-1/16 mm).

Syncline. A fold in which the strata dip inward from both sides toward the axis. The opposite of anticline.

Tephra. A collective term for all clastic volcanic materials which during an eruption are ejected from a crater or from some other type of vent and transported through the air, including volcanic dust, ash, cinders, lapilli, scoria, pumice, bombs, and blocks. Synonymous with volcanic ejecta.

Tephrite. An extrusive rock composed essentially of calcic plagioclase, nepheline, and augite.

Trachyte. An extrusive rock in which less than 25 percent of the feldspar is plagioclase, the rest being anorthoclase. The color index is generally between 10 and 20. Aegirine or aegirine-augite and sodic amphibole are the principal mafic minerals. Trachyte may contain a small amount of nepheline (nepheline trachyte) or quartz (quartz trachyte).

Trachyandesite. A feldspathic extrusive rock in which the color index is less than 40 and plagioclase makes up 25 percent or more of the total feldspar. These vary widely in the feldspar composition and content of mafic minerals, which may include olivine, augite, aegirine, biotite, and amphibole of different types. Sodic trachyandesite refers to trachyandesite in which plagioclase is subordinate to anorthoclase, and in which the pyroxene is aegirine or aegirine-augite.

Tuff. Consolidated, fragmental volcanic rock of primary eruptive origin having a grain size less than 4 mm.

Tuff, eolian. Tuff composed of particles reworked by wind prior to burial and consolidation.

Tuffaceous. Pertaining to tuff. When prefixed to a rock name, it denotes a content of less than 50 percent volcanic ash (for example, tuffaceous sandstone).

Vitroclast. A fragment of volcanic glass.

Vitroclastic. Pertaining to a structure typical of fragmental glassy rocks. The most common structures are pumice and shards with one or more curved surfaces.

Vogesite. An extrusive rock chiefly of olivine and augite, containing alkali feldspar as the only salic mineral.

189

Welded tuff. An ignimbrite which has been indurated by the heat retained by the constituent particles.

Zeolite. A hydrated aluminosilicate mineral with a framework structure enclosing pores and water molecules. Chemically the zeolites can be viewed as hydrated equivalents of the feldspars. More than thirty distinct species of zeolites occur in nature, and about ten of these are common as authigenic constituents in sedimentary rocks.

Zeolitite. A sedimentary deposit consisting largely of zeolites, the bulk of which are not alteration products of volcanic glass.

APPENDIX A.

RADIOCARBON DATES FROM THE OLDUVAI REGION, NGORONGORO, AND THE VICINITY OF OLDOINYO LENGAI

Sample	Laboratory no.	Material dated	Sample location	Date (yr. B.P.)
1	Isotopes I-6038	Calcite grains of silt size	Surface layer of eolian silt on Serengeti Plain 3-4 km southeast of Kelogi.	$1,200 \pm 100$
2	LJ2979	Shell of land snail (*Achatina* sp.)	In fluvial silts 2 m below Namorod Ash, loc. 4*d*.	$1,370 \pm 40$
3	Isotopes I-4984	Calcrete 1.5 cm thick	At land surface near the southwest margin of Olbalbal (1.0 km south of loc. 154).	$2,190 \pm 105$
4	UCLA 1903	Collagen of bone	Eolian tuffs near the top of the upper unit of the Ndutu Beds, loc. 26.	$3,340 \pm 800$
5	Lamont L-1301B	Calcrete 1.5 cm thick	Topmost calcrete at CK (loc. 6); overlies Naisiusiu Beds.	$4,000 \pm 100$
6	Isotopes I-5153	Calcrete 1 cm thick	Over Naisiusiu Beds on plain southeast of Kelogi (0.7 km ENE of loc. 98).	$4,115 \pm 105$
7	Isotopes I-6039	Calcrete 1-2 mm thick underlying hard calcareous tuff 1-2 cm thick	Over top of Naisiusiu Beds at excavated site 110 m east of Second Fault.	$6,135 \pm 140$
8	Isotopes I-4892	Calcrete 1 cm thick	Over top of Naisiusiu Beds at top of trail leading down the Fifth Fault on the north side of the gorge (near loc. 83).	$6,630 \pm 110$
9	Isotopes I-5154	Calcrete 2.5 cm thick	Over top of Naisiusiu Beds, 5 km north of Main Gorge (loc. 205).	$9,130 \pm 130$
10	UCLA 1321	Collagen of bone	Bones collected from eolian tuff of the upper part of the Naisiusiu Beds exposed in the trail down the Fifth Fault (near loc. 83).	$10,400 \pm 600$
11	UCLA 1902B	Collagen of bone	Eolian tuffs of Naisiusiu Beds, same locality as no. 10.	$11,000 \pm 750$
12	Isotopes I-4891	Calcrete 2 cm thick	Near middle of Naisiusiu Beds at location of sample 8.	$14,580 \pm 210$
13	UCLA 1902A	Collagen of bone	Eolian tuffs of Naisiusiu Beds, same locality as nos. 10 and 11.	$15,400 \pm 1,200$
14	Isotopes I-4893	Calcrete 5 mm thick	Between Ndutu and Naisiusiu Beds at north rim of Kestrel K (loc. 84).	$15,850 \pm 270$
15	UCLA 1740	Collagen of bone of *Homo sapiens*	Intrusive burial on north side of gorge near the Third Fault.	$16,920 \pm 920$
16	Lamont	Ostrich eggshell	From bone concentration in the middle part of the Naisiusiu Beds; in excavated site 110 m east of the Second Fault on the north side of the gorge.	17,000
17	UCLA 1695	Collagen of bone	Same bed and locality as sample 16.	$17,500 \pm 1,000$
18	Isotopes I-5152	Calcrete 1 cm thick	Between Ndutu and Naisiusiu Beds, location of sample no. 8.	$20,000 \pm 400$

APPENDIX A

Sample	Laboratory no.	Material dated	Sample location	Date (yr. B.P.)
19	Isotopes I-6767	Calcrete 6-8 mm thick	Between Ndutu and Naisiusiu Beds at head of the trail at the Third Fault on north side of Main Gorge.	$22,470 \pm 420$
20	Lamont	*Limicolaria* shells	From upper part of the upper unit of the Ndutu Beds at several localities to the east of the Second Fault.	$\geqslant 29,000$
21	Lamont L-880	Calcrete 2 cm thick	Between Norkilili Member and Ndutu Beds at CK (loc. 6).	$\geqslant 30,000$
22	Lamont L-901A	Calcrete 2 mm thick	Surface of alluvial fan of dark gray reworked tephra deposits 3-5 km south of Lake Natron.	1,300
23	Lamont L-901B	Calcrete 8 mm thick	Locality of no. 22, 45 cm below top of alluvial fan deposits.	2,050
24	Lamont 1194-B	Calcrete 0.5-1 cm thick	Over top of Naisiusiu Beds on northeast edge of Swallow Crater, southeast of Oldoinyo Lengai.	$3,500 \pm 100$
25	Lamont 1194-A	Calcrete 6 mm thick	Base of Naisiusiu Beds, locality of sample no. 24.	$19,800 \pm 1,000$
26	Isotopes I-6037	Ostracod-rich limestone with authigenic calcite crystals	Bluffs 2 m high at the eastern margin of the crater lake in Ngorongoro.	$24,400 \pm 690$
27	LJ 3086	Ostracod concentration	Layer of ostracods 5 cm thick beneath ostracodal limestone of sample 26. Sample collected about 200 m south of the north end of bluffs along eastern margin of Crater lake in Ngorongoro.	$27,900 \pm 500$

NOTE: Dates published elsewhere are as follows: nos. 4, 11, and 13 (Libby and Berger, in press); no. 10 (L. S. B. Leakey et al., 1968); no. 15 (Protsch, 1973); nos. 16, 17, 20, 21 (M. D. Leakey et al., 1972); nos. 22, 23 (Hay, 1966). Samples 1-21 refer to the Olduvai region and 22-27 to Ngorongoro and the vicinity of Oldoinyo.

APPENDIX B.

ARCHEOLOGIC SITES IN BEDS I AND II

Locality	Archeologic site no.	Description, stratigraphic position	Facies	Published references
		Bed I		
10 (LK)	25a	Artifacts and faunal remains below Tuff IB.	Lake-margin (eastern)	M. D. Leakey, 1971a
11 (MK)	24b	Artifacts and faunal remains below Tuff IB.	Lake-margin (eastern)	M. D. Leakey, 1971a
13 (DK)	22	Occupation site and stone circle 1.5-2.2 m below Tuff IB.	Lake-margin (eastern)	M. D. Leakey, 1967, 1971a
14 (JK)	21	Scattered artifacts and faunal remains in claystone beneath Tuff IB.	Lake-margin (eastern)	M. D. Leakey, 1971a
34	–	Scattered artifacts, displaced basalt blocks, and faunal remains in claystone beneath Tuff IB.	Lake-margin (eastern)	–
45 (FLK)	41a	Occupation floor with skull of "Zinjanthropus" and two hominid limb bones; beneath Tuff IC.	Lake-margin (eastern)	M. D. Leakey, 1967, 1971a
45 (FLK)	41c-41f, 41i	Scattered artifacts and faunal remains in several levels through 2.6 m of tuff and claystone above Tuff ID.	Lake-margin (eastern)	M. D. Leakey, 1967, 1971a
45 (FLK-N)	40a	Butchering site, 1.3 m below Tuff IF.	Lake-margin (eastern)	M. D. Leakey, 1967, 1971a
45a (FLK-N)	40b-e	Occupation floors, faunal remains, and phalange of H-10 in 1 m of claystone beneath Tuff IF.	Lake-margin (eastern)	M. D. Leakey, 1967, 1971a
45b (FLK-NN)	38b	Occupation site and remains of H-7 and H-8; on claystone overlying Tuff IB.	Lake-margin (eastern)	M. D. Leakey, 1967, 1971a
45b (FLK-NN)	38d	Artifacts and faunal remains in claystone beneath Tuff IC (Corresponds to the "Zinjanthropus" level at FLK).	Lake-margin (eastern)	M. D. Leakey, 1967, 1971a
61	–	Several artifacts and faunal remains in conglomeratic sandstone 75 cm below Tuff IF.	Lake-margin (western)	–
63a	–	Numerous artifacts and associated faunal remains in claystone 7.3 m below top of Bed I.	Lake-margin (western)	–
		Bed II		
7 (CK)	27a	Bifaces and faunal remains in conglomeratic sandstone 8 m above Lemuta Member.	Eastern fluvial-lacustrine	L. S. B. Leakey, 1951; also M. D. Leakey, 1971a
12 (EF-HR)	23a	Acheulian occupation site on clay surface buried by conglomeratic sandstone, probably lies above level of brown tuffaceous siltstone marker bed.	Eastern fluvial-lacustrine	M. D. Leakey, 1967, 1971a
12a (DK-EE)	–	Numerous artifacts, principally bifaces, in 65 cm of conglomeratic sandstone 85 cm above Lemuta Member. Nearby, this bed overlies the brown tuffaceous siltstone. This site probably correlative with 23a (loc. 12).	Eastern fluvial-lacustrine	–
14 (JK)	–	Scattered artifacts and bones in 30 cm to 2 m of siliceous earthy claystone overlying Tuff IID.	Eastern fluvial-lacustrine	–

193

Locality	Archeologic site no.	Description, stratigraphic position	Facies	Published references
15 (TK)	–	Concentration of artifacts in sandstone filling channel 1 m above Tuff IIA; sandstone represents lower augitic sandstone.	Eastern fluvial-lacustrine	–
16 (TK)	19a	Artifacts and faunal remains, possibly from adjacent living site; in filling of channel cut into Tuff IID.	Eastern fluvial-lacustrine	M. D. Leakey, 1967, 1971a
16 (TK)	19b	Occupation floor on paleosol above site 19a (loc. 16).	Eastern fluvial-lacustrine	M. D. Leakey, 1967, 1971a
16 (TK)	19c	Numerous artifacts and faunal remains dispersed in rootmarked clayey sandstone 30-60 cm thick above site 19b (loc. 16).	Eastern fluvial-lacustrine	M. D. Leakey, 1967, 1971a
16 (TK)	19d	Occupation floor at base of clayey tuffaceous sandstone above site 19c (loc. 16).	Eastern fluvial-lacustrine	M. D. Leakey, 1967, 1971a
16 (TK)	19e	Numerous artifacts and faunal remains dispersed in clayey tuffaceous sandstone 75 cm thick; overlies site 19d (loc. 16). Sandstone has siliceous root casts.	Eastern fluvial-lacustrine	M. D. Leakey, 1967, 1971a
23 (PLK)	a	Scattered artifacts in sandy claystone, 8 m below top of Bed II.	Eastern fluvial-lacustrine	–
23 (PLK)	b	Scattered artifacts in sandy claystone, 7 m below top of Bed II.	Eastern fluvial-lacustrine	–
23 (PLK)	c	Scattered artifacts and faunal remains in 30 to 45 cm of sandstone and sandy claystone, 5.2 m below top of Bed II.	Eastern fluvial-lacustrine	–
23 (PLK)	d	Scattered artifacts and faunal remains in basal part of ostracodal sandstone 1.5 m thick containing clasts of Tuff IID.	Eastern fluvial-lacustrine	–
23 (PLK)	e	Scattered artifacts and faunal remains at base of 60 cm of sandstone and sandy claystone, 2 m below top of Bed II.	Eastern fluvial-lacustrine	–
23 (PLK)	15	Occupation site with artifacts and catfish remains in claystone at top of Bed II.	Eastern fluvial-lacustrine	M. D. Leakey, 1971a
32 (Elephant K)	55	Artifacts, including bifaces and faunal remains in conglomerate 1.7 m thick that lies 3 m above Lemuta Member.	Eastern fluvial-lacustrine	M. D. Leakey, 1971a
32a	–	Several artifacts in conglomerate 1 m thick that overlies Lemuta Member.	Eastern fluvial-lacustrine	–
33a	–	Scattered artifacts and faunal remains in conglomerate 1.4 m thick beneath Tuff IIC.	Eastern fluvial-lacustrine	–
34	–	Several artifacts in conglomeratic sandstone 60 cm thick that lies 4.3 m below Tuff IID.	Eastern fluvial-lacustrine	–
35a (ISK)	–	Several artifacts in 30 cm of conglomeratic sandstone that lies 60 cm below Tuff IID.	Eastern fluvial-lacustrine	–
38	a	Scattered artifacts in lower 15 cm of lower augitic sandstone; nearby, artifacts lie on paleosol buried by sandstone.	Eastern fluvial-lacustrine	–
38	b	Biface and flake in conglomeratic sandstone 1.5 m thick which lies 75 cm below Tuff IID; numerous artifacts on surface below this level probably come from this bed.	Eastern fluvial-lacustrine	–
42b (HWK-EE)	–	Abundant artifacts and faunal remains in 1.4 to 2 m of conglomeratic sandstone of lower augitic sandstone unit. Most of artifacts are in basal 45 cm.	Eastern fluvial-lacustrine	–
43a (HWK)	–	Numerous artifacts and faunal remains in 40 cm to 1 m of conglomerate filling stream channel; same level as site 48c (loc. 43).	Eastern fluvial-lacustrine	–

194

Locality	Archeologic site no.	Description, stratigraphic position	Facies	Published references
43 (HWK-E)	48a	Occupation site on surface of clay overlying Tuff IF.	Lake-margin (eastern)	M. D. Leakey, 1971a
43 (HWK-E)	48b	Faunal remains and artifacts in claystone approximately 1.5 m above Tuff IF.	Lake-margin (eastern)	M. D. Leakey, 1971a
43 (HWK-E)	48c	Numerous artifacts and faunal remains in 1.2 m of conglomeratic sandstone forming lower part of lower augitic sandstone.	Eastern fluvial-lacustrine	M. D. Leakey, 1971a
44a	–	25 artifacts, including bifaces, at base of sandy claystone 1.2 m thick that overlies bird-print tuff.	Eastern fluvial-lacustrine	–
45 (FLK)	a	Artifacts and faunal remains in conglomeratic sandstone of lower part of lower augitic sandstone, 1.5 m above base of Bed II.	Eastern fluvial-lacustrine	–
45 (FLK)	42	Reworked concentration of artifacts in tuffaceous sandstone representing Tuff IIC.	Eastern fluvial-lacustrine	M. D. Leakey, 1971a
45a (FLK-N)	40f	Scattered artifacts and faunal remains in 1.4 m of claystone above Tuff IF.	Lake-margin (eastern)	M. D. Leakey, 1971a
45a (FLK-N)	40g	Butchering site; artifacts scattered around Deinotherium skeleton in claystone 2 m above Tuff IF.	Lake-margin (eastern)	M. D. Leakey, 1967, 1971a
45a (FLK-N)	40h	Artifacts in 30 cm of clayey sandstone forming lower unit of lower augitic sandstone.	Eastern fluvial-lacustrine	M. D. Leakey, 1971a
45d	–	Several artifacts on paleosol overlying Tuff 11A; may represent paleosol beneath lower augitic sandstone.	Eastern fluvial-lacustrine	–
46b (FLK-S)	44b	Bifaces, other artifacts, and faunal remains in 1.2 m of sandy limestone 5.9 m above bird-print tuff.	Eastern fluvial-lacustrine	M. D. Leakey, 1971a
46c (FLK-S)	44a	Artifacts and faunal remains in sandstone representing lower augitic sandstone.	Eastern fluvial-lacustrine	M. D. Leakey, 1971a
49	35	Scattered artifacts in conglomerate overlying uppermost disconformity in Bed II, which postdates Tuff IID.	Western fluvial-lacustrine	M. D. Leakey, 1971a
52 (MLK-east)	34	Reworked occupation site with bifaces, etc., in conglomeratic sandstone at same level as site 35 (loc. 49).	Western fluvial-lacustrine	M. D. Leakey, 1971a
64	31	Scattered artifacts in reworked tuff of lower 1.2 m of Bed II.	Lake-margin (western)	M. D. Leakey, 1971a
66	–	Approximately one dozen artifacts and faunal remains in base of tuff bed 60 cm thick at base of Bed II.	Lake-margin (western)	–
66b	–	Abundant artifacts, including bifaces, in conglomerate interbedded with limestone 8 to 11 m above base of Bed II. Exact horizon unknown.	Western fluvial-lacustrine	–
75	–	Seven artifacts, including biface and faunal remains in 2-3 m of conglomeratic sandstone that overlies and locally cuts out bird-print tuff.	Western fluvial-lacustrine	–
77 (Kar K)	3	Numerous artifacts, including bifaces and faunal remains, including buffalo skull, in conglomeratic sandstone with clasts of Tuff IID.	Western fluvial-lacustrine	M. D. Leakey, 1971a
79	4	Scattered artifacts in conglomeratic sandstone with clasts of Tuff IID; same horizon as sites 3 (loc. 77) and 35 (loc. 49).	Western fluvial-lacustrine	M. D. Leakey, 1971a
80 (RHC)	5b	Scattered artifacts and faunal remains in conglomerate with blocks of Tuff IID; same horizon as site 3 (loc. 77).	Western fluvial-lacustrine	M. D. Leakey, 1971a
82a	6	Abundant bifaces in conglomerate filling channel of uppermost disconformity of Bed II; same horizon as site 3 (loc. 77).	Western fluvial-lacustrine	–

Locality	Archeologic site no.	Description, stratigraphic position	Facies	Published references
88 (MNK)	—	Chert factory site at level of main chert unit; artifacts in 5-8 cm of sandy limestone overlain by 10-15 cm of sandy claystone.	Eastern fluvial-lacustrine	Stiles *et al.*, 1974
88 (MNK)	71*a*	Artifacts, faunal remains (including H-13 skull fragments) in 1.4 m of sandy tuffaceous clay-stone about 2 m below Tuff IIB.	Eastern fluvial-lacustrine	M. D. Leakey, 1971*a*
88 (MNK)	71*b*	Artifacts, faunal remains (including H-15 teeth) in 1.3 m of clayey sandstone between site 71*a* (loc. 88) and Tuff IIB.	Eastern fluvial-lacustrine	M. D. Leakey, 1971*a*
88 (MNK)	71*c*	Reworked occupation site, faunal remains, and phalange of H-19, in 1.5 m of tuffaceous sand-stone which represents upper augitic sandstone and lies 35 cm above Tuff IIB.	Eastern fluvial-lacustrine	M. D. Leakey, 1971*a*
89 (FC-E)	—	Artifacts and faunal remains in 1.5 m of calcareous sandstone representing upper augitic sandstone.	Eastern fluvial-lacustrine	—
89 (FC-W)	62*a*	Occupation site in topmost claystone of 75-cm bed overlying Tuff IIB.	Eastern fluvial-lacustrine	M. D. Leakey, 1971*a*
89 (FC-W)	62*b*	Artifacts, faunal remains, and broken hominid tooth (H-19), in 1.4-m sandstone bed representing upper augitic sandstone.	Eastern fluvial-lacustrine	M. D. Leakey, 1971*a*
90*a* (SC)	60	Scattered artifacts, one *Homo erectus* ulna, and 3 australopithecine teeth; in siliceous earthy clay-stone above the level of Tuff IID.	Eastern fluvial-lacustrine	—
91 (SHK-west, annexe)	68*a*	Occupation site on brown claystone and buried by siliceous earthy reworked tuff and claystone probably representing Tuff IIC.	Eastern fluvial-lacustrine	L. S. B. Leakey, 1958; M. D. Leakey, 1971*a*
91 (SHK-west, Main site)	68*b*	Abundant artifacts and faunal remains filling channel eroded into claystone and overlain by Tuff IIC.	Eastern fluvial-lacustrine	L. S. B. Leakey, 1958; M. D. Leakey, 1971*a*
91 (SHK-west, Main site)	68*c*	Scattered artifacts and faunal remains in siliceous earthy reworked tuff and sandstone, probably representing Tuff IIC.	Eastern fluvial-lacustrine	L. S. B. Leakey, 1958; M. D. Leakey, 1971*a*
91*a*	—	Scattered artifacts and faunal remains in 60 cm of sandstone probably representing upper augitic sandstone; near base of exposed section, 15 m below top of Bed II.	Eastern fluvial-lacustrine	—
94 (BK)	66	Numerous artifacts and faunal remains including 2 hominid teeth (H-3); in conglomeratic sandstone filling channels eroded into siliceous earthy claystone and sandstone above Tuff IID.	Eastern fluvial-lacustrine	L. S. B. Leakey, 1958; M. D. Leakey, 1971*a*
99	—	Scattered artifacts in bed of siliceous earthy claystone 1 m thick and lying 3.3 m above base of Bed II.	Lake-margin (Kelogi)	—
101	—	Rich concentration of artifacts in 15 cm of con-glomeratic reworked tuff, 2.3 m above base of Bed II.	Lake-margin (Kelogi)	—
201	—	Abundant artifacts and flakes, principally of chert, in augite-rich sandstone filling channel 75 cm deep and 1.1 m wide at base of Bed II.	Lake deposits	—

JOURNAL ABBREVIATIONS

Abst. with Program, geol. Soc. Am.: Abstracts with Program, Geological Society of America

Am. J. Sci. Radiocarbon Suppl.: American Journal of Science, Radiocarbon Supplement

Bull, geol. Soc. Am.: Bulletin of the Geological Society of America

Bull. volcan.: Bulletin Volcanologique

Contr. Mineral. Petrol.: Contributions to Mineralogy and Petrology

Curr. Anthrop.: Current Anthropology

Earth & Planet. Sci. Ltrs.: Earth and Planetary Science Letters

Earth Sci. Rev.: Earth Science Reviews

Geol. Mag.: Geological Magazine

Geol. Surv. Tanganyika: Geological Survey of Tanganyika

Geol. Surv. Tanzania: Geological Survey of Tanzania

Geophys. J. R. Astr. Soc.: Geophysical Journal of the Royal Astronomical Society

J. Geol.: Journal of Geology

J. Sedim. Petrol.: Journal of Sedimentary Petrology

Proc. C.C.T.A. Joint Ctee. Geol.: Proceedings of the Commission for Technical Cooperation in Africa South of the Sahara, East-Central and Southern Regional Committees for Geology

Proc. natn. Acad. Sci. U.S.A.: Proceedings of the National Academy of Science

Prof. Pap. U.S. geol. Surv.: Professional Paper of the U.S. Geological Survey

Quat. Res.: Journal of Quaternary Research

Res. Pap. Dep. Geogr. Univ. Chicago: Research Papers of the Department of Geography, University of Chicago

Rev. Geophys. & Space Phys.: Reviews of Geophysics and Space Physics

Spec. Pap. Geol. Soc. Am.: Special Paper of the Geological Society of America

Spec. Pap. Min. Soc. Am.: Special Paper of the Mineralogical Society of America

Spec. Pap. Soc. Econ. Pal. & Min.: Special Paper of the Society of Economic Paleontologists and Mineralogists

Trans. geol. Soc. S. Afr.: Transactions of the Geological Society of South Africa.

REFERENCES

Bada, J. L., and Protsch, R. (1973). Racemization reaction of aspartic acid and its use in dating fossil bones. *Proc. natn. Acad. Sci. U.S.A.* 70:1331-34.

Baker, B. H., Mohr, P. A., and Williams, L. A. J. (1972). Geology of the eastern rift system of Africa. *Spec. Pap. geol. Soc. Am.* 136:67.

Berger, R., and Libby, W. F. (in press). *Am. J. Sci. Radiocarbon Suppl.* 10.

Boswell, P. G. H. (1932). The Oldoway human skeleton. *Nature, Lond.* 130:237-38.

Bowler, J. M. (1973). Clay dunes: Their occurrence, formation, and environmental significance. *Earth Sci. Rev.* 9:315-38.

Brock, A., Hay, R. L., and Brown, F. H. (1972). Magnetic stratigraphy of Olduvai Gorge and Ngorongoro, Tanzania (abst.). *Abst. with Program, geol. Soc. Am.* 4:457.

Butzer, K. W. (1971). Recent history of an Ethiopean delta. *Res. Pap. Dep. Geogr. Univ. Chicago* 136:184.

Butzer, K. W., Isaac, G. L., Richardson, J. L., and Washburn-Kamau, C. (1972). Radiocarbon dating of East African lake levels. *Science, N.Y.* 175:1069-76.

Cahen, L., and Snelling, N. J. (1966). *The geochronology of equatorial Africa,* Amsterdam: North-Holland.

Chesworth, W. (1971). Laboratory synthesis of dawsonite and its natural occurrences. *Nature, Lond.* 231:40-41.

Cooke, H. B. S. (1957). Observations relating to Quaternary environments in East and Southern Africa. *Trans. geol. Soc. S. Afr., Annexure to* 60:60.

Cox, A. (1969). Geomagnetic reversals. *Science, N.Y.* 163:237-45.

Curtis, G. H., and Hay, R. L. (1972). Further geologic studies and K-Ar dating at Olduvai Gorge and Ngorongoro Crater. In *Calibration of Hominoid Evolution* (ed. W. W. Bishop and J. A. Miller), pp. 289-301. Edinburg: Scottish Academic Press.

Dalrymple, G. B. (1972). Potassium-argon dating of geomagnetic reversals and North American glaciations. In *Calibration of Hominoid Evolution* (ed. W. W. Bishop and J. A. Miller), pp. 107-34. Edinburgh: Scottish Academic Press.

Dawson, J. B. (1962). The geology of Oldoinyo Lengai. *Bull. volcan.* 24:349-87.

Dawson, J. B. (1964). Carbonatitic volcanic ashes in northern Tanganyika. *Bull. volcan* 27:1-11.

Dawson, J. B., and Powell, D. G. (1969). The Natron-Engaruka explosion crater area, northern Tanzania. *Bull. volcan.* 33:781-817.

Day, M. H. (1971). Postcranial remains of *Homo erectus* from Bed IV, Olduvai Gorge, Tanzania. *Nature, Lond.* 232:383-87.

Eugster, H. P. (1967). Hydrous sodium silicates from Lake Magadi, Kenya: precursors of bedded chert. *Science, N.Y.* 157:1177-80.

Eugster, H. P. (1969). Inorganic bedded cherts from the Magadi area, Kenya. *Contr. Mineral. Petrol.* 22:1-31.

Eugster, H. P. (1970). Chemistry and origin of the brines of Lake Magadi, Kenya. In *Spec. Pap. Min. Soc. Am.* 3 (ed. B. A. Morgan), pp. 215-35.

Eugster, H. P., and Jones, B. F. (1968). Gels composed of sodium-aluminum silicate, Lake Magadi, Kenya. *Science, N.Y.* 161: 160-63.

Evernden, J. F., and Curtis, G. H. (1965). The potassium-argon dating of late Cenozoic rocks in East Africa and Italy. *Curr. Anthrop.* 6:343-64.

Fleischer, R. L., Price, P. B., Walker, R. M., and Leakey, L. S. B. (1965). Fission-track dating of Bed I, Olduvai Gorge. *Science, N.Y.* 148:72-74.

Flint, R. F. (1959). On the basis of Pleistocene correlation in East Africa. *Geol. Mag.* 96:265-84.

Friedman, G. M. (1966). Occurrence and origin of Quaternary dolomite of Salt Flat, West Texas. *J. sedim, Petrol.* 36:263-67.

Friedman, G. M., and Sanders, J. E. (1967). Occurrence and origin of dolostone. In *Carbonate Rocks* (ed. G. V. Chilingar, H. J. Bissell, and R. W. Fairbridge), pp. 267-348. Amsterdam and London: Elsevier.

Galehouse, Jon S. (1971). Point counting. In *Procedures in Sedimentary Petrology* (ed. R. E. Carver), pp. 385-407. New York: Wiley-Interscience.

Gentry, A. W., and Gentry, A. (in press). *Fossil Bovidae (Mammalia) of Olduvai Gorge, Tanzania.* London: British Museum (Natural History).

Greenwood, P. H., and Todd, E. J. (1970). Fish remains from Olduvai. In *Fossil Vertebrates in Africa*, vol. 2 (ed. L. S. B. Leakey and R. J. G. Savage), pp. 225-41. London and New York: Academic Press.

Grommé, C. S., and Hay, R. L. (1963). Magnetization of basalt in Bed I, Olduvai Gorge, Tanganyika. *Nature, Lond.* 200:560-61.

Grommé, C. S., and Hay, R. L. (1967). Geomagnetic polarity epochs: New data from Olduvai Gorge, Tanganyika. *Earth & Planet, Sci. Ltrs.* 2:111-15.

Grommé, C. S., and Hay, R. L. (1971). Geomagnetic polarity epochs: Age and duration of the Olduvai normal polarity event. *Earth & Planet. Sci. Ltrs.* 10:179-85.

REFERENCES

Grommé, C. S., Reilly, T. A., Mussett, A. E., and Hay, R. L. (1970). Paleomagnetism and potassium-argon ages of volcanic rocks of Ngorongoro caldera, Tanzania. *Geophys. J. R. Astr. Soc.* 22:101-15.

Guest, N. J., James, T. C., Pickering, R., and Dawson, J. B. (1961). Angata salei. *Geol. Surv. Tanganyika.* Quarter degree sheet, 39, scale 1:125,000.

Harms, J. C., MacKenzie, D. B., and McCubbin, D. G. (1963). Stratification in modern sands of the Red River, Louisiana. *J. Geol.* 71:566-80.

Hay, R. L. (1963a). Stratigraphy of Bed I through IV, Olduvai Gorge. Tanganyika. *Science, N.Y.* 139:829-33.

Hay, R. L. (1963b). Zeolitic weathering in Olduvai Gorge, Tanganyika. *Bull. Geol. Soc. Am.* 74:1281-86.

Hay, R. L. (1965). Discussion and revised stratigraphy. *Curr. Anthrop.* 6:381-83.

Hay, R. L. (1966). Zeolites and zeolitic reactions in sedimentary rocks. *Spec. Pap. geol. Soc. Am.* 85:130.

Hay, R. L. (1967a), Revised stratigraphy of Olduvai Gorge. In *Background to Evolution in Africa* (ed. W. W. Bishop and J. D. Clark), pp. 221-28. Chicago: University of Chicago Press.

Hay, R. L. (1967b). Hominid-bearing deposits of Olduvai Gorge. In *Time and Stratigraphy in the Evolution of Man* (ed. C. B. Hunt, W. L. Straus, and M. G. Wolman), pp. 30-42. Washington, D.C.: National Academy of Sciences and National Research Council U.S.A.

Hay, R. L. (1968). Chert and its sodium-silicate precursors in sodium-carbonate lakes in East Africa. *Contr. Mineral. Petrol.* 17:255-74.

Hay, R. L. (1970a). Silicate reactions in three lithofacies of a semi-arid basin, Olduvai Gorge, Tanzania. In *Spec. Pap. Min. Soc. Am.* 3 (ed. B. A. Morgan), pp. 237-55.

Hay, R. L. (1970b). Pedogenic calcretes of the Serengeti Plain, Tanzania. In *Abst. with Program, geol. Soc. Am.* 2:572.

Hay, R. L. (1971). Geologic background of Beds I and II. In M. D. Leakey (1971a), pp. 9-18.

Hay, R. L. (1973). Lithofacies and environments of Bed I, Olduvai Gorge, Tanzania. *Quat. Res.* 3:541-60.

Hopwood, A. T. (1951). The Olduvai fauna. In L. S. B. Leakey (1951), pp. 20-24, 31-33.

Inman, D. L. (1952). Measures for describing the size distribution of sediments. *J. Sedim, Petrol.* 22:125-45.

Isaac, G. L. (1967). The stratigraphy of the Peninj Group — early Middle Pleistocene formations west of Lake Natron, Tanzania. In *Background to Evolution in Africa* (ed. W. W. Bishop and J. D. Clark), pp. 229-58. Chicago: University of Chicago Press.

Isaac, G. L. (1968). "The Acheulian site complex at Olorgesaile, Kenya: A contribution to the interpretation of middle Pleistocene culture in East Africa." Cambridge University, unpublished PhD thesis, 349 p.

Isaac, G. L., and Curtis, G. H. (1974). Age of early Acheulian industries from the Peninj Group, Tan-

zania. *Nature, Lond.* 249, 624-27.

Jaeger, J. J. (in press). Les Rongeurs (Mammalia, Rodentia) du Pleistocène inférieur d'Olduvai Bed I, Tanzanie. Part 1: General Introduction and Muridae. In *Fossil Vertebrates of Africa,* vol. 4 (ed. R. J. G. Savage and S. C. Coryndon). London: London Academic Press.

Kent, P. E. (1941). The recent history and Pleistocene deposits of the plateau north of Lake Eyasi, Tanganyika. *Geol. Mag.* 78:173-84.

Kleindienst, M. R. (1964). Summary report on excavations at site JK2, Olduvai Gorge, Tanganyika, 1961-1962. *Annual Report, Antiquities Division, Tanganyika, for the year 1962,* pp. 4-6.

Kleindienst, M. R. (1973). Excavation at site JK2, Olduvai Gorge, Tanzania, 1961-1962: The geological setting. *Quaternaria* 17:145-208.

Koenigswald, G. H. R. von, Gentner, W., and Lippolt, H. J. (1961). Age of the basalt flow at Olduvai, East Africa. *Nature, Lond.* 192:720-21.

Leakey, L. S. B. (1951). *Olduvai Gorge.* London: Cambridge University Press.

Leakey, L. S. B. (1958). Recent discoveries at Olduvai Gorge, Tanganyika. *Nature, Lond.* 181:1099-103.

Leakey, L. S. B. (1965). *Olduvai Gorge, 1951-1961.* London: Cambridge University Press.

Leakey, L. S. B. (1969). Age of Bed V, Olduvai Gorge, Tanzania. *Science, N.Y.* 166:532.

Leakey, L. S. B., Boswell, P. G. H., Reck, H., Solomon, J. D., and Hopwood, A. T. (1933). The Oldoway human skeleton. *Nature, Lond.* 131:397.

Leakey, L. S. B., Curtis, G. H., and Evernden, J. F. (1962). Age of basalt underlying Bed I, Olduvai. *Nature, Lond.* 194:610-12.

Leakey, L. S. B., Evernden, J. F., and Curtis, G. H. (1961). The age of Bed I, Olduvai Gorge. Tanganyika. *Nature, Lond.* 191:478-79.

Leakey, L. S. B., Protsch, R., and Berger, R. (1968). Age of Bed V, Olduvai Gorge, Tanzania. *Science, N.Y.* 162:559-60.

Leakey, M. D. (1967). Preliminary survey of the cultural material from Beds I and II, Olduvai Gorge, Tanzania. In *Background to Evolution in Africa* (ed. W. W. Bishop and J. D. Clark), pp. 417-46. Chicago: University of Chicago Press.

Leakey, M. D. (1971a). *Olduvai Gorge, Vol. 3.* London: Cambridge University Press.

Leakey, M. D. (1971b). Discovery of postcranial remains of *Homo erectus* and associated artefacts in Bed IV at Olduvai Gorge, Tanzania. *Nature, Lond.* 232:380-83.

Leakey, M. D. (in press, 1). Cultural patterns in the Olduvai sequence. In *After the Australopithecine* (ed. K. W. Butzer and G. L. Isaac). The Hague: Mouton.

Leakey, M. D. (in press, 2). Olduvai fossil hominids — their stratigraphic positions and associations. In *African hominidae of the Plio-Pleistocene: Evidence, Problems, Strategies* (ed. C. Jolly). London: Duckworth.

Leakey, M. D. (in press, 3). *Valley of the Wild Sisal.* New York: Knopf.

Leakey, M. D., Hay, R. L., Thurber, D. L., Protsch, R.,

and Berger, R. (1972). Stratigraphy, archeology, and age of the Ndutu and Naisiusiu Beds, Olduvai Gorge, Tanzania. *World Archeology* 3:328-41.

Leopold, L. B., and Wolman, M. G. (1957). River channel patterns: Braided, meandering, and straight. *Prof. Pap. U. S. geol. Surv.* 282-B:85.

MacDougall, D., and Price, P. B. (1974). Attempt to date early South African hominids by using fission tracks in calcite. *Science, N.Y.* 185:943-44.

McDougall, I., and Aziz-ur-Rahman (1972). Age of the Gauss-Matuyama boundary and of the Kaena and Mammoth events. *Earth & Planet. Sci. Ltrs.* 14:367-80.

McDougall, I., and Watkins, N. D. (1973). Age and duration of the Réunion geomagnetic polarity event. *Earth & Planet. Sci. Ltrs.* 19:443-52.

Macintyre, R. M., Mitchell, J. G., and Dawson, J. B. (1974). Age of fault movements in Tanzanian sector of East African rift system. *Nature, Lond.* 247:354-56.

Maglio, V. J. (1970). Early Elephantidae of Africa and a tentative correlation of African Plio-Pleistocene deposits. *Nature, Lond.* 225:328-32.

Middleton, G. V., ed. (1965). Primary sedimentary structures and their hydrodynamic interpretation. *Spec. Pap. Econ. Pal. & Min.* 12.

O'Neil, J. R., and Hay, R. L. (1973). O^{18}/O^{16} ratios in cherts associated with the saline lake deposits of East Africa. *Earth & Planet. Sci. Ltrs.* 19:257-66.

Opdyke, N. D. (1972). Paleomagnetism of deep-sea cores. *Rev. Geophys. & Space Phys.* 10:213-49.

Pickering, R. (1958). Oldoinyo Ogol, Serengeti Plain, East. *Geol. Surv. Tanganyika.* Quarter degree sheet, 12 S.W., scale 1:125,000.

Pickering, R. (1960a). A preliminary note on the Quaternary geology of Tanganyika. *Proc. C.C.T.A. Joint Ctee. Geol., Leopoldville, 1958,* pp. 77-89.

Pickering, R. (1960b). Moru, Serengeti Plain, West. *Geol. Surv. Tanganyika.* Quarter degree sheet, 37, scale 1:125,000.

Pickering, R. (1964). Endulen. *Geol. Surv. Tanzania.* Quarter degree sheet, 52, scale 1:125,000.

Pickering, R. (1965). Ngorongoro. *Geol. Surv. Tanzania.* Quarter degree sheet, 53, scale 1:125,000.

Price, W. A. (1963). Physicochemical and environmental factors in clay dune genesis. *J. sedim, Petrol.* 33:766-78.

Protsch, R. (1973). "The dating of upper Pleistocene sub-Saharan fossil hominids and their place in human evolution: With morphological and archeological implications." University of California (Los Angeles), unpublished Ph.D. thesis, 263 p.

Reck, H. (1914a). Erste vorläufige Mitteilung über den Fund eines fossilen Menschenskelets aus Zentralafrika. *Sitzungsberichten der Gesellschaft naturforschender Freunde* 3:81-95, Berlin.

Reck, H. (1914b). Zweite vorläufige Mitteilung über fossile Tier- und Menschenfunde aus Oldoway in Zentralafrika. *Sitzungsberichten der Gesellschaft naturforschender Freunde* 7:305-18, Berlin.

Reck, H. (1933). *Oldoway die Schlucht des Urmenschens.* Leipzig: F. A. Brockhaus.

Reck, H. (1951). A preliminary survey of the tectonics and stratigraphy of Olduvai. In *Olduvai Gorge* (by L. S. B. Leakey), pp. 5-19. London: Cambridge University Press.

Rovner, I. (1971). Potential of opal phytoliths for use in paleo-ecological reconstructions. *Quat. Res.* 1:343-59.

Schroeder, R. A., and Bada, J. L. (1973). Glacial-postglacial temperature difference deduced from aspartic acid racemization in fossil bones. *Science, N.Y.* 182:479-82.

Selley, R. C. (1970). *Ancient sedimentary environments.* Ithaca: Cornell University Press.

Sheppard, R. A., and Gude, A. J. (1968). Distribution and genesis of authigenic silicate minerals in tuffs of Pleistocene Lake Tecopa, Inyo County, California. *Prof. Pap. U.S. geol. Surv.* 597:38.

Sheppard, R. A., and Gude, A. J. (1970). Authigenic fluorite in Pliocene lacustrine rocks near Rome, Malheur County. Oregon. *Prof. Pap. U.S. geol. Surv.* 650D:69-74.

Shuey, R. T., Brown, F. H., and Croes, M. K. (in press). Magnetostratigraphy of the Shungura Formation, southwestern Ethiopia: Fine structure in the lower Matuyama epoch. *Earth & Planet. Sci. Ltrs.*

Stiles, D. N., Hay, R. L., and O'Neil, J. R. (1974). The MNK chert factory site, Olduvai Gorge, Tanzania. *World Archeology* 5:285-308.

Straus, W. L., Jr., and Hunt, C. B. (1962). Age of Zinjanthropus. *Science, N.Y.* 136:293-95.

Verdcourt, B. (1963). The Miocene nonmarine mollusca of Rusinga Island, Lake Victoria, and other localities in Kenya. *Palaeontographica* 121: Abt. A, 1-37.

Visher, G. S. (1965). Fluvial processes as interpreted from ancient and recent fluvial deposits. In *Spec. Pap. Soc. Econ. Pal. & Min.* 12:116-32.

Walker, T. R. (1967). Formation of red beds in modern and ancient deserts. *Bull. geol. Soc. Am.* 78:353-68.

Wayland, E. J. (1932). The Oldoway human skeleton. *Nature, Lond.* 130:578.

Williams, G. E., and Polach, H. A. (1971). Radiocarbon dating of arid-zone calcareous paleosols. *Bull. geol. Soc. Am.* 82:3069-86.

INDEX

INDEX

INDEX